Statistical Process Control

The business, commercial and public-sector world has changed dramatically since John Oakland wrote the first edition of *Statistical Process Control: A Practical Guide* in the mid-eighties. Then people were rediscovering statistical methods of 'quality control' and the book responded to an often desperate need to find out about the techniques and use them on data. Pressure over time from organizations supplying directly to the consumer, typically in the automotive and high technology sectors, forced those in charge of the supplying production and service operations to think more about preventing problems than how to find and fix them. Subsequent editions retained the 'tool kit' approach of the first but included some of the 'philosophy' behind the techniques and their use.

The theme that runs throughout the 7th edition is still *processes*, which require understanding, have variation, must be properly controlled, have a capability, and need improvement – the five sections of this new edition. SPC never has been and never will be simply a 'tool kit' and in this book the authors not only provide the instructional guide for the tools, but communicate the management practices that have become so vital to success in organizations throughout the world. The book is supported by the authors' extensive and latest consulting work within thousands of organizations worldwide.

Fully updated to include real-life case studies, new research based on client work from an array of industries, and integration with the latest computer methods and Minitab software, the book also retains its valued textbook quality through clear learning objectives and end of chapter discussion questions. It can still serve as a textbook for both student and practising engineers, scientists, technologists, managers and for anyone wishing to understand or implement modern statistical process control techniques.

John Oakland is one of the world's top 10 gurus in quality and operational excellence; Chairman, the Oakland Group; Emeritus Professor of Quality & Business Excellence at Leeds University Business School; a Fellow of the Chartered Quality Institute (CQI); and a Member of American Society for Quality (ASQ).

Robert Oakland works across the globe helping complex organizations to unlock the power in their data, using advanced analytical and statistical techniques to improve the quality, cost and delivery of their products and services. He is a partner at the Oakland Group, LLP.

Statistical Process Control

7th Edition

John Oakland and Robert Oakland

Routledge
Taylor & Francis Group

LONDON AND NEW YORK

Seventh edition published 2019
by Routledge
2 Park Square, Milton Park, Abingdon, Oxon OX14 4RN

and by Routledge
52 Vanderbilt Avenue, New York, NY 10017

Routledge is an imprint of the Taylor & Francis Group, an informa business

First edition published by Elsevier, 1999

British Library Cataloguing-in-Publication Data
A catalogue record for this book is available from the British Library

Library of Congress Cataloging-in-Publication Data
Names: Oakland, John S., author. | Oakland, Robert James, editor.
Title: Statistical process control / John Oakland and Robert James Oakland.
Description: 7th edition. | Abingdon, Oxon ; New York, NY : Routledge, 2019. |
    Includes bibliographical references and index. |
Identifiers: LCCN 2018026835 (print) | LCCN 2018027860 (ebook) |
    ISBN 9781315160511 | ISBN 9781138064256 (hardback : alk. paper) |
    ISBN 9781138064263 (pbk. : alk. paper) | ISBN 9781315160511 (ebk)
Subjects: LCSH: Process control—Statistical methods.
Classification: LCC TS156.8 (ebook) | LCC TS156.8 .O23 2019 (print) |
    DDC 658.5/62—dc23
LC record available at https://lccn.loc.gov/2018026835

ISBN: 978-1-138-06425-6 (hbk)
ISBN: 978-1-138-06426-3 (pbk)
ISBN: 978-1-315-16051-1 (ebk)

Typeset in Sabon
by Swales & Willis Ltd, Exeter, Devon, UK

Visit the e-resources: www.routledge.com/9781138064263

For Susan, Jane and Debbie

Contents

Appendices   377

Preface

Stop Producing Chaos – a cry from the heart! When the great guru of quality management and process improvement W. Edwards Deming died at the age of 93 at the end of 1993, the last words on his lips must have been 'Management *still* doesn't understand process variation'.

Despite all his efforts and those of his followers, including the authors of this book, we still find managers in manufacturing, sales, marketing, finance, service and public sector organizations all over the world reacting (badly) to information and data. They often do not understand the processes they are managing, have no knowledge about the extent of their process variation or what causes it, and yet they try to 'control' processes by taking frequent action. This book is written for them and comes with some advice: 'Don't just do something, sit there (and think)!'

The business, commercial and public sector world has changed a lot since John Oakland wrote the first edition of *Statistical Process Control: A Practical Guide* in the mid-eighties. Then people were rediscovering statistical methods of 'quality control' and the book responded to an often desperate need to find out about the techniques and use them on data. Pressure over time from organizations supplying directly to the consumer, typically in the automotive and high technology sectors, forced those in charge of the supplying production and service operations to think more about preventing problems than how to find and fix them. The second edition of *Statistical Process Control* (1990) retained the 'tool kit' approach of the first but included some of the 'philosophy' behind the techniques and their use.

In writing the third, fourth, fifth and sixth editions John found it necessary to completely restructure the book to address the issues found to be most important in those organizations in which Oakland's colleagues work as researchers, teachers and consultants. These increasingly include service and public sector organizations. The theme that runs throughout the book is still PROCESS. Everything we do in any type of organization is a process that:

- requires UNDERSTANDING;
- has VARIATION;
- must be properly CONTROLLED;

- has a CAPABILITY; and
- needs IMPROVEMENT.

Hence the five sections of this edition.

As we move further into the twenty-first century, we know that companies in every market are experiencing fundamental shifts in how customers interact or seek to interact digitally with them and their products and/or services and make purchasing decisions. Understanding the dynamics of this and implementing the necessary changes is the context for the area of information quality engineering and digital transformation.

In this lightning fast environment, from research carried out by the authors, we understand that changing customer behavioural patterns and disruptive technology produce enormous growth opportunities but, at the same time, organizations also face huge challenges when it comes to ensuring the required changes are properly supported in terms of managing quality and the supporting operations.

Organizations need to gain access to approaches, methods and tools to transform and innovate better (higher quality) and faster (more efficient) so that they can generate new growth opportunities and create a significant competitive advantage in the digital age.

Of course, it is still the case that to be successful in today's climate, organizations must be dedicated to continuous improvement. But this requires management – it will not just happen. If more efficient ways to produce goods and services that consistently meet the needs of the customer are to be found, use must be made of appropriate methods to gather information and analyse it, *before* making decisions on any action to be taken.

Part 1 of this edition sets down some of the basic principles of quality and process management to provide a platform for understanding variation and reducing it, if appropriate. The remaining four sections cover the subject of SPC in the basic but comprehensive manner used in the first six editions, with the emphasis on a practical approach throughout. Again a special feature is the use of real-life examples from a number of industries.

John was joined in the second edition by his friend and colleague Roy Followell, who has now passed away. In this edition John has been joined by Robert, now a partner in the Oakland Group and an engineer with extensive experience of bringing about successful change based on improving the process – people links through good data analytics. Rob and John have been helped by their colleagues in the Oakland Group and its research and education division, now Oakland Institute.

The wisdom gained by our colleagues in consultancy work in helping literally thousands of organizations to implement quality management, business excellence, good management systems, lean, six-sigma and SPC has been incorporated, where possible, into this edition. We hope the book continues to provide a comprehensive guide on how to use SPC 'in anger'. Numerous facets of the implementation process, gleaned from many man-years' work in a variety of industries, have been threaded through the book, as the individual techniques are covered. Many new case examples of SPC implementation, including in the service and finance areas, have been added.

SPC never has been and never will be simply a 'tool kit' and in this book we hope to provide not only the instructional guide for the tools, but communicate the philosophy of process understanding and improvement, which has become so vital to success in organizations throughout the world.

The book was never written for the professional statistician or mathematician. As before, attempts have been made to eliminate much of the mathematical jargon that often causes distress, whilst still providing explanations for the basis of the control charts and techniques. Those interested in pursuing the theoretical aspects will find, at the end of each chapter, references to books and papers for further study, together with discussion questions. Several of the chapters end with worked examples taken from a variety of organizational backgrounds.

The book is written, with learning objectives at the front of each chapter, to meet the requirements of students in universities, polytechnics and colleges engaged in courses on science, technology, engineering and management subjects, including quality assurance. It also serves as a textbook for self or group instruction of managers, supervisors, engineers, scientists and technologists. We hope the text continues to offer clear guidance and help to those unfamiliar with either process management or statistical applications.

The book is fully supported by the Routledge website where can be found Excel spreadsheets containing every table of data throughout the book, so that teachers and students may 'drop' these data sets into their preferred statistical software package, such as Minitab, in order to perform the suggested SPC analyses.

We would like to acknowledge the contributions of our colleagues in the Oakland Group. Our collaboration, both in a research/consultancy environment and in a vast array of public and private organizations, has resulted in an understanding of the part to be played by the use of SPC techniques and the recommendations of how to implement them.

John S. Oakland and Robert J. Oakland

Other titles by the same author and publisher

Oakland on Quality Management

Total Organisational Excellence: The Route to World Class Performance

Total Quality Management and Operational Excellence: Text and Cases

Total Quality Management: A Pictorial Guide

Website: www.theoaklandgroup.co.uk

Part I
Process understanding

1   Quality, processes and control

Objectives

- To introduce the subject of statistical process control (SPC) by considering the basic concepts.
- To define terms such as quality, process and control.
- To distinguish between design quality and conformance quality.
- To define the basics of quality-related costs.
- To set down a system for thinking about SPC and introduce some basic tools.

1.1 The basic concepts

Statistical process control (SPC) is not really about statistics or control, it is about competitiveness. Organizations, whatever their nature, compete on three issues: quality, delivery and price. There cannot be many people in the world who remain to be convinced that the reputation attached to an organization for the quality of its products and services is a key to its success and the future of its employees. Moreover, if the quality is right, the chances are the delivery and price performance will be competitive too.

What is quality?

The word 'quality' is often used to signify 'excellence' of a product or service – we hear talk about 'Rolls-Royce quality' and 'top quality'. In some manufacturing companies quality may be used to indicate that a product conforms to certain physical characteristics set down with a particularly 'tight' specification. But if we are to manage quality it must be defined in a way that recognizes the true requirements of the 'customer'.

Quality is defined simply as meeting the requirements of the customer and this has been expressed in many ways by other authors:

> Fitness for purpose or use (Juran).

> The totality of characteristics of an entity that bear upon its ability to satisfy stated and implied needs (ISO 8402, now 9000: 2008).

The total composite product and service characteristics of marketing, engineering, manufacture, and maintenance through which the product and service in use will meet the expectation by the customer (Feigenbaum).

The ability to meet the customer requirements is vital, not only between two separate organizations, but within the same organization. There exists in every factory, every department, every office, a series of suppliers and customers. The PA is a supplier to the boss – is (s)he meeting the requirements consistently? Does the boss receive error-free information set out as (s)he wants it, when (s)he wants it? If so, then we have a quality service. Does the factory receive from its supplier defect-free parts that conform to the requirements of the assembly process every time? If so, then we have a quality supplier.

For industrial and commercial organizations, which are viable only if they provide satisfaction to the consumer, competitiveness in quality is not only central to profitability, but crucial to business survival. The consumer should not be required to make a choice between price and quality, and for manufacturing or service organizations to continue to exist they must learn how to manage quality. In today's tough and challenging business environment, the development and implementation of a comprehensive quality policy is not merely desirable – it is essential.

Every day people in organizations around the world scrutinize together the results of the examination of the previous day's production or operations, and commence the ritual battle over whether the output is suitable for the customer. One may be called the Production Manager, the other the Quality Control Manager. They often argue and debate the evidence before them, the rights and wrongs of the specification, and each tries to convince the other of the validity of their argument. Sometimes they nearly break into fighting.

This ritual is associated with trying to answer the question: 'Have we done the job correctly?' – 'correctly' being a flexible word depending on the interpretation given to the specification on that particular day. This is not quality control, it is post-production/operation detection, wasteful detection of bad output before it hits the customer. There is a belief in some quarters that to achieve quality we must check, test, inspect or measure – the ritual pouring on of quality at the end of the process – and that quality, therefore, is expensive. This is nonsense, but it is frequently still encountered. In the office we find staff checking other people's work before it goes out, validating computer input data, checking invoices, typing, etc. There is also quite a lot of looking for things, chasing things that are late, apologizing to customers for non-delivery and so on – waste, waste and more waste.

The problems are often a symptom of the real, underlying cause of this type of behaviour, the lack of understanding of quality management. The concentration of inspection effort at the output stage merely shifts the failures and their associated costs from outside the organization to inside. To reduce the total costs of quality, control must be at the point of manufacture or operation; quality cannot be inspected into an item or service after it has been produced. It is essential for cost-effective control to ensure that articles are manufactured, documents are produced, or that services are generated correctly the first time. The aim of process control is the prevention of the manufacture of defective products and the generation of errors and waste in non-manufacturing areas.

To get away from the natural tendency to rush into the detection mode, it is necessary to ask different questions in the first place. We should not ask whether the job has been done correctly, we should ask first: 'Can we do the job correctly?' This has wide implications and this book aims to provide some of the tools which should be used to ensure that the answer is 'Yes'. However, we should realize straight away that such an answer will only be obtained using satisfactory methods, materials, equipment, skills and instruction, and a satisfactory or capable 'process'.

What is a process?

A process is the transformation of a set of inputs, which can include materials, actions, methods and operations into desired outputs, in the form of products, information, services or – generally – results. In each area or function of an organization there will be many processes taking place. Each process may be analysed by an examination of the inputs and outputs. This will determine the action necessary to improve quality.

The output from a process is that which is transferred to somewhere or to someone – the customer. Clearly, to produce an output that meets the requirements of the customer, it is necessary to define, monitor and control the inputs to the process, which in turn may have been supplied as output from an earlier process. At every supplier–customer interface there resides a transformation process and every single task throughout an organization must be viewed as a process in this way. To begin to monitor and analyse any process, it is necessary first of all to identify what the process is, and what the inputs and outputs are. Many processes are easily understood and relate to known procedures, e.g. drilling a hole, compressing tablets, filling cans with paint, polymerizing a chemical. Others are less easily identified, e.g. servicing a customer, delivering a lecture, storing a product, inputting to a computer. In some situations it can be difficult to define the process. For example, if the process is making a sales call, it is vital to know whether the scope of the process includes obtaining access to the potential customer or client. Defining the scope of a process is vital, since it will determine both the required inputs and the resultant outputs. A simple 'static' model of a process is shown in Figure 1.1. This describes the boundaries of the process. 'Dynamic' models of processes will be discussed in Chapter 2.

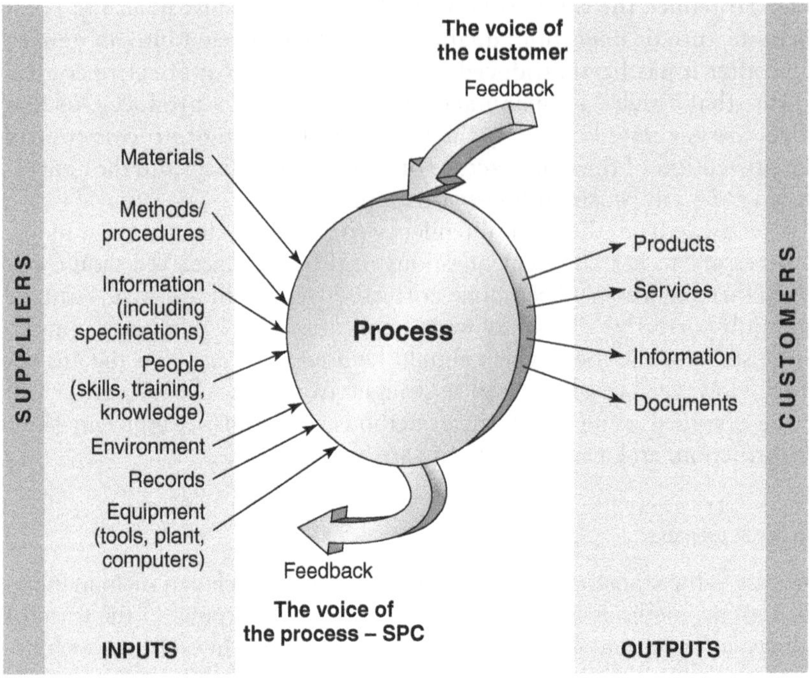

Figure 1.1 A process: SIPOC

Once the process is specified, the suppliers and inputs, outputs and customers (SIPOC) can also be defined, together with the requirements at each of the interfaces (the voice of the customer). Often the most difficult areas in which to do this are in non-manufacturing organization or non-manufacturing parts of manufacturing organizations, but careful use of appropriate questioning methods can release the necessary information. Sometimes this difficulty stems from the previous absence of a precise definition of the requirements and possibilities. Inputs to processes include: equipment, tools, computers or plant required, materials, people (and the inputs they require, such as skills, training, knowledge, etc.); information including the specification for the outputs, methods or procedures instructions and the environment.

Prevention of failure in any transformation is possible only if the process definition, inputs and outputs are properly documented and agreed. This documentation will allow reliable data about the process itself to be collected (the voice of the process), analysis to be performed, and action to be taken to improve the process to prevent failure or non-conformance with the requirements. The target in the operation of any process is the total avoidance of failure. If the objective of no failures or error-free work is not adopted, at least as a target, then certainly it will never be achieved. The key to success is to align the employees of the business, their roles and responsibilities with the organization and its processes. This is the core of process alignment and business process design or re-design (BPR). When an organization focuses on its key processes, that is the value-adding activities and tasks themselves, rather than on abstract issues such as 'culture' and 'participation,' then the change process can begin in earnest.

Process design and particularly re-design challenges managers to rethink their traditional methods of doing work and commit to a customer-focused process. Many outstanding organizations have achieved and maintained their leadership through process re-design or 're-engineering'. Companies using these techniques have reported significant bottom-line results, including better customer relations, reductions in cycle times, time to market, increased productivity, fewer defects/errors and increased profitability. BPR uses recognized techniques for improving business processes and questions the effectiveness of existing structures through 'assumption busting' approaches. Defining, measuring, analysing and re-engineering/designing processes to improve customer satisfaction pays off in many different ways.

What is control?

All processes can be monitored and brought 'under control' by gathering and using data. This refers to measurements of the performance of the process and the feedback required for corrective action, where necessary. Once we have established that our process is 'in control' and capable of meeting the requirement, we can address the next question: 'Are we doing the job correctly?', which brings a requirement to monitor the process and the controls on it. Managers are in control only when they have created a system and climate in which their subordinates can exercise control over their own processes – in other words, the operator of the process has been given the 'tools' to control it.

If we now re-examine the first question: 'Have we done it correctly?', we can see that, if we have been able to answer both of the questions: 'Can we do it correctly?'(capability) and 'Are we doing it correctly?'(control) with a 'yes', we must have done the job correctly – any other outcome would be illogical. By asking the questions in the right order, we have removed the need to ask the 'inspection' question and replaced a strategy of detection with one of prevention. This concentrates attention on the front end of any process – the inputs – and changes the emphasis to making sure the inputs are capable of meeting the requirements of the process. This is a managerial responsibility and these ideas apply to every transformation process, which must be subjected to the same scrutiny of the methods, the people, the skills, the equipment and so on to make sure they are correct for the job.

The control of quality clearly can take place only at the point of transformation of the inputs into the outputs, the point of operation or production, where the letter is typed or the artefact made. The act of inspection is not quality control. When the answer to 'Have we done it correctly?' is given indirectly by answering the questions on capability and control, then we have assured quality and the activity of checking becomes one of quality assurance – making sure that the product or service represents the output from an effective system that ensures capability and control.

1.2 Design, conformance and costs

In any discussion on quality it is necessary to be clear about the purpose of the product or service, in other words, what the customer requirements are. The customer may be inside or outside the organization and his/her satisfaction must be the first and most important ingredient in any plan for success. Clearly, the customer's perception of quality changes with time and an organization's attitude to the product or service, therefore, may have to change with this perception. The skills and attitudes of the people in the organization are also subject to change, and failure to monitor such changes will inevitably lead to dissatisfied customers. The quality of products and services, like all other corporate matters, must be continually reviewed in the light of current circumstances.

The quality of a product or service has two distinct but interrelated aspects:

- quality of design
- quality of conformance to design.

Quality of design

This is a measure of how well the product or service is designed to achieve its stated purpose. If the quality of design is low, either the service or product will not meet the requirements, or it will only meet the requirement at a low level.

A major feature of the design is the specification. This describes and defines the product or service and should be a comprehensive statement of all aspects that must be present to meet the customer's requirements.

A precise specification is vital in the purchase of materials and services for use in any conversion process. All too frequently, the terms 'as previously supplied', or 'as agreed with your representative', are to be found on purchasing orders for bought-out goods and services. The importance of obtaining materials and services of the appropriate quality cannot be overemphasized and it cannot be achieved without proper specifications. Published standards should be incorporated into purchasing documents wherever possible.

There must be a corporate understanding of the company's position in the market place. It is not sufficient that the marketing department specifies a product or service, 'because that is what the customer wants'. There must also be an agreement that the producing departments can produce to the specification. Should 'production' or 'operations' be incapable of achieving this, then one of two things must happen: either the company finds a different position in the market place or substantially changes the operational facilities.

Quality of conformance to design

This is the extent to which the product or service achieves the specified design. What the customer actually receives should conform to the design and operating costs are tied firmly to the level of conformance achieved. The customer satisfaction must be designed into the production system. A high level of inspection or checking at the end is often indicative of attempts to inspect in quality. This may be associated with spiralling costs and decreasing viability. Conformance to a design is concerned largely with the performance of the actual operations. The recording and analysis of information and data play a major role in this aspect of quality and this is where statistical methods must be applied for effective interpretation.

The costs of quality

Obtaining a quality product or service is not enough. The cost of achieving it must be carefully managed so that the long-term effect of 'quality costs' on the business is a desirable one. These costs are a true measure of the quality effort. A competitive product or service based on a balance between quality and cost factors is the principal goal of responsible production/operations management and operators. This objective is best accomplished with the aid of a competent analysis of the costs of quality.

The analysis of quality costs is a significant management tool which provides:

- a method of assessing and monitoring the overall effectiveness of the management of quality;
- a means of determining problem areas and action priorities.

The costs of quality are no different from any other costs in that, like the costs of maintenance, design, sales, distribution, promotion, production and other activities, they can be budgeted, monitored and analysed.

Having specified the quality of design, the producing or operating units have the task of making a product or service that matches the requirement. To do this they add value by incurring costs. These costs include quality-related costs such as prevention costs, appraisal costs and failure costs. Failure costs can be further split into those resulting from internal and external failure.

Prevention costs

These are associated with the design, implementation and maintenance of the quality management system. Prevention costs are planned and are incurred prior to production or operation. Prevention includes:

Product or service requirements: The determination of the requirements and the setting of corresponding specifications, which also take account of capability, for incoming materials, processes, intermediates, finished products and services.

Quality planning: The creation of quality, reliability, production, supervision, process control, inspection and other special plans (e.g. preproduction trials) required to achieve the quality objective.

Quality assurance: The creation and maintenance of the overall quality management system.

Inspection equipment: The design, development and/or purchase of equipment for use in inspection work.

Training: The development, preparation and maintenance of quality training programmes for operators, supervisors and managers to both achieve and maintain capability.

Miscellaneous: Clerical, travel, supply, shipping, communications and other general office management activities associated with quality.

Resources devoted to prevention give rise to the 'costs of getting it right the first time'.

Appraisal costs

These costs are associated with the supplier's and customer's evaluation of purchased materials, processes, intermediates, products and services to assure conformance with the specified requirements. Appraisal includes:

Verification: Of incoming material, process set-up, first-offs, running processes, intermediates and final products or services, and includes product or service performance appraisal against agreed specifications.

Quality audits: To check that the quality management system is functioning satisfactorily.

Inspection equipment: The calibration and maintenance of equipment used in all inspection activities.

Vendor rating: The assessment and approval of all suppliers – of both products and services.

Appraisal activities result in the 'cost of checking it is right'.

Internal failure costs

These costs occur when products or services fail to reach designed standards and are detected before transfer to the consumer takes place. Internal failure includes:

Scrap: Defective product which cannot be repaired, used or sold.

Rework or rectification: The correction of defective material or errors to meet the requirements.

Reinspection: The re-examination of products or work that has been rectified.

Downgrading: Product that is usable but does not meet specifications and may be sold as 'second quality' at a low price.

Waste: The activities associated with doing unnecessary work or holding stocks as the result of errors, poor organization, the wrong materials, exceptional as well as generally accepted losses, etc.

Failure analysis: The activity required to establish the causes of internal product or service failure.

External failure costs

These costs occur when products or services fail to reach design quality standards and are not detected until after transfer to the consumer. External failure includes:

Repair and servicing: Either of returned products or those in the field.

Warranty claims: *Failed products* which are replaced or services redone under guarantee.

Complaints: All work and costs associated with the servicing of customers' complaints.

Returns: The handling and investigation of rejected products, including transport costs.

Liability: The result of product liability litigation and other claims, which may include change of contract.

Loss of goodwill: The impact on reputation and image which impinges directly on future prospects for sales.

External and internal failures produce the 'costs of getting it wrong'.

The relationship between these so-called direct costs of prevention, appraisal and failure (P-A-F) costs, and the ability of the organization to meet the customer requirements is shown in Figure 1.2. Where the ability

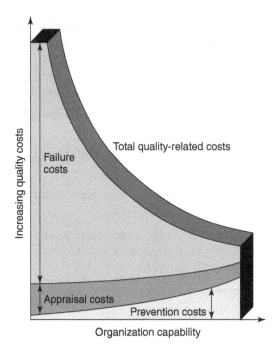

Figure 1.2 Relationship between costs of quality and organizational capability

to produce a quality product or service acceptable to the customer is low, the total direct quality costs are high and the failure costs predominate. As ability is improved by modest investment in prevention, the failure costs and total cost drop very steeply. It is possible to envisage the combination of failure (declining), appraisal (declining less rapidly) and prevention costs (increasing) as leading to a minimum in the combined costs. Such a minimum does not exist because, as it is approached, the requirements become more exacting. The late Frank Price, author of *Right First Time*, also refuted the minimum and called it 'the mathematics of mediocrity'.

So far little has been said about the often intractable indirect quality costs associated with customer dissatisfaction, and loss of reputation or goodwill. These costs reflect the customer attitude towards an organization and may be both considerable and elusive in estimation but not in fact.

The P-A-F model for quality costing has a number of drawbacks, particularly the separation of prevention costs. The so-called 'process cost model' sets out a method for applying quality costing to any process or service. A full discussion of the measurement and management of the cost of quality is outside the scope of this book, but may be found in *Total Quality Management and Operational Excellence*, 5th edn.

Total direct quality costs, and their division between the categories of prevention, appraisal, internal failure and external failure, vary considerably from industry to industry and from site to site. A figure for quality-related costs of less than 10 per cent of sales turnover is seldom quoted when perfection is the goal. This means that in an average organization there exists a 'hidden plant' or 'hidden operation', amounting to perhaps one-tenth of productive capacity. This hidden plant is devoted to producing scrap, rework, correcting errors, replacing or correcting defective goods, services and so on. Thus, a direct link exists between quality and productivity and there is no better way to improve productivity than to convert this hidden resource to truly productive use. A systematic approach to the control of processes provides the only way to accomplish this.

Technologies and market conditions vary between different industries and markets, but the basic concepts of quality management and the financial implications are of general validity. The objective should be to produce, at an acceptable cost, goods and services which conform to the requirements of the customer. The way to accomplish this is to use a systematic approach in the operating departments of: design, manufacturing, quality, purchasing, sales, personnel, administration and all others – nobody is exempt. The statistical approach to process management is not a separate science or a unique theory of quality control – rather a set of valuable tools which becomes an integral part of the 'total' quality approach.

Two of the original and most famous authors on the subject of statistical methods applied to quality management are Shewhart and Deming. In their book, *Statistical Method from the Viewpoint of Quality Control*, they wrote:

> The long-range contribution of statistics depends not so much upon getting a lot of highly trained statisticians into industry as it does on creating a statistically minded generation of physicists, chemists, engineer and others who will in any way have a hand in developing and directing production processes of tomorrow.

This was written in 1939. It is as true today as it was then.

1.3 Quality, processes systems, teams, tools and SPC

The concept of 'total quality' is basically very simple. Each part of an organization has customers, whether within or without, and the need to identify what the customer requirements are, and then set about meeting them, forms the core of the approach. Three hard management necessities are then needed: a good quality management system, the tools and teamwork for improvement. These are complementary in many ways and they share the same requirement for an uncompromising commitment to quality.

This must start with the most senior management and flow down through the organization. Having said that, teamwork, the tools or the management system or all three may be used as a spearhead to drive SPC through an organization. The attention to many aspects of a company's processes – from purchasing through to distribution, from data recording to control chart plotting – which are required for the successful introduction of a good management system, use of tools or the implementation of teamwork, will have a 'Hawthorne effect' concentrating everyone's attention on the customer/supplier interface, both inside and outside the organization.

Good quality management involves consideration of processes in all the major areas: marketing, design, procurement, operations, distribution, etc. Clearly, these each require considerable expansion and thought but if attention is given to all areas using the concept of customer/supplier then very little will be left to chance. A well-operated, documented management system provides the necessary foundation for the successful application of SPC techniques and teamwork. It is not possible simply to 'graft' these onto a poor system.

Much of industry, commerce and the public sector would benefit from the improvements in quality brought about by the approach represented in Figure 1.3. This will ensure the implementation of the management commitment represented in the quality policy, and provide the environment and information base on which teamwork thrives, the culture changes and communications improve.

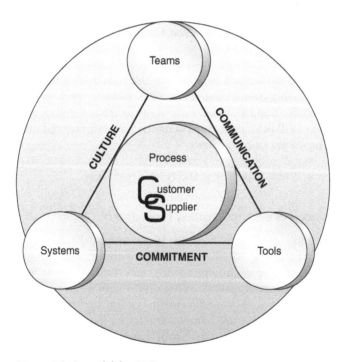

Figure 1.3 A model for SPC

SPC methods, backed by management commitment and good organization, provide objective means of controlling quality in any transformation process, whether used in the manufacture of artefacts, the provision of services, or the transfer of information.

SPC is not only a tool kit. It is a strategy for reducing variability, the cause of most quality problems; variation in products, in times of deliveries, in ways of doing things, in materials, in people's attitudes, in equipment and its use, in maintenance practices, in everything. Control by itself is not sufficient, SPC requires that the process should be improved continually by reducing its variability. This is brought about by studying all aspects of the process using the basic question: 'Could we do the job more consistently and on target (i.e. better)?' the answering of which drives the search for improvements. This significant feature of SPC means that it is not constrained to measuring conformance, and that it is intended to lead to action on processes that are operating within the 'specification' to minimize variability. There must be a willingness to implement changes, even in the ways in which an organization does business, in order to achieve continuous improvement. Innovation and resources will be required to satisfy the long-term requirements of the customer and the organization, and these must be placed before or alongside short-term profitability.

Process control is vital and SPC should form a vital part of the overall corporate strategy. Incapable and inconsistent processes render the best designs impotent and make supplier quality assurance irrelevant. Whatever process is being operated, it must be reliable and consistent. SPC can be used to achieve this objective.

Dr Deming was a statistician who gained fame by helping Japanese companies to improve quality after the Second World War. His basic philosophy was that quality and productivity increase as variability decreases and, because all things vary, statistical methods of quality control must be used to measure and gain understanding of the causes of the variation. Many companies, particularly those in manufacturing industry or its suppliers, have adopted the Deming philosophy and approach to quality. In these companies, attention has been focused on performance improvement through the use of quality management systems and SPC.

In the application of SPC there is sometimes an emphasis on techniques rather than on the implied wider managerial strategies. SPC is not about plotting charts and pinning them to the walls of a plant or office, it must be a component part of a company-wide adoption of 'total quality' and act as the focal point of never-ending improvement in business performance. Changing an organization's environment into one in which SPC can operate properly may take it onto a new plain of performance. For many companies SPC will bring a new approach, a new 'philosophy', but the importance of the statistical techniques should not be disguised. Simple presentation of data using diagrams, graphs and charts should become the means of communication concerning the state of control of processes.

The responsibility for quality in any transformation process must lie with the operators of that process. To fulfil this responsibility, however, people must be provided with the tools necessary to:

- know whether the process is capable of meeting the requirements;
- know whether the process is meeting the requirements at any point in time;
- correct or adjust the process or its inputs when it is not meeting the requirements.

The success of this approach has caused messages to cascade through the supply chains and companies in all industries, including those in the process and service industries which have become aware of the enormous potential of SPC, in terms of cost savings, improvements in quality, productivity and market share. As the authors know from experience, this has created a massive demand for knowledge, education and understanding of SPC and its applications.

A management system, based on the fact that many functions will share the responsibility for any particular process, provides an effective method of acquiring and maintaining desired standards. The 'Quality Department' should not assume direct responsibility for quality but should support, advise and audit the work of the other functions, in much the same way as a financial auditor performs his duty without assuming responsibility for the profitability of the company.

A systematic study of a process through answering the questions below provides knowledge of the process capability and the sources of nonconforming outputs:

Can we do the job correctly? (capability)

Are we doing the job correctly? (control)

Have we done the job correctly? (quality assurance)

Could we do the job better? (improvement)[1]

This information can then be fed back quickly to marketing, design and the 'technology' functions. Knowledge of the current state of a process also enables a more balanced judgement of equipment, both with regard to the tasks within its capability and its rational utilization.

It is worth repeating that SPC procedures exist because there is variation in the characteristics of materials, articles, services and people. The inherent variability in every transformation process causes the output from it to vary over a period of time. If this variability is considerable, it may be impossible to predict the value of a characteristic of any single item or at any point in time. Using statistical methods, however, it is possible to take meagre knowledge of the output and turn it into meaningful statements which may then be used to describe the process itself. Hence, statistically based process control procedures are designed to divert attention from individual pieces of data and focus it on the process as a whole. SPC techniques may be used to measure and understand, and control the degree of variation of any purchased materials, services, processes and products and to compare this, if required, to previously agreed specifications.

The links between customers, digitalization and quality and operational excellence

Companies in every market are experiencing fundamental shifts in how customers interact or seek to interact digitally with them and their products and/or services and make purchasing decisions. Understanding the dynamics of this and implementing the necessary changes is the context for the area of information quality engineering and digital transformation.

In this lightning fast environment, from research carried out by the authors, executives understand that changing customer behavioural patterns and disruptive technology produces enormous growth opportunities but, at the same time, those executives also face huge challenges when it comes to ensuring the required changes are properly supported in terms of managing quality and the supporting operations.

The challenges include ensuring the right quality data is used to feed into capable processes that can in turn operate at the required speed to deliver the desired outputs. With the amount of disruptive technology based changes now taking place, data quality within digital transformation has become a central component of major successful organizations' strategies in the twenty-first century.

Clearly, change must occur rapidly to remain competitive, retain customers and attract new business, but the combination of processes, people, information, and technology can present quality and operational business problems. Organizations face many challenges, including being agile enough to create new processes, form new business models and teams, and keep investing in new technologies and systems to constantly evolve in today's rapidly changing business environment. This entails numerous challenges in maintaining quality and optimizing business processes.

To address this, information quality engineering frameworks are needed that can provide essential support to modern businesses. The ability to quickly change and adapt business processes is a key capability in digitally mature companies. Driven by the need for quality, speed and agility, along with the need for efficiency and optimization, many organizations need to adopt intelligent business process management systems to mitigate the challenges of ongoing digital transformation initiatives.

Customer experience must lie at the heart of the business technology agenda, which means simplifying customer-facing processes to make digital interactions efficient and a 'delight'. This means that many companies need to change focus from cost-cutting approaches to gaining control over all digital touchpoints, both in terms of quality and efficiency. This requires an alignment of customer-facing and operational teams in order to deliver a quality customer experience in the most relevant channel.

With increasing amounts of data, digital interactions, and the number of processes, tools, and applications available on the market, even leading organizations are struggling to keep pace with digital transformation challenges and are increasingly seeking solutions to help them ensure quality and remain competitive. The challenge here is not primarily related to insufficiently skilled employees or limited company resources, it is more about the lack of methods and tools with strong modelling capabilities that can enable them to implement changes faster without compromising on quality.

While many organizations undergoing digital transformation are trying to improve performance, 'breakthrough solutions' are clearly needed to enable them to meet the challenge of innovation, rapid adoption and digital business transformation. Though these are some of the most prevalent challenges cited by businesses maintaining their digital transformation initiatives, countless other bottlenecks exist. To be armed to meet the challenges posed by digital transformation and remain competitive, organizations have to create an environment that enables agility and flexibility of the internal workflows and customer-facing operations that deliver quality. Many organizations need to gain access to approaches, methods and tools to transform and innovate better (higher quality) and faster (more efficient) so that they can generate new growth opportunities and create a significant competitive advantage.

1.4 Some basic tools

In SPC numbers and information will form the basis for decisions and actions, and a thorough data recording system is essential. In addition to the basic elements of a management system, which will provide a framework for recording data, there exists a set of 'tools' that may be applied to interpret fully and derive maximum use of the data. The simple methods listed below will offer any organization a means of collecting, presenting and analysing most of its data:

- Process flowcharting – What is done?
- Check sheets/tally charts – How often is it done?
- Histograms – What does the variation look like?
- Graphs – Can the variation be represented in a time series?
- Pareto analysis – Which are the big problems?
- Cause and effect analysis and brainstorming – What causes the problems?
- Scatter diagrams – What are the relationships between factors?
- Control charts – Which variations to control and how?

A pictorial example of each of these methods is given in Figure 1.4. A full description of the techniques, with many examples, will be given in

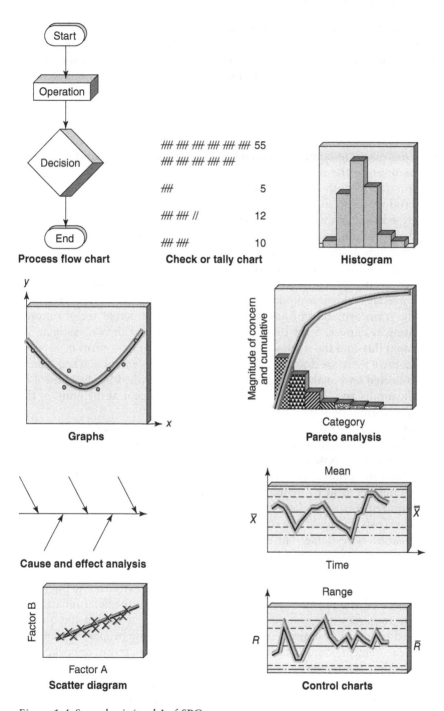

Figure 1.4 Some basic 'tools' of SPC

subsequent chapters. These are written assuming that the reader is neither a mathematician nor a statistician, and the techniques will be introduced through practical examples, where possible, rather than from a theoretical perspective. These techniques are still very applicable, even in our modern world of social media and artificial intelligence.

Chapter highlights

- Organizations compete on quality, delivery and price. Quality is defined as meeting the requirements of the customer. The supplier–customer interface is both internal and external to organizations.

- Product inspection is not the route to good quality management. Start by asking 'Can we do the job correctly?' – and not by asking 'Have we done the job correctly?' – not detection but prevention and control. Detection is costly and neither efficient nor effective. Prevention is the route to successful quality management.

- We need a process to ensure that we can and will continue to do it correctly – this is a model for control. Everything we do is a process – the transformation of any set of inputs into a different set of outputs using resources. Start by defining the process and then investigate its capability and the methods to be used to monitor or control it.

- Control ('Are we doing the job correctly?') is only possible when data is collected and analysed, so the outputs are controlled by the control of the inputs and the process. The latter can only occur at the point of the transformation – then the quality is assured.

- There are two distinct aspects of quality – design and conformance to design. Design is how well the product or service measures against its stated purpose or the specification. Conformance is the extent to which the product or service achieves the specified design. Start quality management by defining the requirement of the customer, keep the requirements up to date.

- The costs of quality need to be managed so that their effect on the business is desirable. The measurement of quality-related costs provides a powerful tool to highlight problem areas and monitor management performance.

- Quality-related costs are made up of failure (both external and internal), appraisal and prevention. Prevention costs include the determination of the requirements, planning, a proper management system for quality and training. Appraisal costs are incurred to allow proper verification, measurement, vendor ratings, etc. Failure includes scrap, rework, reinspection, waste, repair, warranty, complaints, returns and the associated loss of goodwill, among actual and potential customer. Quality-related costs, when measured from perfection, are seldom less than 10 per cent of sales value.

- The route to improved design, increased conformance and reduced costs is the use of statistically based methods in decision making within a framework of 'total quality'.
- SPC includes a set of tools for managing processes, and determining and monitoring the quality of the output of an organization. It is also a strategy for reducing variation in products, deliveries, processes, materials, attitudes and equipment. The question which needs to be asked continually is 'Could we do the job better?'
- SPC exists because there is, and will always be, variation in the characteristics of materials, articles, services, people. Variation has to be understood and assessed in order to be managed.
- Organizations need to gain access to approaches, methods and tools to transform and innovate better (higher quality) and faster (more efficient) so that they can generate new growth opportunities and create a significant competitive advantage in the digital age.
- There are some basic SPC tools. These are: process flowcharting (what is done); check sheets/tally charts (how often it is done); histograms (pictures of variation); graphs (pictures of variation with time); Pareto analysis (prioritizing); cause and effect analysis (what cause the problems); scatter diagrams (exploring relationships); control charts (monitoring variation over time). An understanding of the tools and how to use them requires no prior knowledge of statistics.

Note

1 This system for process capability and control is based on the late Frank Price's very practical framework for thinking about quality in manufacturing: Can we make it OK? Are we making it OK? Have we made it OK? Could we make it better?, which he presented in his excellent book, *Right First Time* (1984).

References and further reading

Beckford, J. (2017) *Quality: A Critical Introduction*, 4th edn, Routledge, London, UK.

Deming, W.E. (1986) *Out of the Crisis*, MIT, Cambridge MA, USA.

Deming, W.E. (1993) *The New Economics*, MIT, Cambridge, MA, USA

Feigenbaum, A.V. (1991) *Total Quality Control*, 3rd edn, McGraw-Hill, New York, USA.

Garvin, D.A. (1988) *Managing Quality*, Free Press, New York, USA.

Hammer, M. and Champy, J. (1993) *Re-engineering the Corporation: A Manifesto for Business Evolution*, Nicholas Brealey, London, UK.

Ishikawa, K. (translated by David J. Lu) (1985) *What Is Total Quality Control?: The Japanese Way*, Prentice Hall, Englewood Cliffs NJ, USA.

Joiner, B.L. (1994) *Fourth Generation Management: the New Business Consciousness*, McGraw-Hill, New York, USA.

Juran, J.M. and De Feo, J.A. (2010) *Quality Handbook*, 6th edn, McGraw-Hill, New York, USA.

Oakland, J.S. (2014) *Total Quality Management and Operational Excellence*, 4th edn, Routledge, Oxford, UK.

Price, F. (1984) *Right First Time*, Gower, Aldershot, UK.

Shewhart, W.A. (1931) *Economic Control of Manufactured Product*, Van Nostrand, New York, USA. (ASQ, 1980).

Shewhart, W.A. and Deeming, W.E. (1939) *Statistical Methods from the Viewpoint of Quality Control*, Van Nostrand, New York, USA.

Discussion questions

1 It has been argued that the definition of product quality as 'fitness for intended purpose' is more likely to lead to commercial success than is a definition such as 'conformance to specification'. Discuss the implication of these alternative definitions for the quality function within a manufacturing enterprise.

2 'Quality' cannot be inspected into a product nor can it be advertised in, it must be designed and built in. Discuss this statement in its application to a service providing organization.

3 Explain the following:
   (a) the difference between quality of design and conformance;
   (b) quality-related costs.

4 E-mail:
   To: Quality Manager;
   From: Managing Director;
   SUBJECT: Quality Costs
   Below are the newly prepared quality costs for the last two quarters:

	Last quarter last year	First quarter this year
Scrap and Rework	$312,000	$624,000
Customer returns/warranty	$524,000	$204,000
Total	$836,000	$828,000

   In spite of agreeing to your request to employ further inspection staff from January to increase finished product inspection to 100 per cent, you will see that overall quality costs have shown no significant change. I look forward to receiving your comments on this.
   Discuss the issues raised by the above memorandum.

5 You are a management consultant and have been asked to assist a manufacturing company in which 15 per cent of the work force are final product inspectors. Currently, 20 per cent of the firm's output has to

be reworked or scrapped. Write a report to the Managing Director of the company explaining, in general terms, how this situation arises and what steps may be taken to improve it.

6   Using a simple model of a process, explain the main features of a process approach to quality management and improvement.

7   Explain a system for SPC which concentrates attention on prevention of problems rather than their detection.

8   What are the basic tools of SPC and their main application areas?

2 Understanding the process

Objectives

- To further examine the concept of process management and improving customer satisfaction.
- To introduce a systematic approach to: defining customer–supplier relationships; defining processes; standardizing processes; designing/modifying processes; improving processes.
- To describe the various techniques of block diagramming and flow-charting and to show their use in process mapping, examination and improvement.
- To position process mapping and analysis in the context of business process re-design/re-engineering (BPR).

2.1 Improving customer satisfaction through process management

An approach to improvement based on process alignment, starting with the organization's vision and mission, analysing its critical success factors (CSFs), and moving on to the key or core processes is the most effective way to engage the people in an enduring change process. In addition to the knowledge of the business as a whole, which will be brought about by an understanding of the mission→CSF→process breakdown links, certain tools, techniques and interpersonal skills will be required for good communication around the processes that are managed by the systems. These are essential for people to identify and solve problems as teams, and form the components of the model for statistical process control (SPC) introduced in Chapter 1 and described in detail in Oakland's *Total Quality Management and Operational Excellence*, 4th edn.

Most organizations have functions: experts of similar backgrounds are grouped together in a pool of knowledge and skills capable of completing any task in that discipline. This focus, however, can foster a 'vertical' view and limits the organization's ability to operate effectively. Barriers to customer satisfaction can evolve, resulting in unnecessary work, restricted

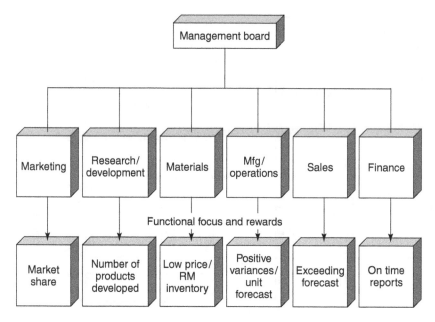

Figure 2.1 Typical functional operation

sharing of resources, limited synergy between functions, delayed development time and no clear understanding of how one department's activities affect the total process of attaining customer satisfaction. Managers can remain tied to managing singular functions, with rewards and incentives for their narrow missions, inhibiting a shared external customer perspective (Figure 2.1).

Concentrating on managing processes breaks down these internal barriers and encourages the entire organization to work as cross-functional teams with a shared horizontal view of the business. It requires shifting the work focus from managing functions to managing processes. Process owners, accountable for the success of major cross-functional processes, are charged with ensuring that employees understand how their individual work processes affect customer satisfaction. The interdependence between one group's work and the next becomes quickly apparent when all understand who the customer is and the value they add to the entire process of satisfying that customer (Figure 2.2).

The core business processes describe what actually is or needs to be done so that the organization meets its CSFs. If the core processes are identified, the questions will come thick and fast: Is the process currently carried out? By whom? When? How frequently? With what performance and how well compared with competitors? The answering of these will

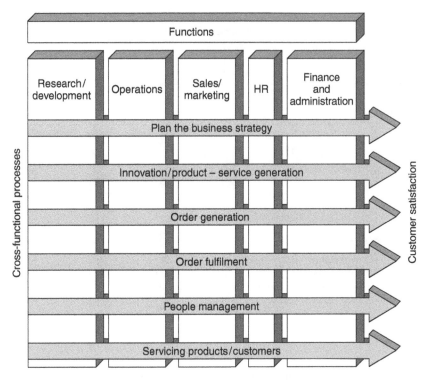

Figure 2.2 Cross-functional approach to managing core processes

force process ownership into the business. The process owners should engage in improvement activities which may lead through process analysis, self-assessment and benchmarking to identifying the improvement opportunities for the business. The processes must then be prioritized into those that require continuous improvement, those that require re-engineering or re-design, and those that require a complete re-think or visioning of the ideal process. The outcome should be a set of 'key processes' that receive priority attention for re-design or re-engineering.

Performance measurement of all processes is necessary to determine progress so that the vision, goals, mission and CSFs may be examined and reconstituted to meet new requirements for the organization and its customers (internal and external). This whole approach forms the basis of a 'Total organizational excellence'[1] implementation framework (Figure 2.3).

Once an organization has defined and mapped out the core processes, people need to develop the skills to understand how the new process structure will be analysed and made to work. The very existence of new process teams with new goals and responsibilities will force the organization into a learning phase. These changes should foster new attitudes and behaviours.

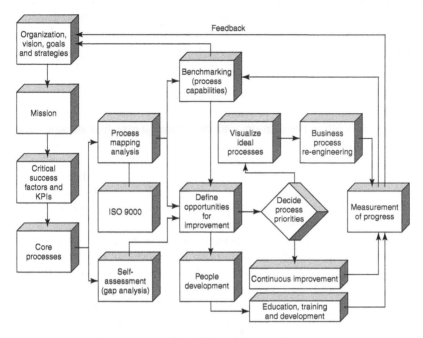

Figure 2.3 Total organizational excellence framework

2.2 Information about the process

One of the initial steps to understand or improve a process is to gather information about the important activities so that a 'dynamic model' – a process map or flowcharts – may be constructed. Process mapping creates a picture of the activities that take place in a process. One of the greatest difficulties here, however, is deciding how many tasks and how much detail should be included. When initially mapping out a process, people often include too much detail or too many tasks. It is important to consider the sources of information about processes and the following aspects should help to identify the key issues:

- defining supplier–customer relationships;
- defining the process;
- standardizing processes;
- designing a new process or modifying an existing one;
- identifying complexity or opportunities for improvement.

Defining supplier–customer relationships

Since quality is defined by the customer, changes to a process are usually made to increase satisfaction of internal and external customers, whether that is to do with quality, timeliness or costs. At many stages in a process, it is necessary for 'customers' to determine their needs or give their reaction to proposed changes in the process. For this it is often useful to describe the edges or boundaries of the process – where does it start and stop? This is accomplished by formally considering the inputs and outputs of the process as well as the suppliers of the inputs and the customers of the outputs – the 'static model' (SIPOC). Figure 2.4 is a form that can be used to provide focus on the boundary of any process and to list the inputs and suppliers to the process, as well as the outputs and customers. These lists do not have to be exhaustive, but should capture the important aspects of the process.

The form asks for some fundamental information about the process itself, such as the name and the 'owner'. The owner of a process is the person at the lowest level in the organization that has the authority to change the process. The owner has the responsibility of organizing and perhaps leading a team to make improvements.

Documentation of the process, perhaps through the use of flowcharts, aids the identification of the customers and suppliers at each stage. It is sometimes surprisingly difficult to define these relationships, especially for internal suppliers and customers. Some customers of an output may also have supplied some of the inputs, and there are usually a number of customers for the same output. For example, information on location and amount of stock or inventory may be used by production planners, material handlers, purchasing staff and accountants.

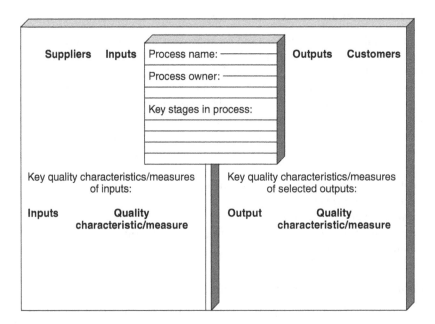

Figure 2.4 Describing the boundary of a process: SIPOC

Defining the process

Many processes in need of improvement are not well defined. A production engineering department may define and document in great detail a manufacturing process, but have little or no documentation on the process of design itself. If the process of design is to be improved, then knowledge of that process will be needed to make it tangible.

The first time any process is examined, the main focus should be to capture everyone's current knowledge of the process. A common mistake is to have a technical process 'expert', usually a technologist, engineer or supervisor, describe the process and then show it to others for their comment. The first information about the process should come instead from a brainstorming session of the people who actually operate or use the process, day in and day out. The technical experts, managers and supervisors should refrain from interjecting their 'ideas' until towards the end of the session. The resulting description will be a reflection of how the process actually works. During this initial stage, the concept of what the process could or should be is distracting to the main purpose of the exercise. These ideas and concepts should be discussed later.

Flowcharts (see Section 2.3) are important to study manufacturing processes, but they are particularly important for non-manufacturing processes. Because of the lack of documentation of administrative and service processes, it is sometimes difficult to reach agreement on the flowcharts for such a process. If this is the case, a first draft of a process map can be circulated to others who are knowledgeable of the process to seek their suggestions. Often, simply putting a team together to define the process using flowcharts will result in some obvious suggestions for improvement. This is especially true for non-manufacturing processes. There are available, of course, many excellent software packages that can aid the collection and presentation of process information but these should not be used in place of the vital work needed by the process team.

Standardizing processes

A significant source of variation in many processes is the use of different methods and procedures by those working in the process. This is caused by the lack of documented, standardized processes, inadequate training or inadequate supervision. Flowcharts are useful for identifying parts of the process where varying procedures are being used. They can also be used to establish a standard process to be followed by all. There have been many cases where standard procedures, developed and followed by operators, with the help of supervisors and technical experts, have resulted in a significant reduction in the variation of the outcomes.

Designing or modifying an existing process

Once process maps have been developed, those knowledgeable in the operation of the process should look for obvious areas of improvement or modification. It may be that steps, once considered necessary, are no longer needed. Time should not be wasted improving an activity that is not worth doing in the first place. Before any team proceeds with its efforts to improve a process, it should consider how the process should be designed from the beginning, and 'assumption or rule-busting' approaches are often required. Flowcharts of the new recommended process, compared to the existing process, will assist in identifying areas for improvement. Flowcharts can also serve as the documentation of a new process, helping those designing the process to identify weaknesses in the design and prevent problems once the new process is put into use.

Identifying complexity or opportunities for improvement

In any process there are many opportunities for things to go wrong and, when they do, what may have been a relatively simple activity can become quite complex. The failure of an airline computer used to document reservations, assign seats and print tickets can make the usually simple task of assigning a seat to a passenger a very difficult one. Documenting the steps in the process, identifying what can go wrong and indicating the increased complexity when things do go wrong will identify opportunities for improving quality and increasing productivity.

2.3 Process mapping and flowcharting

In the systematic planning or examination of any process, whether it is a clerical, manufacturing or managerial activity, it is necessary to record the series of events and activities, stages and decisions in a form that can be easily understood and communicated to all. If improvements are to be made, the facts relating to the existing method should be recorded first. The statements defining the process will lead to its understanding and provide the basis of any critical examination necessary for the development of improvements. It is essential, therefore, that the descriptions of processes are accurate, clear and concise.

Process mapping and flowcharting are very important first steps in improving a process. The flowchart 'pictures' will assist an individual or team in acquiring a better understanding of the system or process under study than would otherwise be possible. Gathering this knowledge provides a graphic definition of the system and of the improvement effort. Process mapping is a communication tool that helps an individual or an improvement team understand a system or process and identify opportunities for improvement.

The usual method of recording and communicating facts is to write them down, but this is not suitable for recording the complicated processes that exist in any organization. This is particularly so when an exact record is required of a long process, and its written description would cover several pages requiring careful study to elicit every detail. To overcome this difficulty certain methods of recording have been developed and the most powerful of these are mapping and flowcharting. There are many different types of maps and flowcharts which serve a variety of uses. The classical form of flowcharting, as used in computer programming, can be used to document current knowledge about a process, but there are other techniques that focus on efforts to improve a process.

Figure 2.5 is a high level process map showing how raw material for a chemical plant was purchased, received, and an invoice for the material was paid. Before an invoice could be paid, there had to be a corresponding receiving report to verify that the material had in fact been received. The accounts department was having trouble matching receiving reports to the invoices because the receiving reports were not available or contained incomplete or incorrect information. A team was formed with members from the accounts, transportation, purchasing and production departments. At the early stage of the project, it was necessary to have a broad overview of the process, including some of the important outputs and some of the problems that could occur at each stage. The process map or block diagram in Figure 2.5 served this purpose. The sub-process activities or tasks are shown under each block.

Figure 2.6 is an example of a process diagram that incorporates another dimension by including the person or group responsible for performing

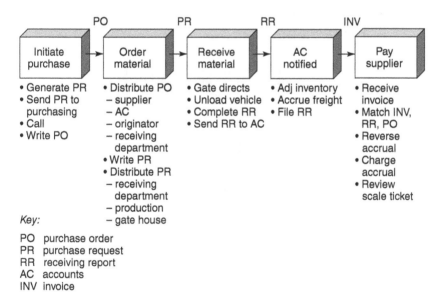

Figure 2.5 Acquisition of raw materials process map

the task in the column headings. This type of flowchart is helpful in determining customer–supplier relationships and is also useful to see where departmental boundaries are crossed and to identify areas where inter-departmental communications are inadequate. The diagram in Figure 2.6 was drawn by a team working on improving the administrative aspects of the 'sales' process. The team had originally drawn a map of the entire sales operation using a form similar to the one in Figure 2.5. After collecting and analysing some data, the team focused on the problem of not being able to locate specific information. Figure 2.6 was then prepared to focus the movement of information, in what are sometimes known as 'swim-lanes'.

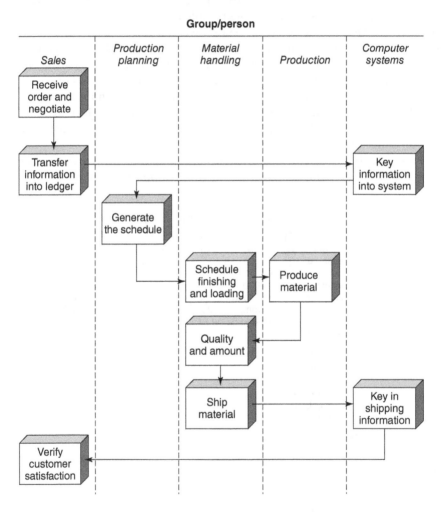

Figure 2.6 Process flowchart for sale of product: 'swim-lane' diagram

Classic flowcharts

Certain standard symbols are used on the 'classic' detailed flowchart and these are shown in Figure 2.7. The starting point of the process is indicated by a circle. Each processing step, indicated by a rectangle, contains a description of the relevant operation, and where the process ends is indicated by an oval. A point where the process branches because of a decision is shown by a diamond. A parallelogram contains useful information but it is not a processing step; a rectangle with a wavy bottom line refers to information or records, including computer files. The arrowed lines are used to connect symbols and to indicate direction of flow. For a complete description of the process, all operation steps (rectangles) and decisions (diamonds) should be connected by pathways from the start circle to the end oval. If the flowchart cannot be drawn in this way, the process is not fully understood.

Flowcharts are frequently used to communicate the components of a system or process to others whose skills and knowledge are needed in the improvement effort. Therefore, the use of standard symbols is necessary to remove any barrier to understanding or communications.

The purpose of the flowchart analysis is to learn why the current system/process operates in the manner it does, and to prepare a method for objective analysis. The team using the flowchart should analyse and document their findings to identify:

1   the problems and weaknesses in the current process system;
2   unnecessary steps or duplication of effort;
3   the objectives of the improvement effort.

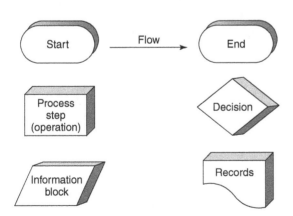

Figure 2.7 Flowcharting symbols

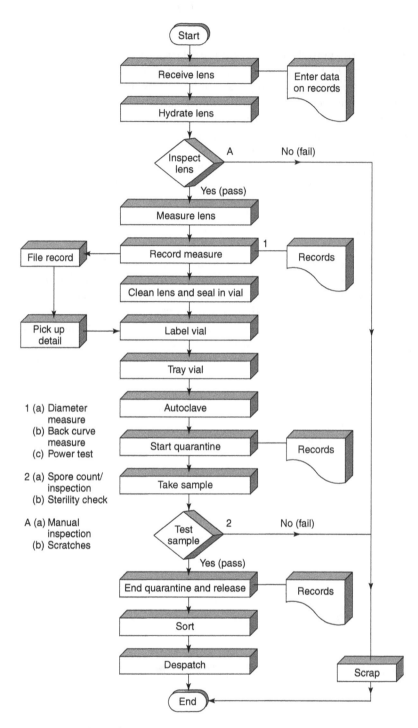

Figure 2.8 'Classic flowchart' for part of a contact lens conversion process

The flowchart techniques can also be used to study a simple system and how it would look if there were no problems. This method has been called 'imagineering' and is a useful aid to visualizing the improvements required.

It is a salutary experience for most people to sit down and try to 'draw' the flowchart for a process in which they are involved every working day. It is often found that:

1   the process flow is not fully understood;
2   a single person is unable to complete the flowchart without help from others.

The very act of flowcharting will improve knowledge of the various levels of the process, and will begin to develop the teamwork necessary to find improvements. In many cases the convoluted flow and octopus-like appearance of the charts will highlight unnecessary movement of people and materials and lead to suggestions for waste elimination.

Flowchart construction features

The boundaries of the process must be clearly defined before the flowcharting begins. This will be relatively easy if the outputs and customers, inputs and suppliers are clearly identified. All work connected with the process to be studied must be included. It is most important to include not only the formal, but also the informal activities. Having said that, it is important to keep the flowcharts as simple as possible.

Every route through a flowchart must lead to an end point and each process step must have one output line. Each decision diamond should have only two outputs which are labelled 'Yes' and 'No', which means that the questions must be phrased so that they may be answered in this way.

An example of a 'classic' flowchart for part of a contact lens conversion process is given in Figure 2.8. Clearly several of the operational steps could be flowcharted in turn to given further detail.

2.4 Process analysis

A process map or flowchart is a picture of the steps used in performing a function. This function can be anything from a chemical process step to accounting procedures, even preparing a meal. Flowcharts provide excellent documentation and are useful troubleshooting tools to determine how each step is related to the others. By reviewing the flowcharts it is often possible to discover inconsistencies and determine potential sources of variation and problems. For this reason, flowcharts are very useful in process improvement when examining an existing process to highlight problem areas. A group of people with knowledge about the process should follow the simple steps:

1   draw maps and flowcharts of the existing process, 'as is';
2   draw charts of the flow the process could or should follow, 'to be';
3   compare the two sets of charts to highlight the sources of the problems or waste, improvements required and changes necessary.

A critical examination of the first set of flowcharts is often required, using a questioning technique, which follows a well-established sequence to examine the:

purpose for which
place at which
sequence in which   }   the activities are undertaken
people by which
method by which

with a view to   {   eliminating,
                                    combining,
                                    rearranging   }   those activities.
                                    or
                                    simplifying

The questions that need to be answered in full are:

Purpose: What is actually done?   } *Eliminate*
             (or What is actually achieved?)   unnecessary parts
             Why is the activity necessary at all?   of the job.
             What else might be or should be done?

Place: Where is it being done?   } *Combine*
             Why is it done at that place?   where possible
             Where else might it or should it be done?   and/or
                                                                 rearrange
Sequence: When is it done?   operations
             Why is it done at that time?   for more
             When might or should it be done?   effective
                                                                 results or
People: Who does it?   waste reduction.
             Why is it done by that person/group?
             Who else might or should do it?

Method: How is it done?   } *Simplify*
             Why is it done in that way?   the
             How else might or should it be done?   operation.

Questions such as these, when applied to any process, will raise many points that will demand explanation and attention.

There is always room for improvement and one does not have to look far to find many real-life examples of what happens when a series of activities is started without being properly planned. Examples of much waste of time and effort can be found in factories, offices, schools and hospitals all over the world.

Development and re-design of the process

Process mapping or flowcharting and analysis is an important component of business process re-design (BPR). As described at the beginning of this chapter, any BPR project should begin with the mission for the organization and an identification of the CSFs and critical processes. Successful practitioners of BPR have made striking improvements in customer satisfaction, productivity and costs, in short periods of time, often by following these simple steps of process analysis using teamwork:

- *Document and map/flowchart the process* – making visible the invisible through mapping/flowcharting is the first crucial step that helps an organization see the way work really is done and not the way one thinks or believes it should be done. Seeing the process 'as is' provides a baseline from which to measure, analyse, test and improve.
- *Identify process customers and their requirements; establish effectiveness measurements* – recognizing that satisfying the external customer is a shared purpose, all internal and external suppliers need to know what customers expect and how well their processes meet customer expectations.
- *Analyse the process; rank problems and opportunities* – collecting supporting data allows an organization to weigh the value each task adds to the total process, to select areas for the greatest improvement and to spot unnecessary work and points of unclear responsibility.
- *Identify root cause of problems; establish control systems* – clarifying the source of errors or defects, particularly those that cross department lines, safeguards against quick-fix remedies and assures proper corrective action.
- *Develop implementation plans for recommended changes* – involving all stakeholders, including senior management, in approval of the action plan commits the organization to implementing change and following through the 'to be' process.
- *Pilot changes and revise the process* – validating the effectiveness of the action steps for the intended effect leads to reinforcement of the 'to be' process strategy and to new levels of performance.
- *Measure performance using appropriate metrics* – once the processes have been analysed in this way, it should be possible to develop metrics for measuring the performance of the 'to be' processes, sub-processes, activities and tasks. These must be meaningful in terms of the inputs and outputs of the processes, and in terms of the customers and suppliers.

Value stream mapping

Value stream mapping (VSM) is a process analysis technique often asso-ciated with so-called 'lean systems'. It studies the set of specific actions required to bring a product family from raw material to finished goods as per customer demand, concentrating on information management and physical transformation tasks.

The outputs of a VSM based study are a current state map, future state map and implementation plan for getting from the current to the future state. Using VSM it should be possible to bring the lead time closer and closer to the actual value added processing time by attacking the identified bottlenecks and constraints. Bottlenecks addressed could include long setup times, unreliable equipment, unacceptable first pass yield, or high work or process inventories.

The VSM technique has been central to the approach advocated by Womack and Jones from their original work with Toyota, and is still the mainstay of lean interventions. The principle is to describe the current pro-cess, looking at both physical/materials flows and information flows, in a highly visual format and to apply measures to each process step to identify the time taken and the cost involved. By also identifying which activities add value and which do not, it is possible to analyse the process from a value creating perspective and determine the potential gains from eliminat-ing non-value adding activity.

VSM may also be used in the generation of a Target Operating Model (TOM), and should come after analysis and investigative work to identify what can be done to reduce non-value adding activities. Figure 2.9 shows a VSM from a manufacturing context where the original time taken has been identified and then altered to show what has been made possible through understanding what can be changed.

In this example, the time taken for information to get from the customer to the manufacturer prior to the start of manufacture was 66 days and amounted to 1360 minutes of actual process time. Manufacturing time took 21 days (1075 minutes actual process time). By identifying these timings and then challenging how much of that time was actually spent adding value and what was not, it was possible to reduce the total lead time from 87 (66 + 21) days to 20 (15 + 5) – a saving of 67 days throughput time. Although this is from a manufacturing environment, the overall process (in terms of process steps and flow) is very similar to many service processes in that much of the non-value added time is spent 'in transit' waiting for things to be done – often on administrative tasks.

Although VSM would appear to be a simplistic tool/technique, it is important that the various activities making up the process are understood. In complex service environments there may be some apparent non-value adding steps that are essential to other processes that are in some way linked to the process being investigated. It is important that these dependencies

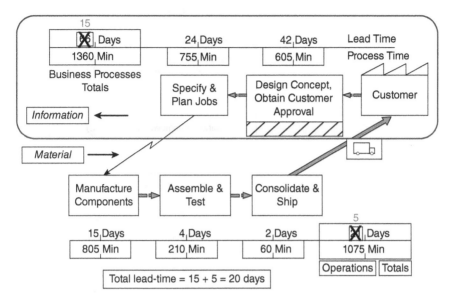

Figure 2.9 Value stream map

are understood and it is therefore essential that value stream maps are not created by individuals but by teams of people working in the process who know what is going on and can challenge each other's perspectives. Lean, as with any other improvement methodology, should be a cross-functional and team-based approach.

2.5 Statistical process control and process understanding

SPC has played a major part in the efforts of many organizations and industries to improve the competitiveness of their products, services, prices and deliveries. But what does SPC mean? A statistician may tell you that SPC is the application of appropriate statistical tools to processes for continuous improvement in quality of products and services, and productivity in the workforce. This is certainly accurate but, at the outset in many organizations, SPC would be better defined as a simple, effective approach to problem solving, and process improvement, or even stop producing chaos!

Every process has problems that need to be solved and SPC tools are universally applicable to everyone's job – manager, operator, secretary, chemist, engineer, whatever. Training in the use of these tools should be available to everyone within an organization, so that each 'worker' can contribute to the improvement of quality in his or her work. Usually, the

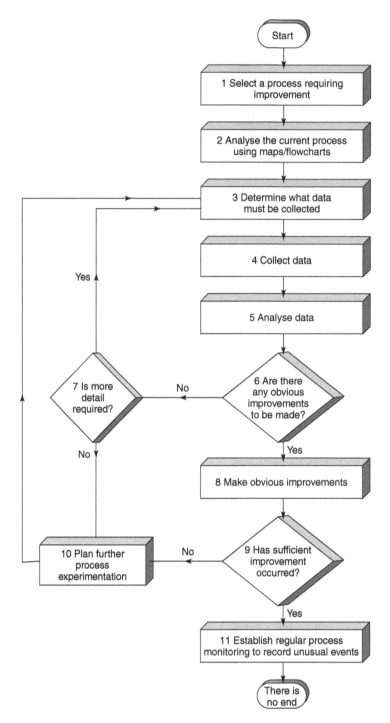

Figure 2.10 Step-by-step approach to developing or improving a process

technical people are the major focus of training in SPC, with concentration on the more technical tools, such as control charts. The other simpler basic tools, such as flowcharts, cause and effect diagrams, check sheets and Pareto charts, however, are well within the capacity of all employees.

Simply teaching individual SPC tools to employees is not enough. Making a successful transition from classroom or on-line examples to on-the-job application is the key to successful SPC implementation and problem solving. With the many tools available, the employee often wonders which one to use when confronted with a quality problem. What is often lacking in SPC training is a simple step-by-step approach to developing or improving a process.

Such an approach is represented in the flowchart of Figure 2.10. This 'road map' for problem solving intuitively makes sense to most people, but its underlying feature is that each step has certain SPC techniques that are appropriate to use in that step. This should reduce the barriers to acceptance of SPC and greatly increase the number of people capable of using it.

The various steps in Figure 2.10 require the use of the basic SPC 'tool kit' introduced in Chapter 1 and which will be described in full in the remaining chapters of this book. This is essential if a systematic approach is to be maintained and satisfactory results are to be achieved. There are several benefits that this approach brings which include:

- There are no restrictions as to the type of problem selected, but the process originally tackled will be improved.
- Decisions are based on facts not opinions – a lot of the 'emotion' is removed from problems by this approach.
- The quality 'awareness' of the workforce increases because they are directly involved in the improvement process.
- The knowledge and experience potential of the people who operate the process is released in a systematic way through the investigative approach. They better understand that their role in problem solving is collecting and communicating the facts with which decisions are made.
- Managers and supervisors solve problems methodically, instead of by using a 'seat-of-the-pants' style. The approach becomes unified, not individual or haphazard.
- Communications across and between all functions are enhanced, due to the excellence of the SPC tools as modes of communication.

The combination of a systematic approach, SPC tools, and outside hand-holding assistance when required helps organizations make the difficult transition from learning SPC in the classroom to applying it in the real world. This concentration on applying the techniques rather than simply learning them will lead to successful problem solving and process improvement.

Chapter highlights

- Improvement should be based on process alignment, starting with the organization's mission statement, its CSFs and core processes.
- Creation of 'dynamic models' through mapping out the core processes will engage the people in an enduring change process.
- A systematic approach to process understanding includes: defining supplier/customer relationships; defining the process; standardizing the procedures; designing a new process or modifying an existing one; identifying complexity or opportunities for improvement. The boundaries of the process must be defined.
- Process mapping and flowcharting allows the systematic planning, description and examination of any process.
- There are various kinds of flowcharts, including block diagrams, person/function based charts, and 'classic' ones used in computer programming. Detailed flowcharts use symbols to provide a picture of the sequential activities and decisions in the process: start, operation (step), decision, information/record block, flow, end. The use of flowcharting to map out processes, combined with a questioning technique based on purpose (what/why?), place (where?), sequence (when?), people (who?) and method (how?) ensures improvements.
- Value stream mapping (VSM) studies the set of specific actions required to bring a product family from raw material to finished goods, as per customer demand, concentrating on information management and physical transformation tasks. The outputs of a VSM based study are a current state map, future state map and implementation plan for getting from the current to the future state.
- Business process re-design (BPR) uses process mapping and flowcharting and teamwork to achieve improvements in customer satisfaction and productivity by moving from the 'as is' to the 'to be' process.
- SPC is above all a simple, effective approach to problem solving and process improvement. Training in the use of the basic tools should be available for everyone in the organization. However, training must be followed up to provide a simple stepwise approach to improvement.
- The SPC approach, correctly introduced, will lead to decisions based on facts, an increase in quality awareness at all levels, a systematic approach to problem solving, release of valuable experience, and all round improvements, especially in communications.

Note

1 Oakland, J.S. (2001) *Total Organisational Excellence*, Butterworth-Heinemann, Oxford.

References and further reading

Becker, J., Kugeler, M. and Rosemann, M. (2011) *Process Management: A Guide for the Design of Business Processes*, Springer, Berlin, Germany.
Harrington, H.J. (1991) *Business Process Improvement*, McGraw-Hill, New York, USA.

Madison, D. (2005) *Process Mapping, Process Improvement, and Process Management*, Paton Professional, Chico CA, USA.

Oakland, J.S. (2014) *Total Quality Management and Operational Excellence*, 4th edn, Taylor & Francis, Oxford, UK.

Pyzdek, T. (1990) *Pyzdek's Guide to SPC, Vol. 1: Fundamentals*, ASQ Press, Milwaukee WI, USA.

Reijers, H.A., Mendling, J., La Rosa, M. and Dumas, M. (2013) *Fundamentals of Business Process Management*, Springer, Berlin, Germany.

Discussion questions

1   Outline the initial steps you would take first to understand and then to improve a process in which you work.

2   Construct a 'static model' or map of a process of your choice, which you know well. Make sure you identify the customer(s) and outputs, suppliers and inputs, how you listen to the 'voice of the customer' and hear the 'voice of the process'.

3   Describe in detail the technique of flowcharting to give a 'dynamic model' of a process. Explain all the symbols and how they are used together to create a picture of events. What improvements would be obtained by using value stream mapping?

4   What are the steps in a critical examination of a process for improvement? Flowchart these into a systematic approach.

3   Process data collection and presentation

Objectives

- To introduce the systematic approach to process improvement.
- To examine the types of data and how data should be recorded.
- To consider various methods of presenting data, in particular bar charts, histograms and graphs.

3.1 The systematic approach

If we adopt the definition of quality as 'meeting the customer requirements', we have already seen the need to consider the quality of design and the quality of conformance to design. To achieve quality therefore requires:

- an appropriate design;
- suitable resources and facilities (equipment, premises, cash, etc.);
- the correct materials;
- people, with their skills, knowledge and training;
- an appropriate process;
- sets of instructions;
- measures for feedback and control.

Already quality management has been broken down into a series of component parts. Basically this process simply entails narrowing down each task until it is of a manageable size. Considering the design stage, it is vital to ensure that the specification for the product or service is realistic. Excessive, unnecessary detail here frequently results in the specification being ignored, at least partially, under the pressures to contain costs. It must be reasonably precise and include some indication of priority areas. Otherwise it will lead to a product or service that is unacceptable to the market. A systematic monitoring of product/service performance should lead to better and more realistic specifications. That is not the same thing as adding to the volume or detail of the documents.

The 'narrowing-down' approach forces attention to be focused on one of the major aspects of quality – the conformance or the ability to provide products or services consistently to the design specification. If all the suppliers in a chain adequately control their processes, then the product/service at each stage will be of the specified quality.

This is a very simple message which cannot be over-stated, but some manufacturing companies still employ a large inspectorate, including many who devote their lives to sorting out the bad from the good, rather than tackling the essential problem of ensuring that the production process remains in control and capable. The role of the 'inspector' should be to check and audit the systems of control, to advise, calibrate and, where appropriate, to undertake complex measurements or assessments. Quality can be controlled only at the point of manufacture or service delivery, it cannot be elsewhere.

In applying a systematic approach to process control and capability, the basic rules are:

- No process without *data collection.*
- No data collection without *analysis.*
- No analysis without *decision.*
- No decision without *action* (which can include no action necessary).

Data collection

If data are not carefully and systematically recorded, especially at the point of manufacture or operation, they cannot be analysed and put to use. Information recorded in a suitable way enables the magnitude of variations and trends to be observed. This allows conclusions to be drawn concerning errors, process capability, vendor ratings, risks, etc. Numerical data are often not recorded, even though measurements have been taken – a simple tick or initials is often used to indicate 'within specifications', but it is almost meaningless. The requirement to record the actual observation (the reading on a measured scale, or the number of times things occurred), can have a marked effect on the reliability of the data. For example, if a result is only just outside a specified tolerance, it is tempting to record another 'tick,' but the actual recording of a false figure is much less likely. The value of this increase in the reliability of the data when recorded properly should not be under-estimated. The practice of recording a result only when it is outside specification is also not recommended, since it ignores the variation going on within the tolerance limits which, hopefully, makes up the largest part of the variation and, therefore, contains the largest amount of information.

Analysis, decision, action

The tools of the 'narrowing-down' approach are a wide range of simple, yet powerful, problem-solving and data-handling techniques, which should form a part of the *analysis–decision–action* chain with all processes. These include:

- process mapping and flowcharting (Chapter 2);
- check sheets/tally charts;
- bar charts/histograms;
- graphs;
- pareto analysis (Chapter 11);
- cause and effect analysis (Chapter 11);
- scatter diagrams (Chapter 11);
- control charts (Chapters 5–9 and 12);
- stratification (Chapter 11).

3.2 Data collection

Data should form the basis for analysis, decision and action, and their form and presentation will obviously differ from process to process. Information is collected to discover the actual situation. It may be used as a part of a product or process control system and it is important to know at the outset what the data are to be used for. For example, if a problem occurs in the amount of impurity present in a product that is manufactured continuously, it is not sufficient to take only one sample per day to find out the variations between – say – different operator shifts. Similarly, in comparing errors produced by two accounting procedures, it is essential to have separate data from the outputs of both. These statements are no more than common sense, but it is not unusual to find that decisions and actions are based on misconceived or biased data. In other words, full consideration must be given to the reasons for collecting data, the correct sampling techniques and stratification. The methods of collecting data and the amount collected must take account of the need for information and not the ease of collection; there should not be a disproportionate amount of a certain kind of data simply because it can be collected easily.

Types of data

Numeric information will arise from both counting and measurement.

Data that arise from counting can occur only in discrete steps. There can be only 0, 1, 2, etc., defectives in a sample of 10 items, there cannot be 2.68 defectives. The number of defects in a length of cloth, the number of typing errors on a page, the presence or absence of a member of staff are all called *attributes*. As there is only a two-way or binary classification, attributes give rise to discrete data, which necessarily varies in steps.

Data that arise from measurements can occur anywhere at all on a continuous scale and are called *variable* data. The weight of a tablet, share prices, time taken for a rail journey, age, efficiency, and most physical dimensions, are all variables, the measurement of which produces continuous data. If variable data were truly continuous, they could take any value within a given range without restriction. However, owing to the limitations of measurement, all data vary in small jumps, the size of which is determined by the instruments in use.

The statistical principles involved in the analysis of whole numbers are not usually the same as those involved in continuous measurement. The theoretical background necessary for the analysis of these different types of data will be presented in later chapters.

Recording data

After data are collected, they are analysed and useful information is extracted through the use of statistical methods. It follows that data should be obtained in a form that will simplify the subsequent analysis. The first basic rule is to plan and construct the pro formas, paperwork or computer systems for data collection. This can avoid the problems of tables of numbers, the origin and relevance of which has been lost or forgotten. It is necessary to record not only the purpose of the observation and its characteristics, but also the date, the sampling plan, the instruments used for measurement, the method, the person or technology collecting the data and so on. Computers play an important role in both establishing and maintaining the format for data collection, of course.

Date	Percentage impurity					Week total	Week average
	15th	16th	17th	18th	19th		
Time							
8 a.m.	0.26	0.24	0.28	0.30	0.26	1.34	0.27
10 a.m.	0.31	0.33	0.33	0.30	0.31	1.58	0.32
12 noon	0.33	0.33	0.34	0.31	0.31	1.62	0.32
2 p.m.	0.32	0.34	0.36	0.32	0.32	1.66	0.33
4 p.m.	0.28	0.24	0.26	0.28	0.27	1.33	0.27
6 p.m.	0.27	0.25	0.24	0.28	0.26	1.30	0.26
Day total	1.77	1.73	1.81	1.79	1.73		
Day average	0.30	0.29	0.30	0.30	0.29	8.83	0.29
Operator	A. Ridgewarth						

Week commencing 15 February

Figure 3.1 Data collection for impurity in a chemical process

Data should be recorded in such a way that they are easy to use. Calculations of totals, averages and ranges are often necessary and the format used for recording the data can make these easier. For example, the format and data recorded in Figure 3.1 have clearly been designed for a situation in which the daily, weekly and grand averages of a percentage impurity are required. Columns and rows have been included for the totals from which the averages are calculated. Fluctuations in the average for a day can be seen by looking down the columns, whilst variations in the percentage impurity at the various sample times can be reviewed by examining the rows.

Careful design of data collection will facilitate easier and more meaningful analysis. A few simple steps in the design are listed below:

- Agree on the exact event to be observed – ensure that everyone is monitoring the same thing(s).
- Decide both how often the events will be observed (the frequency) and over what total period (the duration).
- Design a draft format – keep it simple and leave adequate space for the entry of the observations.
- Tell the observers how to use the format and put it into trial use – be careful to note their initial observations, let them know that it will be reviewed after a period of use and make sure that they accept that there is adequate time for them to record the information required.
- Make sure that the observers record the actual observations and not a 'tick' to show that they made an observation.
- Review the format with the observers to discuss how easy or difficult it has proved to be in use, and also how the data have been of value after analysis.

All that is required is some common sense. Who cannot quote examples of forms or spreadsheets that are almost incomprehensible, including typical forms from government departments and some service organizations? The author recalls a whole improvement programme devoted to the re-design of forms used in a bank – a programme which led to large savings and increased levels of customer satisfaction.

3.3 Bar charts and histograms

Every day, throughout the world, in offices, factories, on public transport, shops, schools and so on, data are being collected and accumulated in various forms: data on prices, quantities, exchange rates, numbers of defective items, lengths of articles, temperatures during treatment, weight, number of absentees, etc. Much of the potential information contained in this data may lie dormant or not be used to the full, and often because it makes little sense in the form presented. Consider, as an example, the data shown in Table 3.1 which refer to the diameter of pistons. It is impossible to visualize

Table 3.1 Diameters of pistons (mm) – raw data

56.1	56.0	55.7	55.4	55.5	55.9	55.7	55.4
55.1	55.8	55.3	55.4	55.5	55.5	55.2	55.8
55.6	55.7	55.1	56.2	55.6	55.7	55.3	55.5
55.0	55.6	55.4	55.9	55.2	56.0	55.7	55.6
55.9	55.8	55.6	55.4	56.1	55.7	55.8	55.3
55.6	56.0	55.8	55.7	55.5	56.0	55.3	55.7
55.9	55.4	55.9	55.5	55.8	55.5	55.6	55.2

the data as a whole. The eye concentrates on individual measurements and, in consequence, a large amount of study will be required to give the general picture represented. A means of visualizing such a set of data is required.

Look again at the data in Table 3.1. Is the average diameter obvious? Can you tell at a glance the highest or the lowest diameter? Can you estimate the range between the highest and lowest values? Given a specification of 55.0 ± 1.00 mm, can you tell whether the process is capable of meeting the specification, and is it doing so? Few people can answer these questions quickly, but given sufficient time to study the data all the questions could be answered. If the observations are placed in sequence or ordered from the highest to the lowest diameters, the problems of estimating the average, the highest and lowest readings, and the range (a measure of the spread of the results) would be simplified. The reordered observations are shown in Table 3.2. After only a brief examination of this table it is apparent that the lowest value is 55.0 mm, that the highest value is 56.2 mm and hence that the range is 1.2 mm (i.e. 55.0–56.2 mm). The average is probably around 55.6 or 55.7 mm and the process is not meeting the specification as three of the observations are greater than 56.0 mm, the upper tolerance.

Tally charts and frequency distributions

The tally chart and frequency distribution are alternative ordered ways of presenting data. To construct a tally chart data may be extracted from the original form given in Table 3.1 or taken from the ordered form of Table 3.2.

Table 3.2 Diameters of pistons ranked in order of size (mm)

55.0	55.1	55.1	55.2	55.2	55.2	55.3	55.3
55.3	55.3	55.4	55.4	55.4	55.4	55.4	55.4
55.5	55.5	55.5	55.5	55.5	55.5	55.5	55.6
55.6	55.6	55.6	55.6	55.6	55.6	55.7	55.7
55.7	55.7	55.7	55.7	55.7	55.7	55.8	55.8
55.8	55.8	55.8	55.8	55.8	55.9	55.9	55.9
55.9	56.0	56.0	56.0	56.0	56.1	56.1	56.2

Table 3.3 Tally sheet and frequency distribution of diameters of pistons (mm)

Diameter	Tally	Frequency
55.0	|	1
55.1	| |	2
55.2	| | |	3
55.3	| | | |	4
55.4	⊞⊞ |	6
55.5	⊞⊞ | |	7
55.6	⊞⊞ | |	7
55.7	⊞⊞ | | |	8
55.8	⊞⊞ |	6
55.9	⊞⊞	5
56.0	| | | |	4
56.1	| |	2
56.2	|	1
		Total 56

A scale over the range of observed values is selected and a tally mark is placed opposite the corresponding value on the scale for each observation. Every fifth tally mark forms a 'five-bar gate' which makes adding the tallies easier and quicker. The totals from such additions form the frequency distribution. A tally chart and frequency distribution for the data in Table 3.1 are illustrated in Table 3.3, which provides a pictorial presentation of the 'central tendency' or the average, and the 'dispersion' or spread or the range of the results. Clearly, all this could be done using appropriate software, such as *Minitab*.

Bar charts and column graphs

Bar charts and column graphs are the most common formats for illustrating comparative data. They are easy to construct and to understand. A bar chart is closely related to a tally chart – with the bars extending horizontally. Column graphs are usually constructed with the measured values on the horizontal axis and the frequency or number of observations on the vertical axis. Above each observed value is drawn a column, the height of which corresponds to the frequency. So the column graph of the data from Table 3.1 will look very much like the tally chart laid on its side (see Figure 3.2).

Like the tally chart, the column graphs show the lowest and highest values, the range, the centring and the fact that the process is not meeting the specification. It is also fairly clear that the process is potentially capable of achieving the tolerances, since the specification range is 2 mm, whilst the spread of the results is only 1.2 mm. Perhaps the idea of capability will be more apparent if you imagine the column graph of Figure 3.2 being moved

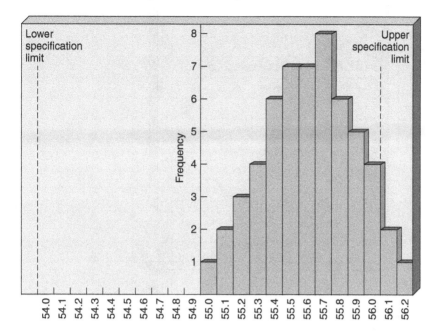

Figure 3.2 Column graph of data in Table 3.1: diameters of pistons

to the left so that it is centred around the mid-specification of 55.0 mm. If a process adjustment could be made to achieve this shift, while retaining the same spread of values, all observations would lie within the specification limits with room to spare.

As mentioned above, bar charts are usually drawn horizontally and can be lines or dots rather than bars, each dot representing a data point. Figure 3.3 shows a dot plot being used to illustrate the difference in a process before and after an operator was trained on the correct procedure to use on a milling machine. In Figure 3.3a the incorrect method of processing caused a 'bimodal' distribution – one with two peaks. After training, the pattern changed to the single peak or 'unimodal' distribution of Figure 3.3b. Notice how the graphical presentation makes the difference so evident.

Group frequency distributions and histograms

In the examples of bar charts given above, the number of values observed was small. When there are a large number of observations, it is often more useful to present data in the condensed form of a grouped frequency distribution. The data shown in Table 3.4 are the thickness measurements of pieces of silicon delivered as one batch. Table 3.5 was prepared by selecting cell boundaries to form equal intervals, called groups or cells, and placing a tally mark in the appropriate group for each observation.

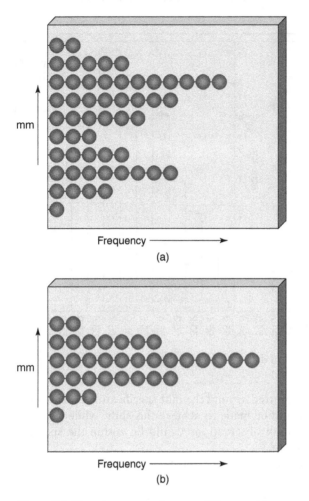

Figure 3.3 Dot plot: output from a milling machine

Table 3.4 Thickness measurements on pieces of silicon (mm × 0.001)

790	1170	970	940	1050	1020	1070	790
1340	710	1010	770	1020	1260	870	1400
1530	1180	1440	1190	1250	940	1380	1320
1190	750	1280	1140	850	600	1020	1230
1010	1040	1050	1240	1040	840	1120	1320
1160	1100	1190	820	1050	1060	880	1100
1260	1450	930	1040	1260	1210	1190	1350
1240	1490	1490	1310	1100	1080	1200	880
820	980	1620	1260	760	1050	1370	950

1220	1300	1330	1590	1310	830	1270	1290
1000	1100	1160	1180	1010	1410	1070	1250
1040	1290	1010	1440	1240	1150	1360	1120
980	1490	1080	1090	1350	1360	1100	1470
1290	990	790	720	1010	1150	1160	850
1360	1560	980	970	1270	510	960	1390
1070	840	870	1380	1320	1510	1550	1030
1170	920	1290	1120	1050	1250	960	1550
1050	1060	970	1520	940	800	1000	1110
1430	1390	1310	1000	1030	1530	1380	1130
1110	950	1220	1160	970	940	880	1270
750	1010	1070	1210	1150	1230	1380	1620
1760	1400	1400	1200	1190	970	1320	1200
1460	1060	1140	1080	1210	1290	1130	1050
1230	1450	1150	1490	980	1160	1520	1160
1160	1700	1520	1220	1680	900	1030	850

Table 3.5 Grouped frequency distribution: measurements on silicon pieces

Cell boundary	Tally						Frequency	Per cent frequency
500–649	||						2	1.0
650–799	╫╫	||||					9	4.5
800–949	╫╫	╫╫	╫╫	╫╫	|		21	10.5
950–1099	╫╫ ╫╫	╫╫ ╫╫	╫╫ ╫╫	╫╫ ╫╫	╫╫	╫╫	50	25.0
1100–1249	╫╫ ╫╫	╫╫ ╫╫	╫╫ ╫╫	╫╫ ╫╫	╫╫	╫╫	50	25.0
1250–1399	╫╫ ╫╫	╫╫ |||	╫╫	╫╫	╫╫	╫╫	38	19.0
1400–1549	╫╫	╫╫	╫╫	╫╫	|		21	10.5
1550–1699	╫╫	||					7	3.5
1700–1849	||						2	1.0

In the preparation of a grouped frequency distribution and the corresponding histogram, it is advisable to:

1 make the cell intervals of equal width;
2 choose the cell boundaries so that they lie between possible observations;
3 if a central target is known in advance, place it in the middle of a cell interval;
4 determine the approximate number of cells from Sturgess rule, which can be represented as a mathematical equation but is much simpler if use is made of Table 3.6.

Table 3.6 Sturgess rule

Number of observations	Number of intervals
0–9	4
10–24	5
25–49	6
50–89	7
90–189	8
190–399	9
400–799	10
800–1599	11
1600–3200	12

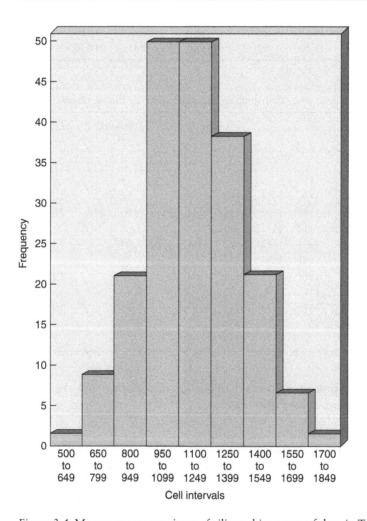

Figure 3.4 Measurements on pieces of silicon: histogram of data in Table 3.4

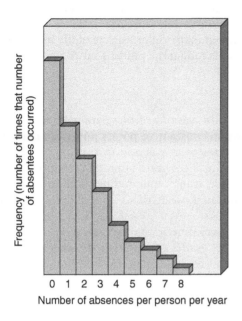

Number of absences per person per year

Figure 3.5 Absenteeism in a small office

The midpoint of a cell is the average of its two boundaries. For example, the midpoint of the cell 475–524 is

$$(475 + 524)/2 = 500$$

The histogram derived from Table 3.5 is shown in Figure 3.4.

All the examples so far have been of histograms showing continuous data. However, numbers of defective parts, accidents, absentees, errors, etc., can be used as data for histogram construction. Figure 3.5 shows absenteeism in a small office which could often be zero. The distribution is skewed to the right – discrete data will often assume an asymmetrical form, so the histogram of absenteeism peaks at zero and shows only positive values.

Other examples of histograms will be discussed along with process capability and Pareto analysis in later chapters.

3.4 Graphs, run charts and other pictures

We have all come across graphs or run charts. Television presenters use them to illustrate the economic situation, newspapers use them to show trends in anything from average rainfall to the sales of mobile phones. Graphs can be drawn in many very different ways. The histogram is one type of graph but graphs also include pie charts, run charts and pictorial graphs. In all cases they are extremely valuable in that they convert tabulated data into a picture, thus revealing what is going on within a process, batches of product, customer returns, scrap, rework and many other aspects of life in manufacturing and service organizations, including the public sector.

Line graphs or run charts

In line graphs or run charts the observations of one parameter are plotted against another parameter and the consecutive points joined by lines. For example, the various defective rates over a period of time of two groups of workers are shown in Figure 3.6. Error rate is plotted against time for the two groups on the same graph, using separate lines and different plot symbols. We can read this picture as showing that Group B performs better than Group A.

Run charts can show changes over time so that we may assess the effects of new equipment, various people, grades of materials or other factors on the process. Graphs are also useful to detect patterns and are an essential part of control charts.

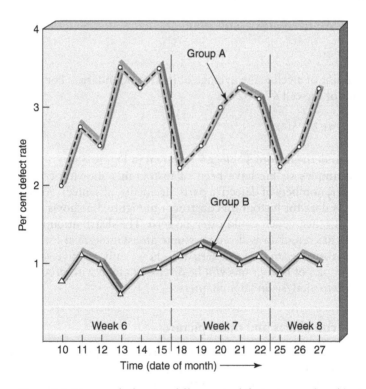

Figure 3.6 Line graph showing difference in defect rates produced by two groups of operatives

Pictorial graphs

Often, when presenting results, it is necessary to catch the eye of the reader. Pictorial graphs usually have high impact, because pictures or symbols of the item under observation are shown. Figure 3.7 shows the number of cars that have been the subject of warranty claims over a 12-month period.

Pie charts

Another type of graph is the pie chart in which much information can be illustrated in a relatively small area. Figure 3.8 illustrates an application of a pie chart in which the types and relative importance of defects in furniture are shown. From this it appears that defect D is the largest contributor. Pie charts applications are limited to the presentation of proportions since the whole 'pie' is normally filled.

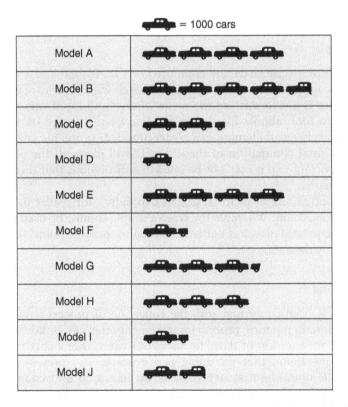

Figure 3.7 Pictorial graph showing the numbers of each model of car which have been repaired under warranty

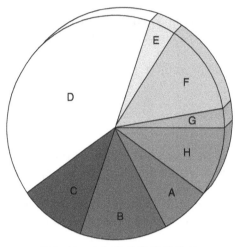

Defect types A, B, C, D, E, F, G, H

Figure 3.8 Pie chart of defects in furniture

The use of graphs

All graphs, except the pie chart, are composed of a horizontal and a vertical axis. The scale for both of these must be chosen with some care if the resultant picture is not to mislead the reader. Large and rapid variations can be made to look almost like a straight line by the choice of scale. Similarly, relatively small changes can be accentuated. In the pie chart of Figure 3.8 the total elimination of the defect D will make all the others look more important and it may not be immediately obvious that the 'pie' will then be smaller.

The inappropriate use of pictorial graphs can induce the reader to leap to the wrong conclusion. Whatever the type of graph, it must be used with care so that the presentation has not been chosen to 'prove a point' that is not supported by the data.

3.5 Conclusions

This chapter has been concerned with the collection of process data and their presentation. In practice, process improvement often can be advanced by the correct presentation of data. In numerous cases, over many years, the authors have found that recording performance, and presenting it appropriately, is often the first step towards an increased understanding of process behaviour by the people involved. The public display of the 'voice of the process' can result in renewed efforts being made by the operators of the processes.

There are many excellent software programmes that can perform data analysis and presentation, with many of the examples in this chapter included. Care must be taken when using IT in analysis and presentation that the quality of the data itself is maintained. The presentation of 'glossy' graphs and pictures may distort or cover up deficiencies in the data or its collection.

Chapter highlights

- Process improvement requires a systematic approach that includes an appropriate design, resources, materials, people, process and operating instructions.
- Narrow quality and process improvement activities to a series of tasks of a manageable size.
- The basic rules of the systematic approach are: no process without data collection, no data collection without analysis, no analysis without decision, no decision without action (which may include no action).
- Without records, analysis is not possible. Ticks and initials cannot be analysed. Record what is observed and not the fact that there was an observation, this makes analysis possible and also improves the reliability of the data recorded.
- The tools of the systematic approach include check sheets/tally charts, histograms, bar charts and graphs.
- There are two types of numeric data: variables that result from measurement, and attributes that result from counting.
- The methods of data collection and the presentation format should be designed to reflect the proposed use of data and the requirements of those charged with its recording. Ease of access is also required.
- Tables of figures are not easily comprehensible but sequencing data reveals the maximum and the minimum values. Tally charts and counts of frequency also reveal the distribution of the data – the central tendency and spread.
- Bar charts and column graphs are in common use and appear in various forms such as vertical and horizontal bars, columns and dots. Grouped frequency distribution or histograms are another type of bar chart of particular value for visualizing large amounts of data. The choice of cell intervals can be aided by the use of Sturgess rule.
- Line graphs or run charts are another way of presenting data as a picture. Graphs include pictorial graphs and pie charts. When reading graphs be aware of the scale chosen, examine them with care, and seek the real meaning – like statistics in general, graphs can be designed to mislead.
- Recording process performance and presenting the results reduce debate and act as a spur to action.
- Collect data, select a good method of presenting the 'voice of the process', and then present it. Use available IT and software for analysing and presenting data with care.

References and further reading

Crossley, M.L. (2000) *The Desk Reference of Statistical Quality Methods*, ASQ Press, Milwaukee WI, USA.

Ishikawa, K. (1982) *Guide to Quality Control*, Asian Productivity Association, Tokyo, Japan.

Oakland, J.S. (2014) *Total Quality Management & Operational Excellence, Text and Cases*, 4th edn, Butterworth-Heinemann, Oxford, UK.

Owen, M. (1993) *SPC and Business Improvement*, IFS Publications, Bedford, UK.

Discussion questions

1   Outline the principles behind a systematic approach to process improvement with respect to the initial collection and presentation of data.

2   Operators on an assembly line are having difficulties when mounting electronic components onto a printed circuit board. The difficulties include: undersized holes in the board, absence of holes in the board, oversized wires on components, component wires snapping on bending, components longer than the corresponding hole spacing, wrong components within a batch, and some other less frequent problems. Design a simple tally chart which the operators could be asked to use in order to keep detailed records. How would you make use of such records? How would you engage the interest of the operators in keeping such records?

3   Describe, with examples, the methods that are available for presenting information by means of charts, graphs, diagrams, etc.

4   The table below shows the recorded thicknesses of steel plates nominally 0.3 cm ± 0.01 cm. Plot a frequency histogram of the plate thicknesses, and comment on the result.

Plate thicknesses (cm)					
.2968	.2921	.2943	.3000	.2935	.3019
.2991	.2969	.2946	.2965	.2917	.3008
.3036	.3004	.2967	.2955	.2959	.2937
.2961	.3037	.2847	.2907	.2986	.2956
.2875	.2950	.2981	.1971	.3009	.2985
.3005	.3127	.2918	.2900	.3029	.3031
.3047	.2901	.2976	.3016	.2975	.2932
.3065	.3006	.3011	.3027	.2909	.2949
.3089	.2997	.3058	.2911	.2993	.2978
.2972	.2919	.2996	.2995	.3014	.2999

5   To establish a manufacturing specification for tablet weight, a sequence of 200 tablets was taken from the production stream and the weight of each tablet was measured. The frequency distribution is shown below.

State and explain the conclusions you would draw from this distribution, assuming the following:

(a) the tablets came from one process,
(b) the tablets came from two processes.

Measured weight of tablets	
Weight (gm)	Number of tablets
.238	2
.239	13
.240	32
.241	29
.242	18
.243	21
.244	20
.245	22
.246	22
.247	13
.248	3
.249	0
.250	1
.251	1
.252	0
.253	1
.254	0
.255	2
	200

Part II
Process variability

4 Variation

Understanding and decision making

Objectives

- To examine the traditional way in which managers look at data.
- To introduce the idea of looking at variation in the data.
- To differentiate between different causes of variation and between accuracy and precision.
- To encourage the evaluation of decision making with regard to process variation.

4.1 How some managers look at data

How do some managers look at data? Imagine the preparations in a production manager's office shortly before the monthly directors' meeting. David, the Production Director, is agitated and not looking forward to the meeting. Figures from the Drying Plant are down again and he is going to have to reprimand John, the Production Manager. David is surprised at the results and John's poor performance. He thought the complete overhaul of the rotary dryer scrubbers would have lifted the output of 2,4 D (the product) and that all that was needed was a weekly chastising of the production middle management to keep them on their toes and the figures up. Still, reprimanding people usually improved things, at least for the following week or so.

If David was not looking forward to the meeting, John was dreading it! He knew he had several good reasons why the drying figures were down but they had each been used a number of times before at similar meetings. He was looking for something new, something more convincing. He listed the old favourites: plant personnel absenteeism, their lack of training (due to never having time to take them off the job), lack of plant maintenance (due to the demand for output, output, output), indifferent material suppliers (the phenol that came in last week was brown instead of white!), late deliveries from suppliers of everything from plant filters to packaging materials (we had 20 tonnes of loose material in sacks in the Spray Dryer for 4 days last week, awaiting re-packing into the correct unavailable cartons). There were a host of other factors that John knew were outside his control, but it would all sound like whinging.

Table 4.1 Sales and production report, Year 6 Month 4

	Month 4 actual	Monthly target	% Difference	% Diff month 4 last year	Actual	YTD target	% Difference	YTD as % Difference (last YTD)
Sales								
Volume	505	530	−4.7	−10.1 (562)	2120	2120	0	+0.7 (2106)
On-time (%)	86	95	−9.5	−4.4 (90)	88	95	−7.4	−3.3 (91)
Rejected (%)	2.5	1.0	+150	+212 (0.8)	1.21	1.0	+21	+2.5 (1.18)
Production								
Volume (1000kg)	341.2	360	−5.2	+5.0 (325)	1385	1440	−3.8	−1.4 (1405)
Material (£/tonne)	453.5	450	+0.8	+13.4 (400)	452	450	+0.4	−0.9 (456)
Man (hours/tonne)	1.34	1.25	+7.2	+3.9 (1.29)	1.21	1.25	−3.2	−2.4 (1.24)
Dryer output (tonnes)	72.5	80	−9.4	−14.7 (85)	295	320	−7.8	−15.7 (350)

John reflected on past occasions when the figures had been high, above target, and everyone had been pleased. But he had been anxious even in those meetings, in case anyone asked him how he had improved the output figures – he didn't really know!

At the directors' meeting David asked John to present and explain the figures to the glum faces around the table. John wondered why it always seemed to be the case that the announcement of low production figures and problems always seemed to coincide with high sales figures. Sheila, the Sales Director, had earlier presented the latest results from her group's efforts. She had proudly listed the actions they had recently taken which had, of course, resulted in the improved sales. Last month a different set of reasons, but recognizable from past usage, had been offered by Sheila in explanation for the poor, below target sales results. Perhaps, John thought, the sales people are like us – they don't know what is going on either!

What John, David and Sheila all knew was that they were all trying to manage their activities in the best interest of the company. So why the anxiety, frustration and conflict?

Let us take a look at some of the figures that were being presented that day. The managers present, like many thousands in industry and the service sector throughout the world every day, were looking at data displayed in tables of variances (Table 4.1). What do managers look for in such tables? Large variances from predicted values are the only things that many managers and directors are interested in. 'Why is that sales figure so low?' 'Why is that cost so high?' 'What is happening to dryer output?' 'What are you doing about it?' Often thrown into the pot are comparisons of this month's figures with last month's or with the same month last year.

4.2 Interpretation of data

The method of 'managing' a company, or part of it, by examining data monthly, in variance tables is analogous to trying to steer a motor car by staring through the off-side wing mirror at what we have just driven past – or hit! It does not take into account the overall performance of the process and the context of the data.

Comparison of only one number with another – say this month's figures compared with last month's or with the same month last year – is also very weak. Consider the figures below for sales of 2,4 D (the product):

	Year 5 Month 4	Year 6 Month 3	Month 4
Sales (tonnes)	562	540	505

What conclusions might be drawn in the typical monthly meeting? 'Sales are down on last month.' 'Even worse they are down on the same month last year!' 'We have a trend here, we're losing market share' (Figure 4.1).

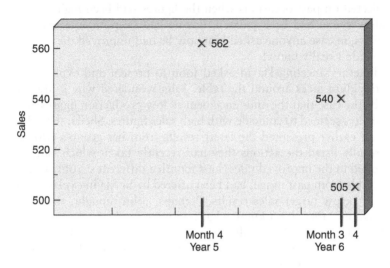

Figure 4.1 Monthly sales data

How can we test these conclusions before reports have to be written, people are reprimanded or fired, the product is re-designed or other possibly futile expensive action is initiated? First, the comparisons made are limited because of the small amount of data used. The conclusions drawn are weak because no account has been taken of the variation in the data being examined.

Let us take a look at a little more sales data on this product – say over the last 24 months (Table 4.2). Tables like this one are also sometimes used in management meeting and attempts are made to interpret the data in them, despite the fact that it is extremely difficult to digest the information contained in such a table of numbers.

Table 4.2 Twenty-four months' sales data

Year/month	Sales
4/5	532
4/6	528
4/7	523
4/8	525
4/9	541
4/10	517
4/11	524
4/12	536
5/1	499
5/2	531

5/3	514
5/4	562
5/5	533
5/6	516
5/7	525
5/8	517
5/9	532
5/10	521
5/11	531
5/12	535
6/1	545
6/2	530
6/3	540
6/4	505

If this information is presented differently, plotted on a simple time series graph or *run chart*, we might be able to begin to see the wood, rather than the trees. Figure 4.2 is such a plot, which allows a visual comparison of the latest value with those of the preceding months, and a decision on whether this value is unusually high or low, whether a trend or cycle is present, or not. Clearly, variation in the figures is present – we expect it, but if we understand that variation and what might be its components or causes, we stand a chance of making better decisions.

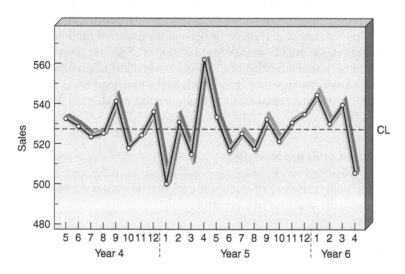

Figure 4.2 Monthly sales data

4.3 Causes of variation

At the basis of the theory of statistical process control is differentiation of the causes of variation during the operation of any process, be it a drying or a sales process. Certain variations belong to the category of chance or random variations, about which little may be done, other than to revise the process. This type of variation is the sum of the multitude of effects of a complex interaction of 'random' or 'common' causes, many of which are slight. When random variations alone exist, it will not be possible to trace their causes. For example, the set of common causes that produces variation in the quality of products may include random variations in the inputs to the process: atmospheric pressure or temperature changes, passing traffic or equipment vibrations, electrical or humidity fluctuations, and changes in operator physical and emotional conditions. This is analogous to the set of forces that cause a coin to land heads or tails when tossed. When only common causes of variations are present in a process, the process is considered to be '*stable*', '*in statistical control*' or '*in control*'.

There is also variation in any test equipment, and inspection/checking procedures, whether used to measure a physical dimension, an electronic or a chemical characteristic or a property of an information system. The inherent variations in checking and testing contribute to the overall process variability. In a similar way, processes whose output is not an artefact but a service will be subject to common causes of variation, e.g. traffic problems, electricity supply, operator performance and the weather all affect the time likely to complete an insurance estimate, the efficiency with which a claim is handled, etc. Sales figures are similarly affected by common causes of variation.

Causes of variation that are relatively large in magnitude and readily identified are classified as 'assignable' or '*special*' causes. When special causes of variation are present, variation will be excessive and the process is classified as '*unstable*', '*out of statistical control*' or beyond the expected random variations. For brevity this is usually written '*out-of-control*'. Special causes include tampering or unnecessary adjusting of the process when it is inherently stable, and structural variations caused by things such as the four seasons.

In Chapter 1 it was suggested that the first question which must be asked of any process is:

'CAN WE DO this job correctly?'

Following our understanding of common and special causes of variation, this must now be divided into two questions:

1   'Is the process stable, or in control?' In other words, are there present any special causes of variation, or is the process variability due to common causes only?

2 'What is the extent of the process variability?' or what is the natural capability of the process when only common causes of variation are present?

This approach may be applied to both variables and attribute data, and provides a systematic methodology for process examination, control and investigation.

It is important to determine the extent of variability when a process is supposed to be stable or 'in control', so that systems may be set up to detect the presence of special causes. A systematic study of a process then provides knowledge of the variability and capability of the process, and the special causes which are potential sources of changes in the outputs. Knowledge of the current state of a process also enables a more balanced judgement of the demands made of all types of resources, both with regard to the tasks within their capability and their rational utilization.

Changes in behaviour

So back to the directors' meeting and what should David, John and Sheila be doing differently? Firstly, they must recognize that variation is present and part of everything: suppliers' products and delivery performance, the dryer temperature, the plant and people's performance, the market. Secondly, they must understand something about the theory of variation and its causes: common versus special. Thirdly, they must use the data appropriately so that they can recognize, interpret and react appropriately to the variation in the data; that is they must be able to distinguish between the presence of common and special causes of variation in their processes. Finally, they must develop a strategy for dealing with special causes.

How much variation and its nature, in terms of common and special causes, may be determined by carrying out simple statistical calculations on the process data. From these, control limits may be set for use with the simple run chart shown in Figure 4.2. These describe the extent of the variation that is being seen in the process due to all the common causes, and indicate the presence of any special causes. If or when the special causes have been identified, accounted for or eliminated, the control limits will allow the managers to predict the future performance of the process with some confidence. The calculations involved and the setting up of 'control charts' with limits are described in Chapters 5 and 6.

A control chart is a device intended to be used at the point of operation, where the process is carried out, and by the operators of that process. Results are plotted on a chart that reflects the variation in the process. As shown in Figure 4.3, the control chart has three zones and the action required depends on the zone in which the results fall. The possibilities are:

1   Carry on or do nothing (stable zone – common causes of variation only).
2   Be careful and seek more information, since the process maybe showing special causes of variation (warning zone).
3   Take action, investigate or, where appropriate, adjust the process (action zone – special causes of variation present).

This is rather like a set of traffic lights that signal 'stop', 'caution' or 'go'.

Look again at the sales data now plotted with control limits in Figure 4.4. We can see that this process was stable and it is unwise to ask, 'Why were sales so low in Year 5 Month 1?' or 'Why were sales so high in Year 5 Month 4?' Trying to find the answers to these questions could waste much time and effort, but would not change or improve the process. It would be useful, however to ask, 'Why was the sales *average* so low and how can we increase it?'

Consider now a different set of sales data (Figure 4.5). This process was unstable and it is wise to ask, 'Why did the average sales increase after week 18?' Trying to find an answer to this question may help to identify a special cause of variation. This in turn may lead to action that ensures

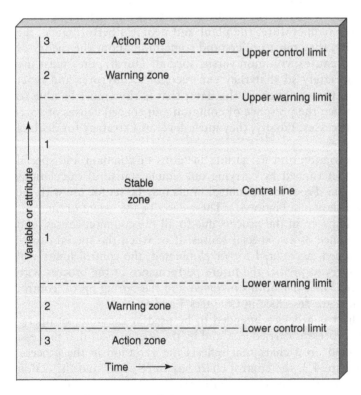

Figure 4.3 Schematic control chart

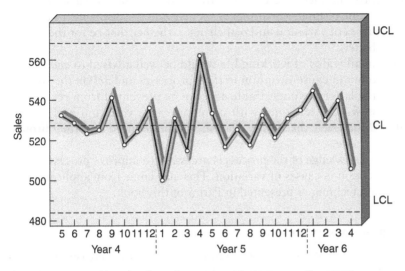

Figure 4.4 Monthly sales data (for years 4–6). CL: centre line; UCL: upper control limit; LCL: lower control limit

that the sales do not fall back to the previous average. If the cause of this beneficial change is not identified, the managers may be powerless to act if the process changes back to its previous state.

The use of run charts and control limits can help managers and process operators to ask useful questions that lead to better process management and improvements. They also discourage the asking of questions that lead to

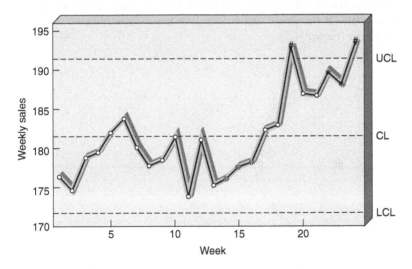

Figure 4.5 Monthly sales data (in weeks)

wasted efforts and increased cost. Control charts (in this case a simple run chart with control limits) help managers generally to distinguish between common causes of variation and real change, whether that be for the worse or for the better.

People in all walks of working life would be well advised to *accept* the inherent common cause variation in their processes and *act* on the special causes. If the latter are undesirable and can be prevented from recurring, the process will be left only with common cause variation and it will be stable. Moreover, the total variation will be reduced and the outputs more predictable.

In-depth knowledge of the process is necessary to improve processes that show only common causes of variation. This may come from application of the ideas and techniques presented in Part 5 of this book.

4.4 Accuracy and precision

In the examination of process data, confusion often exists between the *accuracy* and *precision* of a process. An analogy may help to clarify the meaning of these terms.

Two men with rifles each shoot one bullet at a target, both having aimed at the bull's eye. By a highly improbable coincidence, each marksman hits exactly the same spot on the target, away from the bull's eye (Figure 4.6). What instructions should be given to the men in order to improve their performance? Some may feel that each man should be told to alter his gunsights to adjust the aim: 'down a little and to the right'. Those who have done some shooting, however, will realize that this is premature, and that a more sensible instruction is to ask the men to fire again – perhaps using four more bullets, without altering the aim, to establish the nature of each man's shooting process. If this were to be done, we might observe two different types of pattern (Figure 4.7).

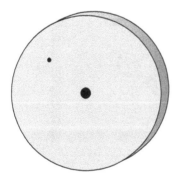

Figure 4.6 The first coincidental shot from each of two marksmen

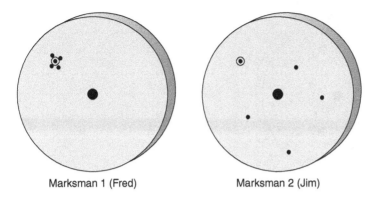

Marksman 1 (Fred)   Marksman 2 (Jim)

Figure 4.7 The results of five shots each for Fred and Jim – their first identical
shots are ringed

Clearly, marksman 1 (Fred) is *precise* because all the bullet holes are clustered together – there is little spread, but he is not *accurate* since on average his shots have missed the bull's eye. It should be a simple job to make the adjustment for accuracy – perhaps to the gun-sight – and improve his performance to that shown in Figure 4.8. Marksman 2 (Jim) has a completely different problem. We now see that the reason for his first wayward shot was completely different to the reason for Fred's. If we had adjusted Jim's gun-sights after just one shot, 'down a little and to the right', Jim's whole process would have shifted, and things would have been worse (Figure 4.9). Jim's next shot would then have been even further away from the bull's eye, as the adjustment affects only the accuracy and not the precision.

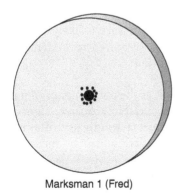

Marksman 1 (Fred)

Figure 4.8 Shooting process, after adjustment of the gun-sight

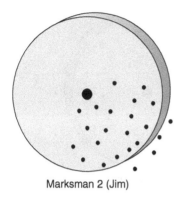

Marksman 2 (Jim)

Figure 4.9 Marksman 2 (Jim) after incorrect adjustment of gun-sight

Jim's problem of spread or lack of precision is likely to be a much more complex problem than Fred's lack of accuracy. The latter can usually be amended by a simple adjustment, whereas problems of wide scatter require a deeper investigation into the causes of the variation.

Several points are worth making from this simple analogy:

- There is a difference between the accuracy and the precision of a process.
- The accuracy of a process relates to its ability to hit the target value.
- The precision of a process relates to the degree of spread of the values (variation).
- The distinction between accuracy and precision may be assessed only by looking at a number of results or values, not by looking at individual ones.
- Making decisions about adjustments to be made to a process, on the basis of one individual result, may give an undesirable outcome, owing to lack of information about process accuracy and precision.
- The adjustment to correct lack of process accuracy is likely to be 'simpler' than the larger investigation usually required to understand or correct problems of spread or large variation.

The shooting analogy is useful when we look at the performance of a manufacturing process producing goods with a variable property. Consider a steel rod cutting process that has as its target a length of 150mm. The overall variability of such a process may be determined by measuring a large sample – say 100 rods – from the process (Table 4.3), and shown graphically as a histogram (Figure 4.10). Another method of illustration is a frequency polygon which is obtained by connecting the mid-points of the tops of each column (Figure 4.11).

Table 4.3 Lengths of 100 steel rods (mm)

144	146	154	146
151	150	134	153
145	139	143	152
154	146	152	148
157	153	155	157
157	150	145	147
149	144	137	155
141	147	149	155
158	150	149	156
145	148	152	154
151	150	154	153
155	145	152	148
152	146	152	142
144	160	150	149
150	146	148	157
147	144	148	149
155	150	153	148
157	148	149	153
153	155	149	151
155	142	150	150
146	156	148	160
152	147	158	154
143	156	151	151
151	152	157	149
154	140	157	151

When the number of rods measured is very large and the class intervals small, the polygon approximates to a curve, called the frequency curve (Figure 4.12). In many cases, the pattern would take the symmetrical form shown – the bell-shaped curve typical of the 'normal distribution'. The greatest number of rods would have the target value, but there would be appreciable numbers either larger or smaller than the target length. Rods with dimensions further from the central value would occur progressively less frequently.

It is possible to imagine four different types of process frequency curve, which correspond to the four different performances of the two marksmen (see Figure 4.13). Hence, process 4 is accurate and relatively precise, as the average of the lengths of steel rod produced is on target, and all the lengths are reasonably close to the mean. If only common causes of variation are present, the output from a process forms a distribution that is stable

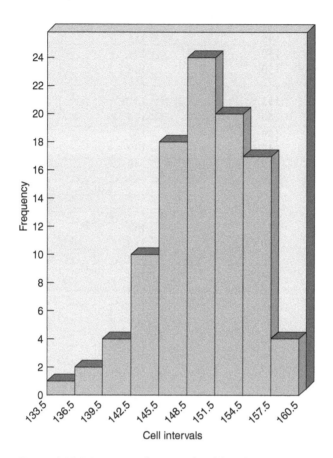

Figure 4.10 Histogram of 100 steel rod lengths

over time and is, therefore, predictable (Figure 4.14a). Conversely, if special causes of variation are present, the process output is not stable over time and is not predictable (Figure 4.14b). For a detailed interpretation of the data, and before the design of a process control system can take place, this intuitive analysis must be replaced by more objective and quantitative methods of summarizing the histogram or frequency curve. In particular, some measure of both the location of the central value and of the spread must be found. These are introduced in Chapter 5.

4.5 Variation and management

So how should John, David and Sheila, whom we met at the beginning of this chapter, manage their respective processes? First of all, basing each decision on just one result is dangerous. They all need to get the 'big picture', and see the context of their data/information. This is best achieved by

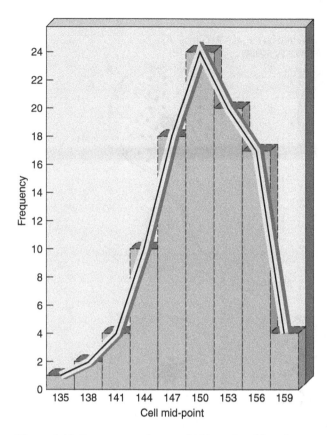

Figure 4.11 Frequency polygon of 100 steel rod lengths

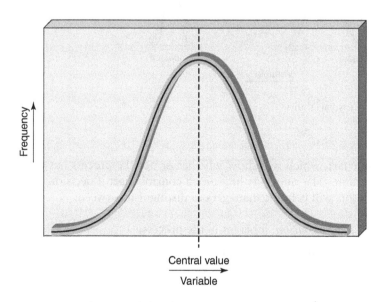

Figure 4.12 The normal distribution of a continuous variable

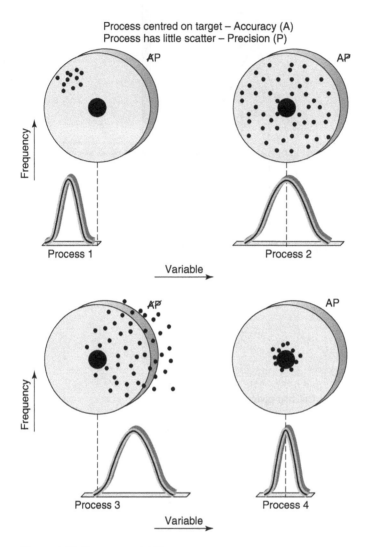

Figure 4.13 Process variability

plotting a run chart, which will show whether or not the process has or is changing over time. The run chart becomes a control chart if decision lines are added and this will help the managers to distinguish between:

Common cause variation: inherent in the process.

Special cause variation: due to real changes.

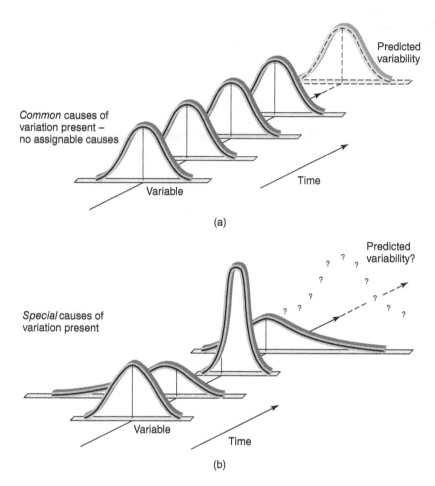

Figure 4.14 Common and special causes of variation

These managers should stop blaming people and start examining processes and the causes of variation.

The purpose of a control chart is to detect change in the performance of a process. A control chart illustrates the dynamic performance of the process, whereas a histogram gives a static picture of variations around a mean or average. Ideally these should be used together to detect:

Changes in absolute level (centring/accuracy).

Changes in variability (spread/precision).

Generally pictures are more meaningful than tables of results. It is easier to detect relatively large changes, with respect to the underlying variation, than small changes and control limits help the detection of change.

Chapter highlights

- Managers tend to look at data presented in tables of variances from predicted or target values, reacting to individual values. This does not take into account the overall performance of the process, the context of the data and its variation.
- Data plotted on simple time series graphs or run charts enable the easy comparison of individual values with the remainder of the data set.
- It is important to differentiate between the random or 'common' causes of variation and the assignable or 'special' causes. When only common causes of variation are present, the process is said to be stable or 'in statistical control'. Special causes lead to an unstable or 'out of statistical control' process.
- Following an understanding of common and special causes of variation, the 'Can we do the job correctly?' question may be split into two questions: 'Is the process in control?' followed by 'What is the extent of the process variability?' (or 'What is the natural process capability?').
- It is important to know the extent of the variation (capability) when the process is stable, so that systems may be set up to detect the presence of special causes.
- Managers must: (i) recognize that process variation is present; (ii) understand the theory of variation and its causes (common and special); (iii) use data appropriately so they can recognize, interpret and react properly to variation and (iv) develop a strategy for dealing with special causes.
- Control charts with limits may be used to assist in the interpretation of data. Results are plotted onto the charts and fall into three zones: one in which no action should be taken (common causes only present); one that suggests more information should be obtained and one that requires some action to be taken (special causes present) – such as a set of stop, caution, go traffic lights.
- In the examination of process data a distinction should be made between accuracy (with respect to a target value) and precision (with respect to the spread of data). This can be achieved only by looking at a number of results, not at individual values.
- The overall variability of any process may be determined from a reasonable size sample of results. This may be presented as a histogram, or a frequency polygon or curve. In many cases, a symmetrical bell-shaped curve, typical of the 'normal distribution' is obtained.
- A run chart or control chart illustrates the dynamic performance of the process, whereas a histogram/frequency curve gives a static picture

of variations around an average value. Ideally these should be used together to detect special causes of changes in absolute level (accuracy) or in variability (precision).

- It can generally be said that: (i) pictures are more meaningful than tables of results; (ii) it is easier to detect relatively large changes and (iii) control chart limits help the detection of change.

References and further reading

Deming, W.E. (1993) *The New Economics: For Industry, Government and Education*, MIT, Cambridge MA, USA.

Joiner, B.L. (1994) *Fourth Generation Management: The New Business Consciousness*, McGraw-Hill, New York, USA.

Shewhart, W.A. (edited and new foreword by Deming, W.E.) (1986) *Statistical Method from the Viewpoint of Quality Control*, Dover Publications, New York, USA.

Wheeler, D.J. (1993) *Understanding Variation: The Key to Managing Chaos*, SPC Press, Knoxville TN, USA.

Wheeler, D.J. (2005) *Making Sense of Data: SPC for the Service Sector*, SPC Press, Knoxville TN, USA.

Discussion questions

1 Design a classroom 'experience', with the aid of computers if necessary, for a team of senior managers who do not appear to understand the concepts of variation. Explain how this will help them understand the need for better decision-making processes.

2 (a) Explain why managers tend to look at individual values – perhaps monthly results, rather than obtain an overall picture of data.

(b) Which simple techniques would you recommend to managers for improving their understanding of process and the variation in them?

3 (a) What is meant by the inherent variability of a process?

(b) Distinguish between common (or random) and special (or assignable) causes of variation, and explain how the presence of special causes may be detected by simple techniques.

4 'In the examination of process data, a distinction should be made between accuracy and precision.' Explain fully the meaning of this statement, illustrating with simple everyday examples, and suggesting which techniques may be helpful.

5 How could the overall variability of a process be determined? What does the term 'capability' mean in this context?

5   Variables and process variation

Objectives

- To introduce measures for accuracy (centring) and precision (spread).
- To describe the properties of the normal distribution and its use in understanding process variation and capability.
- To consider some theory for sampling and subgrouping of data and see the value in grouping data for analysis.

5.1 Measures of accuracy or centring

In Chapter 4 we saw how objective and quantitative methods of summarizing variable data were needed to help the intuitive analysis used so far. In particular a measure of the central value is necessary, so that the accuracy or centring of a process may be estimated. There are various ways of doing this, such as follows.

Mean (or arithmetic average)

This is simply the average of the observations, the sum of all the measurements divided by the number of the observations. For example, the mean of the first row of four measurements of rod lengths in Table 4.3: 144 mm, 146 mm, 154 mm and 146 mm is obtained:

$$
\begin{array}{r}
144 \text{ mm} \\
146 \text{ mm} \\
154 \text{ mm} \\
\underline{146 \text{ mm}} \\
\end{array}
$$

Sum   590 mm

$$
\text{Sample Mean} = \frac{590 \text{ mm}}{4} = 147.5 \text{ mm}
$$

The 100 results in Table 4.3 are shown as 25 different groups or samples of four rods and we may calculate a sample mean $\overline{X}$ for each group. The 25 sample means are shown in Table 5.1.

The mean of a whole population, i.e. the total output from a process rather than a sample, is represented by the Greek letter μ. We can never know μ, the true mean, but the 'Grand' or 'Process Mean', $\overline{\overline{X}}$, the average of all the sample means, is a good estimate of the population mean.

The value of $\overline{\overline{X}}$ for the steel rods is 150.1 mm.

Median

If the measurements are arranged in order of magnitude, the median is simply the value of the middle item. This applies directly if the number in the series

Table 5.1 100 steel rod lengths as 25 samples of size 4

Sample number	Rod lengths (mm)				Sample mean (mm)	Sample range (mm)
	(i)	*(ii)*	*(iii)*	*(iv)*		
1	144	146	154	146	147.50	10
2	151	150	134	153	147.00	19
3	145	139	143	152	144.75	13
4	154	146	152	148	150.00	8
5	157	153	155	157	155.50	4
6	157	150	145	147	149.75	12
7	149	144	137	155	146.25	18
8	141	147	149	155	148.00	14
9	158	150	149	156	153.25	9
10	145	148	152	154	149.75	9
11	151	150	154	153	152.00	4
12	155	145	152	148	150.00	10
13	152	146	152	142	148.00	10
14	144	160	150	149	150.75	16
15	150	146	148	157	150.25	11
16	147	144	148	149	147.00	5
17	155	150	153	148	151.50	7
18	157	148	149	153	151.75	9
19	153	155	149	151	152.00	6
20	155	142	150	150	149.25	13
21	146	156	148	160	152.50	14
22	152	147	158	154	152.75	11
23	143	156	151	151	150.25	13
24	151	152	157	149	152.25	8
25	154	140	157	151	150.50	17

is odd. When the number in the series is even, as in our example of the first four rod lengths in Table 4.1, the median lies between the two middle numbers. Thus, the four measurements arranged in order of magnitude are:

144,   146,   146,   154.

The median is the 'middle item'; in this case 146. In general, about half the values will be less than the median value, and half will be more than it. An advantage of using the median is the simplicity with which it may be determined, particularly when the number of items is odd.

Mode

A third method of obtaining a measure of central tendency is the most commonly occurring value, or mode. In our example of four, the value 146 occurs twice and is the modal value. It is possible for the mode to be non-existent in a series of numbers or to have more than one value. When data are grouped into a frequency distribution, the mid-point of the cell with the highest frequency is the modal value. During many operations of recording data, the mode is often not easily recognized or assessed.

Relationship between mean, median and mode

Some distributions, as we have seen, are symmetrical about their central value. In these cases, the values for the mean, median and mode are identical. Other distributions have marked asymmetry and are said to be skewed. Skewed distributions are divided into two types. If the 'tail' of the distribution stretches to the right – the higher values, the distribution is said to be positively skewed; conversely in negatively skewed distributions the tail extends towards the left – the smaller values.

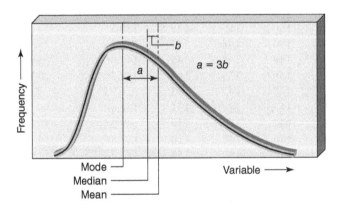

Figure 5.1 Mode, median and mean in skew distributions

Figure 5.1 illustrates the relationship between the mean, median and mode of moderately skew distributions. An approximate relationship is:

Mean – mode = 3(mean – median).

Thus, knowing two of the parameters enables the third to be estimated.

5.2 Measures of precision or spread

Measures of the extent of variation in process data are also needed. Again there are a number of methods:

Range

The range is the difference between the highest and the lowest observations and is the simplest possible measure of scatter. For example, the range of the first four rod lengths is the difference between the longest (154 mm) and the shortest (144 mm), that is 10 mm. The range is usually given the symbol R_i. The ranges of the 25 samples of four rods are given in Table 5.1. The mean range $\overline{R}$, the average of all the sample ranges, may also be calculated at 10.8.

The range offers a measure of scatter which can be used widely, owing to its simplicity. There are, however, two major problems in its use:

(i) The value of the range depends on the number of observations in the sample. The range will tend to increase as the sample size increases. This can be shown by considering again the data on steel rod lengths in Table 4.3:

   The range of the first two observations is 2 mm.
   The range of the first four observations is 10 mm.
   The range of the first six observations is also 10 mm.
   The range of the first eight observations is 20 mm.

(ii) Calculation of the range uses only a portion of the data obtained. The range remains the same despite changes in the values lying between the lowest and the highest values.

   It would seem desirable to obtain a measure of spread that is free from these two disadvantages.

Standard deviation

The standard deviation takes all the data into account and is a measure of the 'deviation' of the values from the mean. It is best illustrated by an example. Consider the deviations of the first four steel rod lengths from the mean:

Value x_i (mm)		Deviation $(x_i - \overline{X})$
144		–3.5 mm
146		–1.5 mm
154	+6.5 mm	
146		–1.5 mm
Mean $\overline{X}$ = 147.5 m	Total =	0

Measurements above the mean have a positive deviation and measurements below the mean have a negative deviation. Hence, the total deviation from the mean is zero, which is obviously a useless measure of spread. If, however, each deviation is multiplied by itself, or squared, since a negative number multiplied by a negative number is positive, the squared deviations will always be positive:

Value x_i (mm)	Deviation $(x_i - \overline{X})$	$(x_i - \overline{X})^2$
144	–3.5	12.25
146	–1.5	2.25
154	+6.5	42.25
146	–1.5	2.25
Sample	Total: $\Sigma(x_i - \overline{X})^2 =$	59.00
Mean $\overline{X}$ = 147.5		

The average of the squared deviations may now be calculated and this value is known as the *variance* of the sample. In the above example, the variance or mean squared variation is 14.75.

The *standard deviation*, normally denoted by the Greek letter sigma (σ), is the square root of the variance, which then measures the spread in the same units as the variables, i.e. in the case of the steel rods, in millimetres:

$$\sigma = \sqrt{14.75} = 3.84 \text{ mm.}$$

The true standard deviation σ, like μ, can never be known, but for simplicity, the conventional symbol σ will be used throughout this book to represent the process standard deviation. If a sample is being used to estimate the spread of the process, then the sample standard deviation will tend to underestimate the standard deviation of the whole process. This bias is particularly marked in small samples. To correct for the bias, the sum of the squared deviations is divided by the sample size minus one. In the above example, the *estimated process standard deviation s* is:

$$s = \sqrt{\frac{59.00}{3}} = \sqrt{19.67} = 4.43 \text{ mm.}$$

Whilst the standard deviation gives an accurate measure of spread, it is laborious to calculate. Software capable of statistical calculations may be used, of course. A much greater problem is that, unlike range, standard deviation takes a little more effort to understand.

5.3 The normal distribution

The meaning of the standard deviation is perhaps most easily explained in terms of the *normal* distribution. If a continuous variable is monitored, such as the lengths of rod from the cutting process, the volume of paint in tins from a filling process, the weights of tablets from a pelletizing process, or the monthly sales of a product, that variable will usually be distributed normally about a mean μ. The spread of values may be measured in terms of the population standard deviation, σ, which defines the width of the bell-shaped curve. Figure 5.2 shows the proportion of the output expected to be found between the values of $\mu \pm \sigma$, $\mu \pm 2\sigma$ and $\mu \pm 3\sigma$.

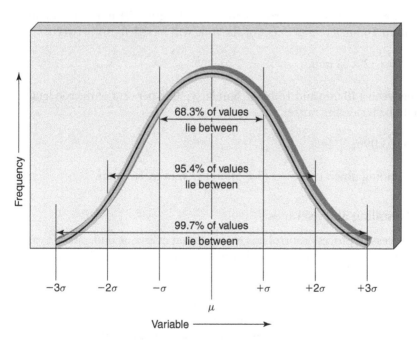

Figure 5.2 Normal distribution

Suppose the process mean of the steel rod cutting process is 150 mm and that the standard deviation is 5 mm, then from a knowledge of the shape of the curve and the properties of the normal distribution, the following facts would emerge:

a   68.3 per cent of the steel rods produced will lie within ±5 mm of the mean, i.e. $\mu \pm \sigma$,
b   95.4 per cent of the rods will lie within ±10 mm ($\mu \pm 2\sigma$),
c   99.7 per cent of the rods will lie within ±15 mm ($\mu \pm 3\sigma$).

We may be confident then that almost all the steel rods produced will have lengths between 135 mm and 165 mm. The approximate distance between the two extremes of the distribution, therefore, is 30 mm, which is equivalent to six standard deviations or 6σ.

The mathematical equation and further theories behind the normal distribution are given in Appendix A. This appendix includes a table on page 381 which gives the probability that any item chosen at random from a normal distribution will fall outside a given number of standard deviations from the mean. The table shows that, at the value $\mu + 1.96\sigma$, only 0.025 or 2.5 per cent of the population will exceed this length. The same proportion will be less than $\mu - 1.96\sigma$. Hence ca 95 per cent of the population will lie within $\mu \pm 2\sigma$.

In the case of the steel rods with mean length 150 mm and standard deviation 5 mm, ca 95 per cent of the rods will have lengths between:

$$150 \pm (2 \times 5) \text{ mm},$$

i.e. between 140 mm and 160 mm. Similarly, 99.8 per cent of the rod lengths should be inside the range:

$$\mu \pm 3.09\sigma,$$

i.e. rounding down to: $150 \pm (3 \times 5)$ or 135 mm to 165 mm.

5.4 Sampling and averages

For successful process control it is essential that everyone understands variation, and how and why it arises. The absence of such knowledge will lead to action being taken to adjust or interfere with processes which, if left alone, would be quite capable of achieving the requirements. Many processes are found to be out-of-statistical-control or unstable, when first examined using statistical process control techniques. It is frequently observed that this is due to an excessive number of adjustments being made to the process based on individual tests or measurements. This behaviour, commonly known as tampering or hunting, causes an overall increase in variability of results from the process, as shown in Figure 5.3. The process is initially set at the target value: $\mu = T$, but a single measurement at A results in the process being adjusted downwards to a new mean μ_A. Subsequently, another single measurement at B results in an upwards adjustment of the process to a new mean μ_B. Clearly if this tampering continues throughout the operation of the process, its variability will be greatly and unnecessarily increased, with a detrimental effect on the ability of the process to meet the specified requirements. Indeed it is not uncommon for such behaviour to lead to a call for even tighter tolerances and for the process to be 'controlled' *very* carefully. This in turn leads to even more frequent adjustment, further increases in variability and more failure to meet the requirements.

To improve this situation and to understand the logic behind process control methods for variables, it is necessary to give some thought to the behaviour of sampling and of averages. If the length of a single steel rod is measured, it is clear that occasionally a length will be found which is towards one end of the tails of the process's normal distribution. This occurrence, if taken on its own, may lead to the wrong conclusion that the cutting process requires adjustment. If, on the other hand, a sample of four or five is taken, it is extremely unlikely that all four or five lengths will lie towards one extreme end of the distribution.

If, therefore, we take the average or mean length of four or five rods, we shall have a much more reliable indicator of the state of the process.

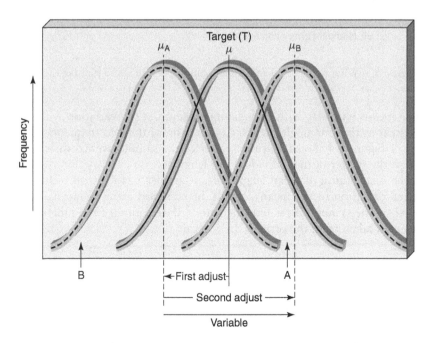

Figure 5.3 Increase in process variability due to frequent adjustment

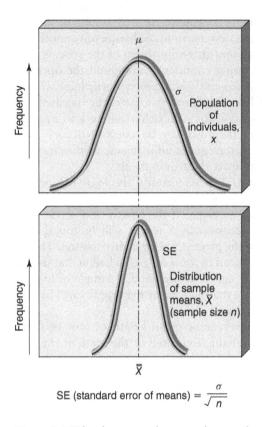

SE (standard error of means) $= \dfrac{\sigma}{\sqrt{n}}$

Figure 5.4 What happens when we take samples of size *n* and plot the means

Sample means will vary with each sample taken, but the variation will not be as great as that for single pieces. Comparison of the two frequency diagrams of Figure 5.4 shows that the scatter of the sample averages is much less than the scatter of the individual rod lengths.

In the distribution of mean lengths from samples of four steel rods, the standard deviation of the means, called the standard error of means, and denoted by the symbol SE, is half the standard deviation of the individual rod lengths taken from the process. In general:

Standard error of means, $SE = \sigma / \sqrt{n}$.

and when $n = 4$, $SE = \sigma/2$, i.e. half the spread of the parent distribution of individual items. SE has the same characteristics as any standard deviation, and normal tables may be used to evaluate probabilities related to the distribution of sample averages. We call it by a different name to avoid confusion with the population standard deviation.

The smaller spread of the distribution of sample averages provides the basis for a useful means of detecting changes in processes. Any change in the process mean, unless it is extremely large, will be difficult to detect from individual results alone. The reason can be seen in Figure 5.5a, which shows the parent distributions for two periods in a paint filling process between which the average has risen from 1000 ml to 1012 ml. The shaded portion is common to both process distributions and, if a volume estimate occurs in the shaded portion, say at 1010 ml, it could suggest either a volume above the average from the distribution centred at 1000 ml, or one slightly below the average from the distribution centred at 1012 ml. A large number of individual readings would, therefore, be necessary before such a change was confirmed.

The distribution of sample means reveals the change much more quickly, the overlap of the distributions for such a change being much smaller (Figure 5.5b). A sample mean of 1010 ml would almost certainly not come from the distribution centred at 1000 ml. Therefore, on a chart for sample means, plotted against time, the change in level would be revealed almost immediately. For this reason sample means rather than individual values are used, where possible and appropriate, to control the centring of processes.

The Central Limit Theorem

What happens when the measurements of the individual items are not distributed normally? A very important piece of theory in SPC is the *Central Limit Theorem*. This states that if we draw samples of size n, from a population with a mean μ and a standard deviation σ, then as n increases in size, the distribution of sample means approaches a normal distribution with a mean μ and a standard error of the means of $\sigma / \sqrt{n}$. This tells us that, even if the individual values are not normally distributed, the distribution of the means will tend to have a normal distribution, and the larger the sample size the greater will be this tendency. It also tells us that the Grand or Process Mean $\overline{\overline{X}}$ will be a very good estimate of the true mean of the population μ.

Even if n is as small as 4 and the population is not normally distributed, the distribution of sample means will be very close to normal. This may be illustrated by sketching the distributions of averages of 1000 samples of size four taken from each of two boxes of strips of paper, one box containing a rectangular distribution of lengths, and the other a triangular distribution (Figure 5.6). The mathematical proof of the Central Limit Theorem is beyond the scope of this book. The reader may perform the appropriate experimental work if (s)he requires further evidence. The main point is that, when samples of size $n = 4$ or more are taken from a process that is stable, we can assume that the distribution of the sample means $\overline{X}$ will be very nearly normal, even if the parent population is not normally distributed. This provides a sound basis for the Mean Control Chart which, as mentioned in Chapter 4, has decision 'zones' based on predetermined *control limits*. The setting of these will be explained in the next chapter.

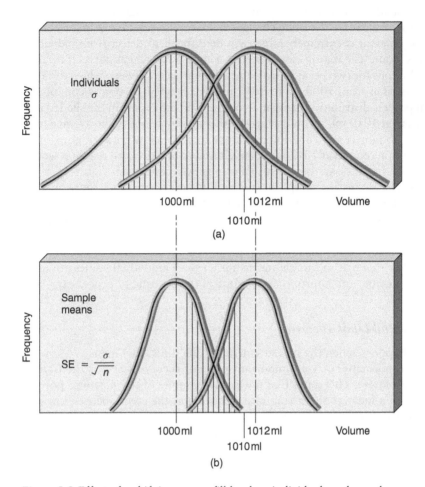

Figure 5.5 Effect of a shift in average fill level on individuals and sample means. Spread of sample means is much less than spread of individuals

   The Range Chart is very similar to the mean chart, the range of each sample being plotted over time and compared to predetermined limits. The development of a more serious fault than incorrect or changed centring can lead to the situation illustrated in Figure 5.7, where the process collapses from form A to form B, perhaps due to a change in the variation of material. The ranges of the samples from B will have higher values than ranges in samples taken from A. A range chart should be plotted in conjunction with the mean chart.

Rational subgrouping of data

We have seen that a subgroup or sample is a small set of observations on a process parameter or its output, taken together in time. The two major problems

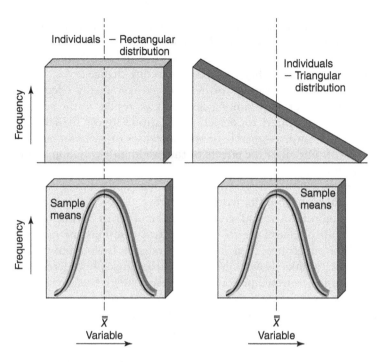

Figure 5.6 The distribution of sample means from rectangular and triangular universes

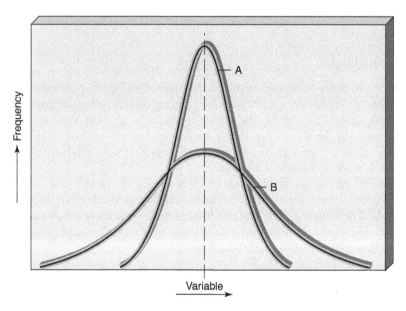

Figure 5.7 Increase in spread of a process

with regard to choosing a subgroup relate to its size and the frequency of sampling. The smaller the subgroup, the less opportunity there is for variation within it, but the larger the sample size the narrower the distribution of the means, and the more sensitive they become to detecting change.

A rational subgroup is a sample of items or measurements selected in a way that minimizes variation among the items or results in the sample, and maximizes the opportunity for detecting variation between the samples. With a rational subgroup, assignable or special causes of variation are not likely to be present, but all of the effects of the random or common causes are likely to be shown. Generally, subgroups should be selected to keep the chance for differences within the group to a minimum, and yet maximize the chance for the subgroups to differ from one another.

The most common basis for subgrouping is the order of output or production. When control charts are to be used, great care must be taken in the selection of the subgroups, their frequency and size. It would not make sense, for example, to take as a subgroup the chronologically ordered output from an arbitrarily selected period of time, especially if this overlapped two or more shifts, or a change over from one grade of products to another, or four different machines. A difference in shifts, grades or machines may be an assignable cause that may not be detected by the variation between samples, if *irrational* subgrouping has been used.

An important consideration in the selection of subgroups is the type of process – one-off, short run, batch or continuous flow – and the type of data available. This will be considered further in Chapter 7, but at this stage it is clear that, in any type of process control charting system, the careful selection of subgroups is a key factor.

Chapter highlights

- There are three main measures of the central value of a distribution (accuracy). These are the mean μ (the average value), the median (the middle value), the mode (the most common value). For symmetrical distributions the values for mean, median and mode are identical. For asymmetric or skewed distributions, the approximate relationship is mean – mode = 3 (mean – median).

- There are two main measures of the spread of a distribution of values (precision). These are the range (the highest minus the lowest) and the standard deviation σ. The range is limited in use but it is easy to understand. The standard deviation gives a more accurate measure of spread, but takes a little more effort to understand.

- Continuous variables usually form a normal or symmetrical distribution. The normal distribution is explained by using the scale of the standard deviation around the mean. Using the normal distribution, the proportion falling in the 'tail' may be used to assess process capability or the amount out-of-specification or to set targets.

- A failure to understand and manage variation often leads to unjustified changes to the centring of processes, which results in an unnecessary increase in the amount of variation.
- Variation of the mean values of samples will show less scatter than individual results. The Central Limit Theorem gives the relationship between standard deviation (σ), sample size (n) and standard error of means (SE) as $SE = \sigma / \sqrt{n}$.
- The grouping of data results in an increased sensitivity to the detection of change, which is the basis of the mean chart.
- The range chart may be used to check and control variation.
- The choice of sample size is important in the control chart system and depends on the process under consideration.

References and further reading

Besterfield, D. (2013) *Quality Improvement*, 9th edn, Pearson, Englewood Cliffs NJ, USA.

Pyzdek, T. (1990) *Pyzdek's Guide to SPC, Vol. 1: Fundamentals*, ASQC Quality Press, Milwaukee WI, USA.

Shewart, W.A. (1931 – 50th Anniversary Commemorative Reissue 1980) *Economic Control of Quality of Manufactured Product*, D. Van Nostrand, New York, USA.

Wheeler, D.J. and Chambers, D.S. (1992) *Understanding Statistical Process Control*, 2nd edn, SPC Press, Knoxville TN, USA.

Discussion questions

1 Calculate the mean and standard deviation of the melt flow rate data below (g/10 minutes):

3.2	3.3	3.2	3.3	3.2
3.5	3.0	3.4	3.3	3.7
3.0	3.4	3.5	3.4	3.3
3.2	3.1	3.0	3.4	3.1
3.3	3.5	3.4	3.3	3.2
3.2	3.1	3.5	3.2	
3.3	3.2	3.6	3.4	
2.7	3.5	3.0	3.3	
3.3	2.4	3.1	3.6	
3.6	3.5	3.4	3.1	
3.2	3.3	3.1	3.4	
2.9	3.6	3.6	3.5	

If the specification is 3.0–3.8 g/10 minutes, comment on the capability of the process.

2 Describe the characteristics of the normal distribution and construct an example to show how these may be used in answering questions that arise from discussions of specification limits for a product.

3   A bottle filling machine is being used to fill 150 ml bottles of a shampoo. The actual bottles will hold 156 ml. The machine has been set to discharge an average of 152 ml. It is known that the actual amounts discharged follow a normal distribution with a standard deviation of 2 ml.

(a) What proportion of the bottles overflow?

(b) The overflow of bottles causes considerable problems and it has therefore been suggested that the average discharge should be reduced to 151 ml. In order to meet the weights and measures regulations, however, not more than 1 in 40 bottles, on average, must contain less than 146 ml. Will the weights and measures regulations be contravened by the proposed changes?

You will need to consult Appendix A to answer these questions.

4   State the Central Limit Theorem and explain how it is used in SPC.

5   **To:**   International Chemicals Supplier
    From: Senior Buyer, Perplexed Plastics Ltd
    SUBJECT: *MFR Values of Polyglyptalene*
    As promised, I have now completed the examination of our delivery records and have verified that the values we discussed were not in fact in chronological order. They were simply recorded from a bundle of certificates of analysis held in our quality records file. I have checked, however, that the bundle did represent all the daily deliveries made by ICS since you started to supply in October last year.
    Using your own lot identification system I have put them into sequence as manufactured:

(1) 4.1	(13) 3.2	(25) 3.3	(37) 3.2	(49) 3.3	(61) 3.2
(2) 4.0	(14) 3.5	(26) 3.0	(38) 3.4	(50) 3.3	(62) 3.7
(3) 4.2	(15) 3.0	(27) 3.4	(39) 3.5	(51) 3.4	(63) 3.3
(4) 4.2	(16) 3.2	(28) 3.1	(40) 3.0	(52) 3.4	(64) 3.1
(5) 4.4	(17) 3.3	(29) 3.5	(41) 3.4	(53) 3.3	
(6) 4.2	(18) 3.2	(30) 3.1	(42) 3.5	(54) 3.2	
(7) 4.3	(19) 3.3	(31) 3.2	(43) 3.6	(55) 3.4	
(8) 4.2	(20) 2.7	(32) 3.5	(44) 3.0	(56) 3.3	
(9) 4.2	(21) 3.3	(33) 2.4	(45) 3.1	(57) 3.6	
(10) 4.1	(22) 3.6	(34) 3.5	(46) 3.4	(58) 3.1	
(11) 4.3	(23) 3.2	(35) 3.3	(47) 3.1	(59) 3.4	
(12) 4.1	(24) 2.9	(36) 3.6	(48) 3.6	(60) 3.5	

I hope you can make use of this information.

Quickly analyse the above data and report on what information can be derived.

Worked examples using the normal distribution

1 Estimating proportion defective produced

In manufacturing it is frequently necessary to estimate the proportion of product produced outside the tolerance limits, when a process is not capable of meeting the requirements. The method to be used is illustrated in the following example: 100 units were taken from a margarine packaging unit which was 'in statistical control' or stable. The packets of margarine were weighed and the mean weight, $\overline{X} = 255$ g, the estimated standard deviation, $s = 4.73$ g. If the product specification demanded a weight of 250 ± 10 g, how much of the output of the packaging process would lie outside the tolerance zone?

This situation is represented in Figure 5.8. Since the characteristics of the normal distribution are measured in units of standard deviations, we must first convert the distance between the process mean and the Upper Specification Limit into s units. This is done as follows:

$$Z = (\mathrm{USL} - \overline{X}) / s,$$

where USL = Upper Specification Limit

$\overline{X}$ = Estimated Process Mean

s = Estimated Process Standard Deviation

Z = Number of standard deviations between USL and $\overline{X}$

(termed the standardized normal variate).

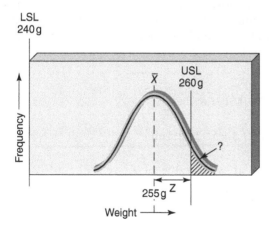

Figure 5.8 Determination of proportion defective produced

Hence, $Z = (260 - 255)/4.73 = 1.057$. Using the Table of Proportion Under the Normal Curve in Appendix A, it is possible to determine that the proportion of packages lying outside the USL was 0.145 or 14.5 per cent. There are two contributory causes for this high level of rejects:

(i) the setting of the process, which should be centred at 250 g and not 255 g, and
(ii) the spread of the process.

If the process was centred at 250 g, and with the same spread, one may calculate using the above method the proportion of product which would then lie outside the tolerance band. With a properly centred process, the distance between both the specification limits and the process mean would be 10 g. So:

$$Z = (USL - \overline{X}) / s = (\overline{X} - LSL) / s = 10 / 4.73 = 2.11.$$

Using this value of Z and the table in Appendix A the proportion lying outside each specification limit would be 0.0175. Therefore, a total of 3.5 per cent of product would be outside the tolerance band, even if the process mean was adjusted to the correct target weight. Statistical software may be used to perform the same analysis, of course.

2 Setting targets

(a) It is still common in some industries to specify an acceptable quality level (AQL) – this is the proportion or percentage of product that the producer/customer is prepared to accept outside the tolerance band. The characteristics of the normal distribution may be used to determine the target maximum standard deviation, when the target mean and AQL are specified. For example, if the tolerance band for a filling process is 5 ml and an AQL of 2.5 per cent is specified, then for a centred process:

$$Z = (USL - \overline{X}) / s = (\overline{X} - LSL) / s \text{ and}$$

$$(USL - \overline{X}) = (\overline{X} - LSL) / s = 5 / 2 = 2.5 \, ml.$$

We now need to know at what value of Z we will find (2.5%/2) under the tail – this is a proportion of 0.0125, and from Appendix A this is the proportion when $Z = 2.24$. So rewriting the above equation we have:

$$S_{msx} = (USL - \overline{X}) / Z = 2.5 / 2.24 = 1.12 \, ml.$$

In order to meet the specified tolerance band of 5 ml and an AQL of 2.5 per cent, we need an estimated standard deviation, measured on the products, of at most 1.12 ml.

(b) Consider a paint manufacturer who is filling nominal 1-litre cans with paint. The quantity of paint in the cans varies according to the normal distribution with an estimated standard deviation of 2 ml. If the stated minimum quality in any can is 1000 ml, what quantity must be put into the cans on average in order to ensure that the risk of underfill is 1 in 40?

1 in 40 in this case is the same as an AQL of 2.5 per cent or a probability of non-conforming output of 0.025 – the specification is one sided. The 1 in 40 line must be set at 1000 ml. From Appendix A this probability occurs at a value for Z of ca $2s$. So 1000 ml must be $2s$ below the average quantity. The process mean must be set at:

$$(1000 + 2s) \text{ ml} = 1000 + (2 \times 2) \text{ ml}$$

$$= 1004 \text{ ml}$$

This is illustrated in Figure 5.9.

A special type of graph paper, normal probability paper, which is also described in Appendix A, can be of great assistance to the specialist in handling normally distributed data. Again the analysis may be performed using statistical software.

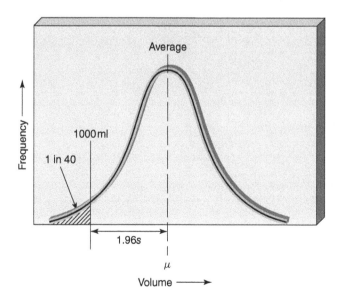

Figure 5.9 Setting target fill quantity in paint process

3 Setting targets

A bagging line fills plastic bags with polyethylene pellets which are automatically heat sealed and packed in layers on a pallet. SPC charting of the bag weights by packaging personnel has shown an estimated standard deviation of 20 g. Assume the weights vary according to a normal distribution. If the stated minimum quantity in one bag is 25 kg what must be average quantity of resin put in a bag be if the risk for underfilling is to be about one chance in 250?

The 1 in 250 (4 out of 1000 = 0.0040) line must be set at 25,000 g. From Appendix A, Average $- 2.65s = 25,000$ g. Thus, the average target should be $25,000 + (2.65 \times 20)$ g $= 25, 053$ g $= 25,053$ kg (see Figure 5.10).

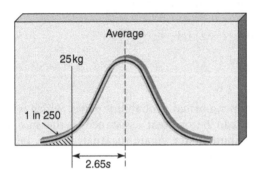

Figure 5.10 Target setting for the pellet bagging process

Part III
Process control

6 Process control using variables

Objectives

- To introduce the use of mean and range charts for the control of process accuracy and precision for variables.
- To provide the method by which process control limits may be calculated.
- To set out the steps in assessing process stability and capability.
- To examine the use of mean and range charts in the real-time control of processes.
- To look at alternative ways of calculating and using control charts limits.

6.1 Means, ranges and charts

To control a process using variable data, it is necessary to keep a check on the current state of the accuracy (central tendency) and precision (spread) of the distribution of the data. This may be achieved with the aid of control charts.

All too often processes are adjusted on the basis of a single result or measurement ($n = 1$), a practice that can increase the apparent variability. As pointed out in Chapter 4, a control chart is like a traffic signal, the operation of which is based on evidence from process samples taken at random intervals. A green light is given when the process should be allowed to run without adjustment, only random or common causes of variation being present. The equivalent of an amber light appears when trouble is possible. The red light shows that there is practically no doubt that assignable or special causes of variation have been introduced; the process has wandered.

Clearly, such a scheme should be introduced only when the process is 'in statistical control', i.e. is not changing its characteristics of average and spread. When interpreting the behaviour of a whole population from a sample, often small and typically less than 10, there is a risk of error. It is important to know the size of such a risk.

The American Shewhart was credited with the invention of control charts for variable and attribute data in the 1920s, at the Bell Telephone

Laboratories, and the term 'Shewhart charts' is in common use. The most frequently used charts for variables are mean and range charts which are used together. There are, however, other control charts for special applications to variables data. These are dealt with in Chapter 7. Control charts for attributes data are to be found in Chapter 8.

We have seen in Chapter 5 that with variable parameters, to distinguish between and control for accuracy and precision, it is advisable to group results, and a sample size of $n = 4$ or more is preferred. This provides an increased sensitivity with which we can detect changes of the mean of the process and take suitable corrective action.

Is the process in control?

The operation of control charts for sample mean and range to detect the state of control of a process proceeds as follows. Periodically, samples of a given size (e.g. four steel rods, five tins of paint, eight tablets, four delivery times) are taken from the process at reasonable intervals, when it is believed to be stable or in control and adjustments are not being made. The variable (length, volume, weight, time, etc.) is measured for each item of the sample and the sample mean and range recorded on a chart, the layout of which resembles Figure 6.1. The layout of the chart makes sure the following information is presented:

- chart identification;
- any specification;
- statistical data;
- data collected or observed;
- sample means and ranges;
- plot of the sample mean values;
- plot of the sample range values.

The grouped data on steel rod lengths from Table 5.1 have been plotted on mean and range charts, without any statistical calculations being performed, in Figure 6.2. Such a chart should be examined for any 'fliers', for which, at this stage, only the data itself and the calculations should be checked. The sample means and ranges are not constant; they vary a little about an average value. Is this amount of variation acceptable or not? Clearly we need an indication of what is acceptable, against which to judge the sample results.

Mean chart

We have seen in Chapter 5 that if the process is stable, we expect most of the individual results to lie within the range $\bar{X} \pm 3\sigma$ Moreover, if we are sampling from a stable process most of the sample means will lie within the range $\bar{X} \pm 3SE$. Figure 6.3 shows the principle of the mean control chart

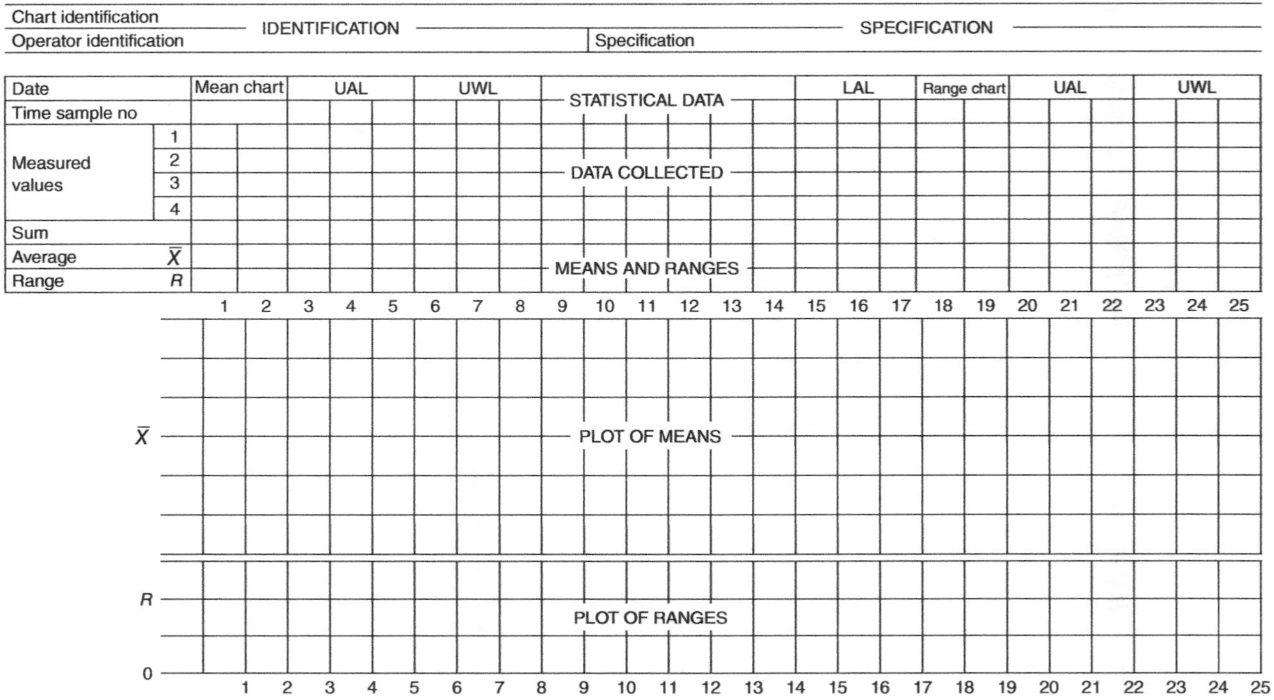

Figure 6.1 Layout of mean and range charts

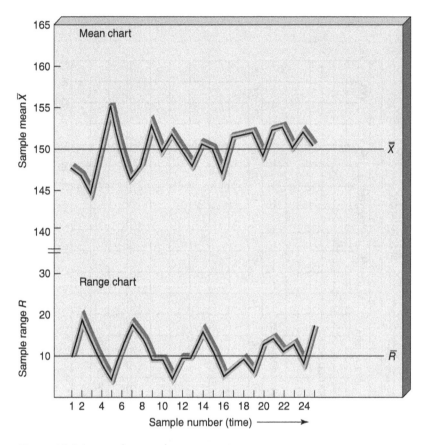

Figure 6.2 Mean and range chart

where we have turned the distribution 'bell' onto its side and extrapolated the ±2SE and ±3SE lines as well as the Grand or Process Mean line. We can use this to assess the degree of variation of the 25 estimates of the mean rod lengths, taken over a period of supposed stability. This can be used as the 'template' to decide whether the means are varying by an expected or unexpected amount, judged against the known degree of random variation. We can also plan to use this in a control sense to estimate whether the means have moved by an amount sufficient to require us to make a change to the process.

If the process is running satisfactorily, we expect from our knowledge of the normal distribution that more than 99 per cent of the means of successive samples will lie between the lines marked Upper Action and Lower Action. These are set at a distance equal to 3SE on either side of the mean. The chance of a point falling outside either of these lines is approximately 1 in 1000, unless the process has altered during the sampling period.

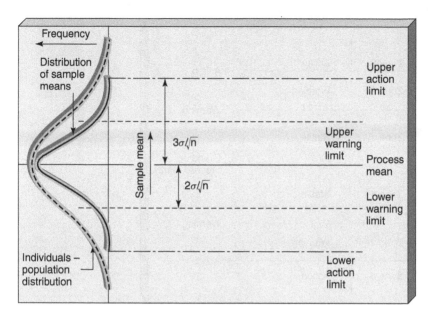

Figure 6.3 Principle of mean control chart

Figure 6.3 also shows warning limits which have been set 2SE each side of the process mean. The chance of a sample mean plotting outside either of these limits is about 1 in 40, i.e. it is expected to happen but only once in approximately 40 samples, if the process has remained stable.

So, as indicated in Chapter 4, there are three zones on the mean chart (Figure 6.4). If the mean value based on four results lies in zone 1 – and remember it is only an estimate of the actual mean position of the whole family – this is a very likely place to find the estimate, if the true mean of the population has not moved.

If the mean is plotted in zone 2 – there is, at most, a 1 in 40 chance that this arises from a process that is still set at the calculated process mean value, $\bar{X}$.

If the result of the mean of four lies in zone 3 there is only about a 1 in 1000 chance that this can occur without the population having moved, which suggests that the process must be unstable or 'out of control'. The chance of two consecutive sample means plotting in zone 2 is approximately $1/40 \times 1/40 = 1/1600$, which is even lower than the chance of a point in zone 3. Hence, two consecutive warning signals suggest that the process is out of control.

The presence of unusual patterns, such as runs or trends, even when all sample means and ranges are within zone 1, can be evidence of changes in process average or spread. This may be the first warning of unfavourable conditions which should be corrected even before points

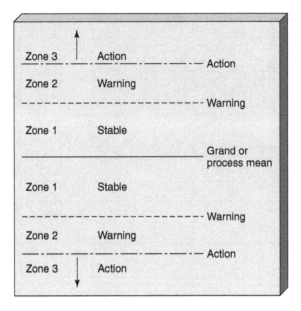

Figure 6.4 The three zones on the mean chart

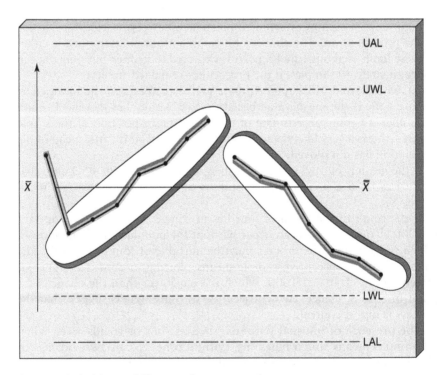

Figure 6.5 A rising or falling trend on a mean chart

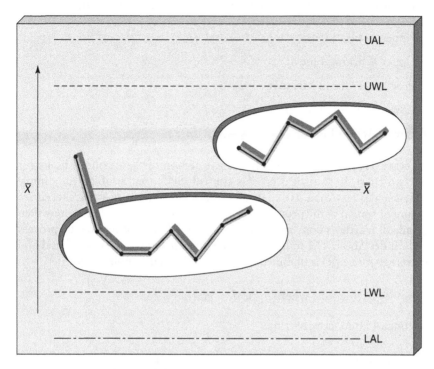

Figure 6.6 A run above or below the process mean value

occur outside the warning or action lines. Conversely, certain patterns or trends could be favourable and should be studied for possible improvement of the process.

Runs are often signs that a process shift has taken place or has begun. A run is defined as a succession of points that are above or below the average. A trend is a succession of points on the chart that are rising or falling, and may indicate gradual changes, such as tool wear. The rules concerning the detection of runs and trends are based on finding a series of seven points in a rising or falling trend (Figure 6.5), or in a run above or below the mean value (Figure 6.6). These are treated as out of control signals.

The reason for choosing seven is associated with the risk of finding one point above the average, but below the warning line being *ca*. 0.475. The probability of finding seven point in such a series will be (0.475)7 = *ca*. 0.005. This indicates how a run or trend of seven has approximately the same probability of occurring as a point outside an action line (zone 3). Similarly, a warning signal is given by five consecutive points rising of falling, or in a run above or below the mean value.

The formulae for setting the action and warning lines on mean charts are:

Upper Action Line at	$\overline{\overline{X}} + 3\sigma / \sqrt{n}$
Upper Warning Line at	$\overline{\overline{X}} + 2\sigma / \sqrt{n}$
Process or Grand Mean at	$\overline{\overline{X}}$
Lower Warning Line at	$\overline{\overline{X}} - 2\sigma / \sqrt{n}$
Lower Action Line at	$\overline{\overline{X}} - 3\sigma / \sqrt{n}$.

It is, however, possible to simplify the calculation of these control limits for the mean chart. In statistical process control (SPC) for variables, the sample size is usually less than 10, and it becomes possible to use the alternative measure of spread of the process – the mean range of samples $\overline{R}$. Use may then be made of Hartley's conversion constant (d_n or d_2) for estimating the process standard deviation. The individual range of each sample R_i is calculated and the average range ($\overline{R}$) is obtained from the individual sample ranges.

$$\sigma = \overline{R} / d_n \text{ or } \overline{R} / d_2, \text{ where } d_n \text{ or } d_2 = \text{Hartley's constant.}$$

The control limits now become:

Action Lines at $\quad \overline{\overline{X}} \quad \pm \quad A_2 \quad \times \quad \overline{R}$
$\qquad\qquad\qquad\quad | \qquad\qquad | \qquad\qquad |$
$\qquad\qquad$ Grand or Process $\qquad$ A constant $\quad$ Mean of sample ranges
$\qquad\qquad$ Mean of sample means

Warning Lines at $\quad \overline{\overline{X}} \pm 2/3\, A_2 R$

The constants d_n, A_2 and 2/3 A_2 for sample sizes $n = 2$ to $n = 12$ have been calculated and appear in Appendix B. For sample sizes up to $n = 12$, the range method of estimating σ is relatively efficient. For values of n greater than 12, the range loses efficiency rapidly as it ignores all the information in the sample between the highest and lowest values. For the small samples sizes ($n = 4$ or 5) often employed on variables control charts, it is entirely satisfactory.

Using the data on lengths of steel rods in Table 5.1, we may now calculate the action and warning limits for the mean chart for that process:

$$\text{Process Mean, } \overline{\overline{X}} = \frac{147.5 + 147.0 + 144.75 + \dots + 150.5}{25}$$

$$= 150.1\,\text{mm.}$$

$$\text{Mean Range, } \overline{R} = \frac{10 + 19 + 13 + 8 + \dots + 17}{25}$$

$$= 10.8\,\text{mm.}$$

From Appendix B, for a sample size $n = 4$; d_n or $d_2 = 2.059$

Therefore, $\sigma = \dfrac{\bar{R}}{d_n} = \dfrac{10.8}{2.059} = 5.25\,\text{mm}$,

and

$$\text{Upper Action Line} = 150.1 + (3 \times 5.25 / \sqrt{4})$$
$$= 157.98\text{mm}$$
$$\text{Upper Warning Line} = 150.1 + (2 \times 5.25 / \sqrt{4})$$
$$= 155.35\text{mm}$$
$$\text{Lower Warning Line} = 150.1 - (2 \times 5.25 / \sqrt{4})$$
$$= 144.85\text{mm}$$
$$\text{Lower Action Line} = 150.1 - (3 \times 5.25 / \sqrt{4})$$
$$= 142.23\text{mm}.$$

Alternatively the values of 0.73 for A_2 and 0.49 for 2/3 A_2 may be derived directly from Appendix B.
Now,
Action Lines at

Action Lines at $\bar{\bar{X}} \pm A_2\bar{R}$

therefore, Upper Action Line $= 150.1 + (0.73 \times 10.8)\ mm$
$$= 157.98\ mm,$$

and

Lower Action Line $= 150.1 - (0.73 \times 10.8)\ mm$
$$= 142.22\ mm.$$

Similarly,

Warning Lines $\bar{\bar{X}} \pm 2/3\ A_2\bar{R}$

therefore, Upper Warning Line $= 150.1 + (0.49 \times 10.8)\ mm$
$$= 155.40\ mm,$$

and

Lower Warning Line $= 150.1 - (0.49 \times 10.8)\ mm$
$$= 144.81\ mm.$$

These days most people will use an SPC software package to set up mean charts, of course, but these derivations are provided to ensure a complete understanding for those that require it.

Range chart

The control limits on the range chart are asymmetrical about the mean range since the distribution of sample ranges is a positively skewed distribution (Figure 6.7). The table in Appendix C provides four constants $D'_{0.001}$, $D'_{0.025}$, $D'_{0.975}$ and $D'_{0.999}$ which may be used to calculate the control limits for a range chart. Thus:

Upper Action Line at $\quad D'_{0.001}\overline{R}$

Upper Warning Line at $\quad D'_{0.025}\overline{R}$

Lower Warning Line at $\quad D'_{0.975}\overline{R}$

Lower Action Line at $\quad D'_{0.999}\overline{R}$.

For the steel rods, the sample size is four and the constants are thus:

$D'_{0.001} = 2.57, D'_{0.025} = 1.93,$

$D'_{0.999} = 0.10, D'_{0.975} = 0.29.$

As the mean range $\overline{R}$ is 10.8 mm the control limits for range are:

Action Lines at $\quad 2.57 \times 10.8 = 27.8$ mm

and $\qquad\qquad\quad 0.10 \times 10.8 = 1.1$ mm,

Warning Lines at $1.93 \times 10.8 = 10.8$ mm

and $\qquad\qquad\quad 0.29 \times 10.8 = 3.1$ mm.

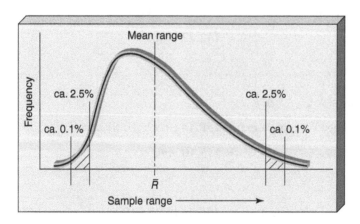

Figure 6.7 Distribution of sample ranges

The action and warning limits for the mean and range charts for the steel rod cutting process have been added to the data plots in Figure 6.8. Although the statistical concepts behind control charts for mean and range may seem complex to the non-mathematically inclined, the steps in setting up the charts are remarkably simple:

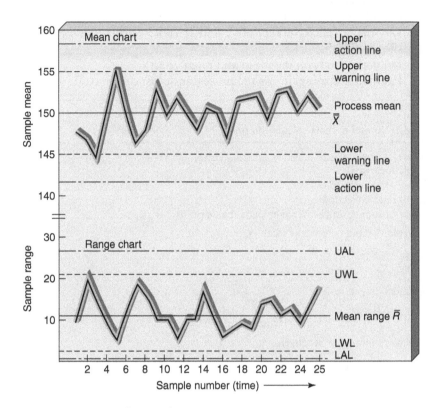

Figure 6.8 Mean and range chart

Steps in assessing process stability

1   Select a series of random samples of size *n* (greater than 4 but less than 12) to give a total number of individual results between 50 and 100.
2   Measure the variable *x* for each individual item.
3   Calculate $\overline{X}$, the sample mean and R, the sample range for each sample.
4   Calculate the Process Mean $\overline{\overline{X}}$ – the average value of $\overline{X}$ and the Mean Range $\overline{R}$ – the average value of R.

(continued)

(continued)

5   Plot all the values of $\overline{X}$ and R and examine the charts for any possible miscalculations.

6   Look up: d_n, A_2, $2/3A_2$, $D'_{0.999}$, $D'_{0.975}$, $D'_{0.025}$ and $D'_{0.001}$ (see Appendices B and C).

7   Calculate the values for the action and warning lines for the mean and range charts. A typical $\overline{X}$ and R chart calculation form is shown in Table 6.1.

8   Draw the limits on the mean and range charts.

9   Examine charts again – is the process in statistical control?

Table 6.1 $\overline{X}$ and R chart calculation form

Process: _____   Date:
Variable measured:
Number of subgroups (K):
Dates of data collection:
Number of samples/measurements per subgroup: (n)

1   Calculate grand or process mean $\overline{\overline{X}}$:

$$\overline{\overline{X}} = \frac{\Sigma\overline{X}}{K} = \underline{\quad} =$$

2   Calculate mean range:

$$\overline{R} = \frac{\Sigma R}{K} = \underline{\quad} =$$

3   Calculate limits for $\overline{X}$ chart:

UAL/LAL $= \overline{\overline{X}} \pm (A_2 \times \overline{R})$
UAL/LAL $=\quad \pm$
UAL $=\quad$ LAL $=$
UWL/LWL $= \overline{\overline{X}} \pm (2/3\ A_2 \times \overline{R})$
UWL/LWL $=\quad \pm$
UWL $=\quad$ LWL $=$

4   Calculate limits for R chart:

UAL $= D'_{0.001} \times \overline{R}$ LAL $= D'_{0.999} \times \overline{R}$
UAL $=$ LAL $=$
UWL $= D'_{0.025} \times \overline{R}$ LWL $= D'_{0.975} \times \overline{R}$
UWL $=$ LWL $=$

There are many computer packages available that will perform these calculations and plot data on control charts. One of the easiest to use and highly aligned with the methods of SPC presented in this book is Minitab.

6.2 Are we in control?

At the beginning of the section on mean charts it was stated that samples should be taken to set up control charts when it is believed that the process is in statistical control. Before the control charts are put into use or the process capability is assessed, it is important to confirm that when the samples were taken the process was indeed 'in statistical control', i.e. the distribution of individual items was reasonably stable.

Assessing the state of control

A process is in statistical control when all the variations have been shown to arise from random or common causes. The randomness of the variations can best be illustrated by collecting at least 30 observations of data and grouping these into samples or sets of at least four observations; presenting the results in the form of both mean and range control charts – the limits of which are worked out from the data. If the process from which the data was collected is in statistical control there will be:

- NO Mean or Range values that lie outside the Action Limits (zone 3 Figure 6.4);
- NO more than about 1 in 40 values between the Warning and Action Limits (zone 2);
- NO incidence of two consecutive Mean or Range values that lie outside the same Warning Limit on either the mean or the range chart (zone 2);
- NO run or trend of five or more that also infringes a warning or action limit (zone 2 or 3);
- NO runs of more than six sample Means that lie either above or below the Grand Mean (zone 1);
- NO trends of more than six values of the sample Means that are either rising or falling (zone 1).

If a process is 'out of control', the special causes will be located in time and must now be identified and eliminated. The process can then be re-examined to see whether it is in statistical control. If the process is shown to be in statistical control the next task is to compare the limits of this control with the tolerance sought.

The means and ranges of the 25 samples of four lengths of steel rods, which were plotted in Figure 6.2, may be compared with the calculated control limits in this way, using Figure 6.8.

We start by examining the range chart in all cases, because it is the range that determines the position of the range chart limits *and* the 'separation' of the limits on the mean chart. The range is in control – all the points lie inside the warning limits, which means that the spread of the distribution remained constant – the process is in control with respect to range or spread.

For the mean chart there are two points that fall in the warning zone – they are not consecutive and of the total points plotted on the charts we are expecting 1 in 40 to be in each warning zone when the process is stable. There are not 40 results available but we have to make a decision. It is reasonable to assume that the two plots in the warning zone have arisen from the random variation of the process and do not indicate an out of control situation.

There are no runs or trends of seven or more points on the charts and, from Figure 6.8, the process is judged to be in statistical control; the mean and range charts may now be used to control the process.

During this check on process stability, should any sample points plot outside the action lines, or several points appear between the warning and action lines, or any of the trend and run rules be contravened, then the assignable causes of variation should be investigated. When the special causes of variation have been identified and eliminated, either another set of samples from the process is taken and the control chart limits recalculated, or approximate control chart limits are recalculated by simply excluding the out of control results for which special causes have been found and corrected. The exclusion of samples representing unstable conditions is not just throwing away bad data. By excluding the points affected by known causes, we have a better estimate of variation due to common causes only. Most industrial processes are not in control when first examined using control chart methods and the special causes of the out of control periods should be found and corrected.

A clear distinction must be made between the tolerance limits set down in the product specification and the limits on the control charts. The former should be based on the functional requirements of the products, the latter are based on the stability and actual capability of the process. The process may be unable to meet the specification requirements but still be in a state of statistical control (Figure 6.9). A comparison of process capability and tolerance can take place with confidence when it has been established that the process is in control statistically.

Capability of the process

So with both the mean and the range charts in statistical control, we have shown that the process was stable for the period during which samples were taken. We now know that the variations were due to common causes only, but how much scatter is present, and is the process capable of meeting the requirements? We know that, during this period of stable running,

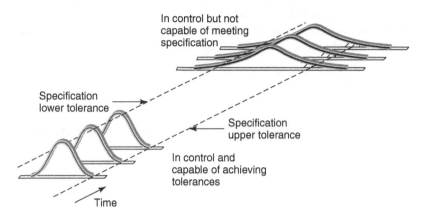

Figure 6.9 Process capability

the results were scattered around a Process Mean of $\overline{\overline{X}}=150.1$ mm, and that, during this period, the Mean Range $\overline{R}=10.8$ mm. From this we have calculated that the standard deviation was 5.25 mm, and it is possible to say that more than 99 per cent of the output from the process will lie within three standard deviations on either side of the mean, i.e. between $150.1 \pm 3 \times 5.25$ mm or 134.35 to 165.85 mm.

If a specification for the rod-cutting process had been set, it would be possible at this stage to compare the capability of the process with the requirements. It is important to recognize that the information about capability and the requirements come from different sources – they are totally independent, one being the 'voice of the process,' the other being the 'voice of the customer'. The specification does not determine the capability of the process and the process capability does not determine the requirement, but they do need to be known, compared and found to be compatible. The quantitative assessment of capability with respect to the specified requirements is the subject of Chapter 10.

6.3 Do we continue to be in control?

When the process has been shown to be in control, the mean and range charts may be used to make decisions about the state of the process during its operation. Just as for testing whether a process was in control, we can use the three zones on the charts for *controlling* or managing the process:

Zone 1 – If the points plot in this zone it indicates that the process has remained stable and actions/adjustments are unnecessary, indeed they may increase the amount of variability.

Zone 3 – Any points plotted in this zone indicate that the process should be investigated and that, if action is taken, the latest estimate of the mean and its difference from the original process mean or target value should be used to assess the size of any 'correction'.

Zone 2 – A point plotted in this zone suggests there may have been an assignable change and that another sample should be taken in order to check.

Such a second sample can lie in only one of the three zones as shown in Figure 6.10.

- If it lies in zone 1 – then the previous result was a statistical event which has approximately a 1 in 40 chance of occurring every time we estimate the position of the mean.
- If it lies in zone 3 – there is only approximately a 1 in 1000 chance that it can get there without the process mean having moved, so the latest estimate of the value of the mean may be used to correct it.
- If it again lies in zone 2 – then there is approximately a $1/40 \times 1/40 = 1/1600$ chance that this is a random event arising from an unchanged mean, so we can again use the latest estimate of the position of the mean to decide on the corrective action to be taken.

This is a simple list of instructions to give to an 'operator' of any process. The first three options corresponding to points in zones 1, 2, 3, respectively are: 'do nothing', 'take another sample', 'investigate or adjust the process'. If a second sample is taken following a point in zone 2, it is done in the certain knowledge that this time there will be one of two conclusions: either

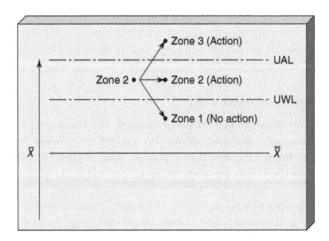

Figure 6.10 The second sample following a warning signal in zone 2

'do nothing', or 'investigate/adjust'. In addition, when the instruction is to adjust the process, it is accompanied by an estimate of by how much, and this is based on four observations not one. The rules given on page 117 for detecting runs and trends should also be used in controlling the process.

Figure 6.11 provides an example of this scheme in operation. It shows mean and range charts for the next 30 samples taken from the steel rod cutting process. The process is well under control, i.e. within the action lines, until sample 11, when the mean almost reaches the Upper Warning Line. A cautions person may be tempted to take a repeat sample here although, strictly speaking, this is not called for if the technique is applied rigidly. This decision depends on the time and cost of sampling, among other factors.

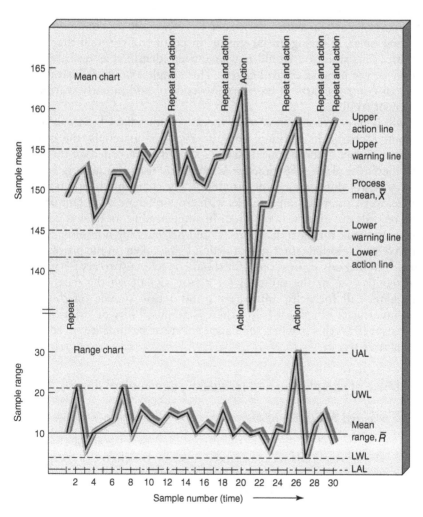

Figure 6.11 Mean and range chart in process control

Sample 12 shows that the cautious approach was justified for its mean has plotted above the Upper Action Line and corrective action should be taken. This action brings the process back into control again until sample 18 which is the fifth point in a run above the mean – another sample should be taken immediately, rather than wait for the next sampling period. The mean of sample 19 is in the warning zone and these two consecutive 'warning' signals indicate that corrective action should be taken. However, sample 20 gives a mean well above the action line, indicating that the corrective action caused the process to move in the wrong direction. The action following sample 20 results in over-correction and sample mean 21 is below the lower action line.

The process continues to drift upwards out of control between samples 21 and 26 and from 28 to 30. The process equipment was investigated as a result of this – a worn adjustment screw was slowly and continually vibrating open, allowing an increasing speed of rod through the cutting machine. This situation would not have been identified as quickly in the absence of the process control charts. This simple example illustrates the power of control charts in both process control and in early warning of equipment trouble.

It will be noted that 'action' and 'repeat' samples have been marked on the control charts. In addition, any alterations in materials, the process, operators or any other technical changes should be recorded on the charts when they take place. This practice is extremely useful in helping to track down causes of shifts in mean or variability. The chart should not, however, become over-cluttered; simple marks with cross-references to plant or operators' records are all that is required. In some organizations it is common practice to break the pattern on the $\overline{X}$ and R charts, by not joining points that have been plotted either side of action being taken on the process.

It is vital that any process operator should be told how to act for warning zone signals (repeat the sample), for action signals on the mean (stop, investigate, call for help, adjust, etc.) and action signals on the range (stop, investigate or call for help – there is no possibility of 'adjusting' the process spread – this is where management must become involved in the investigative work).

6.4 Choice of sample size and frequency, and control limits

Sample size and frequency of sampling

In the example used to illustrate the design and use of control charts, 25 samples of four steel rods were measured to set up the charts. Subsequently, further samples of size four were taken at regular intervals to control the process. This is a common sample size, but there may be justification for taking other sample sizes. Some guidelines may be helpful:

1 The sample size should be at least 2 to give an estimate of residual variability, but a minimum of 4 is preferred, unless the infrequency of sampling limits the available data to 'one at a time'.

2 As the sample size increases, the mean control chart limits become closer to the process mean. This makes the control chart more sensitive to the detection of small variations in the process average.

3 As the sample size increases, the inspection costs per sample may increase. One should question whether the greater sensitivity justifies any increase in cost.

4 The sample size should not exceed 12 if the range is to be used to measure process variability. With larger samples the resulting mean range ($\bar{R}$) does not give a good estimate of the standard deviation and sample standard deviation charts should be used.

5 When each item has a high monetary value and destructive testing is being used, a small sample size is desirable and satisfactory for control purposes.

6 A sample size of $n = 5$ is often used because of the ease of calculation of the sample mean (multiply sum of values by 2 and divide result by 10 or move decimal point 1 digit to left). However, with the advent of inexpensive computers and calculators, this is no longer necessary.

7 The technology of the process may indicate a suitable sample size. For example, in the control of a paint filling process the filling head may be designed to discharge paint through six nozzles into six cans simultaneously. In this case, it is obviously sensible to use a sample size of six – one can from each identified filling nozzle, so that a check on the whole process and the individual nozzles may be maintained.

There are no general rules for the frequency of taking samples. It is very much a function of the product being made and the process used. It is recommended that samples are taken quite often at the beginning of a process capability assessment and process control. When it has been confirmed that the process is in control, the frequency of sampling may be reduced. It is important to ensure that the frequency of sampling is determined in such a way that ensures no bias exists and that, if autocorrelation (see Appendix I) is a problem, it does not give false indications on the control charts. The problem of how to handle additional variation is dealt with in the next section.

In certain types of operation, measurements are made on samples taken at different stages of the process, when the results from such samples are expected to follow a predetermined pattern. Examples of this are to be found in chemical manufacturing, where process parameters change as the starting materials are converted into products or intermediates. It may be desirable to plot the sample means against time to observe the process profile or progress of the reaction, and show warning and action control limits on these graphs, in the usual way. Alternatively, a chart of means of differences from a target value, at a particular point in time, may be plotted with a range chart.

Control chart limits

Instead of calculating upper and lower warning lines at two standard errors, the American automotive and other industries use simplified control charts and set an 'Upper Control Limit' (UCL) and a 'Lower Control Limit' (LCL) at three standard errors either side of the process mean. To allow for the use of only one set of control limits, the UCL and LCL on the corresponding range charts are set in between the 'action' and 'warning' lines. The general formulae are:

$$\text{Upper Control Limit} = D_4\overline{R},$$
$$\text{Lower Control Limit} = D_2\overline{R},$$

where n is 6 or less, the LCL will turn out to be less than 0 but, because the range cannot be less than 0, the lower limit is not used. The constants D_2 and D_4 may be found directly in Appendix C for sample sizes of 2 to 12. A sample size of 5 is commonly used in the automotive industry.

Such control charts are used in a very similar fashion to those designed with action and warning lines. Hence, the presence of any points beyond either UCL or LCL is evidence of an out of control situation and provides a signal for an immediate investigation of the special cause. Because there are no warning limits on these charts, some additional guidance is usually offered to assist the process control operation. This guidance is more complex and may be summarized as:

1   Approximately two-thirds of the data points should be within the middle third region of each chart – for mean and for range. If substantially more or less than two-thirds of the points lie close to $\overline{\overline{X}}$ or $\overline{R}$, then the process should be checked for possible changes.
2   If common causes of variation only are present, the control charts should not display any evidence of runs or trends in the data. The following are taken to be signs that a process shift or trend has been initiated:

   •   seven points in a row on one side of the average;
   •   seven lines between successive points which are continually increasing or decreasing.

3   There should be no occurrences of two mean points out of three consecutive points on the *same* side of the centreline in the zone corresponding to one standard error (SE) from the process mean $\overline{\overline{X}}$.

4   There should be no occurrences of four mean points out of five consecutive points on the *same* side of the centreline in the zone between one and two standard errors away from the process mean $\overline{\overline{X}}$.

It is useful practice for those using the control chart system with warning lines to also apply the simple checks described above. The control charts with warning lines, however, offer a less stop or go situation than the UCL/LCL system, so there is less need for these additional checks. The more complex the control chart system rules, the less likely they will be adhered to. The temptation to adjust the process when a point plots near to a UCL or an LCL is real. If it falls in a warning zone, there is a clear signal to check, not to panic and above all not to adjust. It is authors' experience that the use of warning limits and zones give process operators and managers clearer rules and quicker understanding of variation and its management.

The precise points on the normal distribution at which 1 in 40 and 1 in 1000 probabilities occur are at 1.96 and 3.09 standard deviation from the process mean, respectively. Using these refinements instead of the simpler 2 and 3 standard deviations makes no significant difference to the control system. The original British Standards on control charts quoted the 1.96 and 3.09 values. Appendix F gives confidence limits and tests of significance and Appendix G gives operating characteristics (OC) and average run lengths (ARL) curves for mean and range charts.

There are clearly some differences between the various types of control charts for mean and range. Far more important than any operating discrepancies is the need to understand and adhere to whichever system has been chosen.

6.5 Short-, medium- and long-term variation: a change in the standard practice

In their excellent paper on control chart design, Caulcutt and Porter (1992) pointed out that, owing to the relative complexity of control charts and the lack of understanding of variability at all levels, many texts on SPC (including this one!) offer simple rules for setting up such charts. As we have seen earlier in this chapter, these rules specify how the values for the centreline and the control lines, or action lines, should be calculated from data. The rules work very well in many situations but they do not produce useful charts in all situations. Indeed, the failure to implement SPC in many organizations may be due to following rules that are based on an over-simplistic model of process variability.

Caulcutt and Porter examined the widely used procedures for setting up control charts and illustrated how these may fail when the process variability has certain characteristics. They suggested an alternative, more robust, procedure that involves taking a closer look at variability and the many ways in which it can be quantified.

Caulcutt and Porter's survey of texts on SPC revealed a consensus view that data should be subgrouped and that the ranges of these groups (or perhaps the standard deviations of the groups) should be used to calculate values for positioning the control lines. In practice there may be a natural subgrouping of the data or there may be a number of arbitrary groupings that are possible, including groups of one, i.e. 'one-at-a-time' data.

They pointed out that, regardless of the ease or difficulty of grouping the data from a particular process, the forming of subgroups is an essential step in the investigation of stability and in the setting up of control charts. Furthermore, the use of group ranges to estimate process variability is so widely accepted that 'the mean of subgroup ranges' $\bar{R}$ may be regarded as the central pillar of a standard procedure.

Many people follow the standard procedure given on page 116 and achieve great success with their SPC charts. The short-term benefits of the method include fast reliable detection of change which enables early corrective action to be taken. Even greater gains may be achieved in the longer term, however, if charting is carried out within the context of the process itself, to facilitate greater process understanding and reduction in variability.

In many processes, including many in the chemical and so-called process industries, there is a tendency for observations that are made over a relatively short time period to be more alike than those taken over a longer period. In such instances the additional 'between group' or 'medium-term' variability may be comparable with or greater than the 'within group' or 'short-term' variability. If this extra component of variability is random there may be no obvious way that it can be eliminated and the within group variability will be a poor estimate of the natural random longer-term variation of the process. It should not then be used to control the process.

Caulcutt and Porter observed many cases in which sampling schemes based on the order of output or production gave unrepresentative estimates of the random variation of the process, if $\bar{R} / d_n$ was used to calculate σ. Use of the standard practice in these cases gave control lines for the mean chart which were too 'narrow', and resulted in the process being over-controlled. Unfortunately, not only do many people use bad estimates of the process variability, but in many instances sampling regimes are chosen on an arbitrary basis. It was not uncommon for them to find very different sampling regimes being used in the preliminary process investigation/chart design phase and the subsequent process monitoring phase.

Caulcutt and Porter showed an example of this (Figure 6.12) in which mean and range charts were used to control can heights on can-making production lines (the measurements are expressed as the difference from a nominal value and are in units of 0.001 cm.) It can be seen that 13 of the 50 points lie outside the action lines and the fluctuations in the mean can height result in the process appearing to be 'out-of-statistical control'. There is,

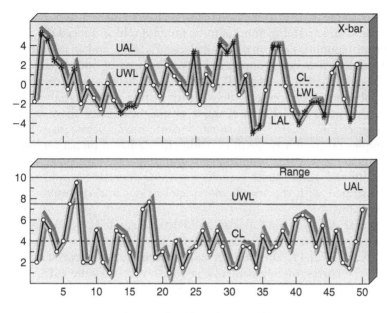

Figure 6.12 Mean and range chart based on standard practice

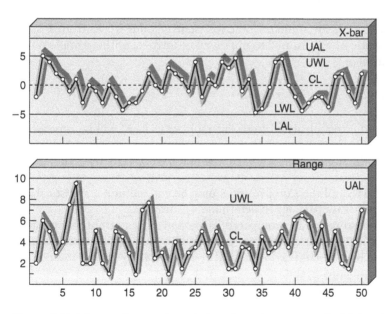

Figure 6.13 Mean and range chart designed to take account of additional random variation

however, no simple pattern to these changes, such as trend or a step change, and the additional variability appears to be random. This is indeed the case for the process contains random *within* group variability, and an additional source of random *between* group variability. This type of additional variability is frequently found in can-making, filling and many other processes.

A control chart design based solely on the within group variability is inappropriate in this case. In the example given, the control chart would mislead its user into seeking an assignable cause on 22 occasions out of the 50 samples taken, if a range of decision criteria based on action lines, repeat points in the warning zone and runs and trends are used (page 117). As this additional variation is actually random, operators would soon become frustrated with the search for special causes and corresponding corrective actions.

To overcome this problem Caulcutt and Porter suggested calculating the standard error of the means directly from the sample means to obtain, in this case, a value of 2.45. This takes account of within and between group variability. The corresponding control chart is shown in Figure 6.13. The process appears to be in statistical control and the chart provides a basis for effective control of the process.

Stages in assessing additional variability

1 Test for additional variability

As we have seen, the standard practice yields a value of $\overline{R}$ from k small samples of size n. This is used to obtain an estimate of within sample standard deviation σ:

$$\sigma = \overline{R} / d_n.$$

The standard error calculated from this estimate ($\sigma / \sqrt{n}$) will be appropriate if σ describes all the natural random variation of the process. A different estimate of the standard error, σ_e, can be obtained directly from the sample means, $\overline{X}$.

Alternatively, all the sample means may be entered into a statistical calculator or software package to determine σ_{n-1} directly.

The two estimates are compared: if σ_e and $\sigma / \sqrt{n}$ are approximately equal there is no extra component of variability and the standard practice for control chart design may be used. If σ_e is appreciably greater than $\sigma / \sqrt{n}$ there is additional variability.

In the can-making example previously considered, the two estimates are:

$$\sigma / \sqrt{n} = 0.94$$
$$\sigma_e = 2.45.$$

This is a clear indication that additional medium-term variation is present.

(A formal significance test for the additional variability can be carried out by comparing $n\sigma^2/\sigma^2$ with a required or critical value from tables of the F distribution with $(k-1)$ and $k(n-1)$ degrees of freedom. A 5 per cent level of significance is usually used. See Appendix G.)

2 Calculate the control lines

If stage 1 has identified additional between group variation, then the mean chart action and warning lines are calculated from σ_e:

Action lines $\overline{\overline{X}} \pm 3\sigma_e$

Warning lines $\overline{\overline{X}} \pm 2\sigma_e$.

These formulae can be safely used as an *alternative to the standard practice* even if there is no additional medium-term variability, i.e. even when $\sigma = \overline{R}/d_n$ is a good estimate of the natural random variation of the process.

(The standard procedure is used for the range chart as the range is unaffected by the additional variability. The range chart monitors the within sample variability only.)

In the can-making example the alternative procedure gives the following control lines for the mean chart:

Upper Action Line	7.39
Lower Action Line	−7.31
Upper Warning Line	4.94
Lower Warning Line	−4.86.

These values provide a sound basis for detecting any systematic variation without over-reacting to the inherent medium-term variation of the process.

The use of σ_e to calculate action and warning lines has important implications for the sampling regime used. Clearly a fixed sample size, n, is required but the sampling frequency must also remain fixed as σ_e takes account of any random variation over time. It would not be correct to use different sampling frequencies in the control chart design phase and subsequent process monitoring phase.

6.6 Summary of SPC for variables using $\overline{X}$ and R charts

If data is recorded on a regular basis, SPC for variables proceeds in three main stages:

1   An examination of the 'State of Control' of the process (Are we in control?)
     A series of measurements are carried out and the results plotted on $\overline{X}$
     and R control charts to discover whether the process is changing due to
     assignable causes. Once any such causes have been found and removed,
     the process is said to be 'in statistical control' and the variations then
     result only from the random or common causes.
2   A 'Process Capability' Study (Are we capable?). It is never possible to
     remove all random or common causes – some variations will remain.
     A process capability study shows whether the remaining variations are
     acceptable and whether the process will generate products or services
     that match the specified requirements.
3   Process Control Using Charts (Do we continue to be in control?). The
     $\overline{X}$ and R charts carry 'control limits' which form traffic light signals or
     decision rules and give operators information about the process and its
     state of control.

Control charts are an essential tool of continuous improvement and great
improvements in quality can be gained if well-designed control charts are
used by those who operate processes. Badly designed control charts lead to
confusion and disillusionment among process operators and management.
They can impede the improvement process as process workers and management rapidly lose faith in SPC techniques. Unfortunately, the authors and
their colleagues have observed too many examples of this across a range of
industries, when SPC charting can rapidly degenerate into a paper or computer exercise. A well-designed control chart can result only if the nature of
the process variation is thoroughly investigated.

In this chapter an attempt has been made to address the setting up of
mean and range control charts and procedures for designing the charts
have been outlined. For mean charts the SE estimate σ_e calculated directly
from the sample means, rather than the estimate based on $\overline{R} / d_n$, provides
a sound basis for designing charts that take account of complex patterns of
random variation as well as simple short-term or inter-group random variation. It is always sound practice to use pictorial evidence to test the validity
of summary statistics used.

In using SPC software packages, such as Minitab, it is important to know
the bases of the setting up of the control chart parameters.

Chapter highlights

- Control charts can be used to monitor and control processes based on
  means ($\overline{X}$) and ranges (R).
- There is a recommended method of collecting data for a process capability study and prescribed layouts for $\overline{X}$ and R control charts which
  include warning and action lines (limits). The control limits on the mean
  and range charts are based on simple calculations from the data.

- Mean chart limits are derived using the process mean $\overline{\overline{X}}$, the mean range $\overline{R}$, and either A_2 constants or by calculating the standard error (SE) from $\overline{R}$. The range chart limits are derived from $\overline{R}$ and D_1 constants.
- The interpretation of the plots are based on rules for action, warning and trend signals. Mean and range charts are used together to control the process.
- A set of rules is required to assess the stability of a process and to establish the state of statistical control. The capability of the process can be measured in terms of σ, and its spread compared with the specified tolerances.
- Mean and range charts may be used to monitor the performance of a process. There are three zones on the charts which are associated with rules for determining what action, if any, is to be taken.
- There are various forms of the charts originally proposed by Shewhart. These include charts without warning limits, which require slightly more complex guidance in use.
- Caulcutt and Porter's procedure is recommended when short- and medium-term random variation is suspected, in which case the standard procedure may lead to over-control of the process.
- SPC for variables is in three stages:

1 Examination of the 'state of control' of the process using mean and range/standard deviation charts.
2 A process capability study, comparing spread with specifications.
3 Process control using the charts.

References and further reading

Bissell, A.F. (1991) 'Getting More from Control Chart Data: Part 1', *Total Quality Management*, Vol. 2, No. 1, pp. 45–55.

Box, G.E.P., Hunter, W.G. and Hunter, J.S. (1978) *Statistics for Experimenters*, John Wiley & Sons, New York, USA.

Caulcutt, R. (1995) 'The Rights and Wrongs of Control Charts', *Applied Statistics*, Vol. 44, No. 3, pp. 279–88.

Caulcutt, R. and Coates, J. (1991) 'Statistical Process Control with Chemical Batch Processes', *Total Quality Management*, Vol. 2, No. 2, pp. 191–200.

Caulcutt, R. and Porter, L.J. (1992) 'Control Chart Design – A Review of Standard Practice', *Quality and Reliability Engineering International*, Vol. 8, pp. 113–122.

Duncan, A.J. (1986) *Quality Control and Industrial Statistics*, 5th edn, Richard D. Irwin IL, USA.

Grant, E.L. and Leavenworth, R.W. (1996) *Statistical Quality Control*, 7th edn, McGraw-Hill, New York, USA.

Montgomery, D. (2008) *Statistical Process Control: A Modern Introduction*, ASQ Press, Milwaukee WI, USA.

Pyzdek, T. (1990) Pyzdek's Guide to SPC, Vol. 1: Fundamentals, ASQC Quality Press, Milwaukee WI, USA.

Shewhart, W.A. (1931) *Economic Control of Quality of Manufactured Product*, Van Nostrand, New York, USA.
Wheeler, D.J. and Chambers, D.S. (1992) *Understanding Statistical Process Control*, 2nd edn, SPC Press, Knoxville TN, USA.

Discussion questions

1   (a) Explain the principles of Shewhart control charts for sample mean and sample range.
     (b) State the Central Limit Theorem and explain its importance in SPC.
2   A machine is operated so as to produce ball bearings having a mean diameter of 0.55 cm and with a standard deviation of 0.01 cm. To determine whether the machine is in proper working order a sample of six ball bearings is taken every half-hour and the mean diameter of the six is computed.

     (a) Design a decision rule whereby one can be fairly certain that the ball bearings constantly meet the requirements.
     (b) Show how to represent the decision rule graphically.
     (c) How could even better control of the process be maintained?

3   The following are measures of the impurity, iron, in a fine chemical that is to be used in pharmaceutical products. The data is given in parts per million (ppm).

Sample	X_1	X_2	X_3	X_4	X_5
1	15	11	8	15	6
2	14	16	11	14	7
3	13	6	9	5	10
4	15	15	9	15	7
5	9	12	9	8	8
6	11	14	11	12	5
7	13	12	9	6	10
8	10	15	12	4	6
9	8	12	14	9	10
10	10	10	9	14	14
11	13	16	12	15	18
12	7	10	9	11	16
13	11	7	16	10	14
14	11	7	10	10	7
15	13	9	12	13	17
16	17	10	11	9	8
17	4	14	5	11	11
18	8	9	6	13	9
19	9	10	7	10	13
20	15	10	10	12	16

Set up mean and range charts and comment on the possibility of using them for future control of the iron content.

4   You are responsible for a small plant that manufacturers and packs jollytots, a children's sweet. The average contents of each packet should be 35 sugar-coated balls of candy that melt in your mouth. Every half-hour a random sample of five packets is taken and the contents counted. These figures are shown below:

Sample	Packet contents				
	1	2	3	4	5
1	33	36	37	38	36
2	35	35	32	37	35
3	31	38	35	36	38
4	37	35	36	36	34
5	34	35	36	36	37
6	34	33	38	35	38
7	34	36	37	35	34
8	36	37	35	32	31
9	34	34	32	34	36
10	34	35	37	34	32
11	34	34	35	36	32
12	35	35	41	38	35
13	36	36	37	31	34
14	35	35	32	32	39
15	35	35	34	34	34
16	33	33	35	35	34
17	34	40	36	32	37
18	33	35	33	34	40
19	34	33	37	34	34
20	37	32	34	35	34

Use the data to set up mean and range charts, and briefly outline their usage.

5   Plot the following data on mean and range charts and interpret the results. The sample size is 4 and the specification is 60.0 ± 2.0.

Sample number	Mean	Range	Sample number	Mean	Range
1	60.0	5	26	59.6	3
2	60.0	3	27	60.0	4
3	61.8	4	28	61.2	3
4	59.2	3	29	60.8	5
5	60.4	4	30	60.8	5

(continued)

(continued)

Sample number	Mean	Range	Sample number	Mean	Range
6	59.6	4	31	60.6	4
7	60.0	2	32	60.6	3
8	60.2	1	33	63.6	3
9	60.6	2	34	61.2	2
10	59.6	5	35	61.0	7
11	59.0	2	36	61.0	3
12	61.0	1	37	61.4	5
13	60.4	5	38	60.2	4
14	59.8	2	39	60.2	4
15	60.8	2	40	60.0	7
16	60.4	2	41	61.2	4
17	59.6	1	42	60.6	5
18	59.6	5	43	61.4	5
19	59.4	3	44	60.4	5
20	61.8	4	45	62.4	6
21	60.0	4	46	63.2	5
22	60.0	5	47	63.6	7
23	60.4	7	48	63.8	5
24	60.0	5	49	62.0	6
25	61.2	2	50	64.6	4

(See also Chapter 10, Discussion question 2)

6   You are a Sales Representative of International Chemicals. Your Manager has received the following letter of complaint from Perplexed Plastics, now one of your largest customers.

> To: Sales Manager, International Chemicals
>
> From: Senior Buyer, Perplexed Plastics
>
> Subject: *MFR Values of Polymax*

We have been experiencing line feed problems recently which we suspect are due to high MFR values on your Polymax. We believe about 30 per cent of your product is out of specification.

As agreed in our telephone conversation, I have extracted from our records some MFR values on approximately 60 recent lots. As you can see, the values are generally on the high side. It is vital that you take urgent action to reduce the MFR so that we can get out lines back to correct operating speed.

MFR Values					
4.4	3.3	3.2	3.5	3.3	4.3
3.2	3.6	3.5	3.6	4.2	3.7
3.5	3.2	2.4	3.0	3.2	3.3
4.1	2.9	3.5	3.1	3.4	3.1

3.0	4.2	3.3	3.4	3.3
3.2	3.3	3.6	3.1	3.6
4.3	3.0	3.2	3.6	3.1
3.3	3.4	3.4	4.2	3.4
3.2	3.1	3.5	3.3	4.1
3.3	4.1	3.0	3.3	3.5
4.0	3.5	3.4	3.4	3.2
2.7	3.1	4.2	3.4	4.2

Specification 3.0 to 3.8 g/10 minute.

Subsequent to the letter, you have received a telephone call advising you that they are now approaching a stock-out position. They are threatening to terminate the contract and seek alternative supplies unless the problem is solved quickly.

- Do you agree that their complaint is justified?
- Discuss what action you are going to take. (See also Chapter 10, Discussion question 3)

7   You are a trader in foreign currencies. The spot exchange rates of all currencies are available to you at all times. The following data for one currency were collected at intervals of 1 minute for a total period of 100 minutes, five consecutive results are shown as one sample.

Sample	Spot exchange rates				
1	1333	1336	1337	1338	1339
2	1335	1335	1332	1337	1335
3	1331	1338	1335	1336	1338
4	1337	1335	1336	1336	1334
5	1334	1335	1336	1336	1337
6	1334	1333	1338	1335	1338
7	1334	1336	1337	1335	1334
8	1336	1337	1335	1332	1331
9	1334	1334	1332	1334	1336
10	1334	1335	1337	1334	1332
11	1334	1334	1335	1336	1332
12	1335	1335	1341	1338	1335
13	1336	1336	1337	1331	1334
14	1335	1335	1332	1332	1339
15	1335	1335	1334	1334	1334
16	1333	1333	1335	1335	1334
17	1334	1340	1336	1338	1342
18	1338	1336	1337	1337	1337
19	1335	1339	1341	1338	1338
20	1339	1340	1342	1339	1339

Use the data to set up mean and range charts, interpret the charts and discuss the use that could be made of this form of presentation of the data.

8   The following data were obtained when measurements of the zinc concentration (measured as percentage of zinc sulphate on sodium sulphate) were made in a viscose rayon spin-bath. The mean and range values of 20 samples of size 5 are given in the table.

Sample	Zn conc. (%)	Range (%)	Sample	Zn conc. (%)	Range (%)
1	6.97	0.38	11	7.05	0.23
2	6.93	0.20	12	6.92	0.21
3	7.02	0.36	13	7.00	0.28
4	6.93	0.31	14	6.99	0.20
5	6.94	0.28	15	7.08	0.16
6	7.04	0.20	16	7.04	0.17
7	7.03	0.38	17	6.97	0.25
8	7.04	0.25	18	7.00	0.23
9	7.01	0.18	19	7.07	0.19
10	6.99	0.29	20	6.96	0.25

If the data are to be used to initiate mean and range charts for controlling the process, determine the action and warning lines for the charts. What would your reaction be to the development chemist setting a tolerance of 7.00 ± 0.25 per cent on the zinc concentration in the spin-bath? (See also Chapter 10, Discussion question 4)

9   Conventional control charts are to be used on a process manufacturing small components with a specified length of 60 ± 1.5 mm. Two identical machines are involved in making the components and process capability studies carried out on them reveal the following data: Sample size, $n = 5$

Sample number	Machine I		Machine II	
	Mean	Range	Mean	Range
1	60.10	2.5	60.86	0.5
2	59.92	2.2	59.10	0.4
3	60.37	3.0	60.32	0.6
4	59.91	2.2	60.05	0.2
5	60.01	2.4	58.95	0.3
6	60.18	2.7	59.12	0.7
7	59.67	1.7	58.80	0.5

8	60.57	3.4	59.68	0.4
9	59.68	1.7	60.14	0.6
10	59.55	1.5	60.96	0.3
11	59.98	2.3	61.05	0.2
12	60.22	2.7	60.84	0.2
13	60.54	3.3	61.01	0.5
14	60.68	3.6	60.82	0.4
15	59.24	0.9	59.14	0.6
16	59.48	1.4	59.01	0.5
17	60.20	2.7	59.08	0.1
18	60.27	2.8	59.25	0.2
19	59.57	1.5	61.50	0.3
20	60.49	3.2	61.42	0.4

Calculate the control limits to be used on a mean and range chart for each machine and give the reasons for any differences between them. Compare the results from each machine with the appropriate control chart limits and the specification tolerances.

(See also Chapter 10, Discussion question 5)

10 The following table gives the average width in millimetres for each of 20 samples of five panels used in the manufacture of a domestic appliance. The range of each sample is also given.

Sample number	Mean	Range	Sample number	Mean	Range
1	550.8	4.2	11	553.1	3.8
2	552.7	4.2	12	551.7	3.1
3	553.8	6.7	13	561.2	3.5
4	555.8	4.7	14	554.2	3.4
5	553.8	3.2	15	552.3	5.8
6	547.5	5.8	16	552.9	1.6
7	550.9	0.7	17	562.9	2.7
8	552.0	5.9	18	559.4	5.4
9	553.7	9.5	19	555.8	1.7
10	557.3	1.9	20	547.6	6.7

Calculate the control chart limits for the control charts and plot the values on the charts. Interpret the results. Given a specification of 540 ± 5 mm, comment on the capability of the process.

(See also Chapter 9, Discussion question 4, and Chapter 10, Discussion question 6)

Worked examples

1 Lathe operation

A component used as a part of a power transmission unit is manufactured using a lathe. Twenty samples, each of five components, are taken at half-hourly intervals. For the most critical dimension, the process mean ($\overline{X}$) is found to be 3.5000 cm, with a normal distribution of the results about the mean, and a mean sample range ($\overline{R}$) of 0.0007 cm.

(a) Use this information to set up suitable control charts.
(b) If the specified tolerance is 3.498–3.502 cm, what is your reaction? Would you consider any action necessary?
(See also Chapter 10, Worked example 1)
(c) The following table shows the operator's results over the day. The measurements were taken using a comparitor set to 3.500 cm and are shown in units of 0.001 cm. The means and ranges have been added to the results. What is your interpretation of these results? Do you have any comments on the process and/or the operator?

Record of results recorded from the lathe operation

Time	1	2	3	4	5	Mean	Range
7.30	0.2	0.5	0.4	0.3	0.2	0.32	0.3
7.35	0.2	0.1	0.3	0.2	0.2	0.20	0.2
8.00	0.2	−0.2	−0.3	−0.1	0.1	−0.06	0.5
8.30	−0.2	0.3	0.4	−0.2	−0.2	0.02	0.6
9.00	−0.3	0.1	−0.4	−0.6	−0.1	−0.26	0.7
9.05	−0.1	−0.5	−0.5	−0.2	−0.5	−0.36	0.4
Machine stopped tool clamp readjusted							
10.30	−0.2	−0.2	0.4	−0.6	−0.2	−0.16	1.0
11.00	0.6	0.2	−0.2	0.0	0.1	0.14	0.8
11.30	0.4	0.1	−0.2	0.5	0.3	0.22	0.7
12.00	0.3	−0.1	−0.3	0.2	0.0	0.02	0.6
Lunch							
12.45	−0.5	−0.1	0.6	0.2	0.3	0.10	1.1
13.15	0.3	0.4	−0.1	−0.2	0.0	0.08	0.6
Reset tool by 0.15 cm							
13.20	−0.6	0.2	−0.2	0.1	−0.2	−0.14	0.8
13.50	0.4	−0.1	−0.5	−0.1	−0.2	−0.10	0.9
14.20	0.0	−0.3	0.2	0.2	0.4	0.10	0.7
14.35	*Batch finished – machine reset*						
16.15	1.3	1.7	2.1	1.4	1.6	1.62	0.8

Solution

(a) Since the distribution is known and the process is in statistical control with:

Process mean $\overline{\overline{X}} = 3.5000$ *cm*
Mean sample range $\overline{R} = 0.0007$ *cm*
Sample size $n = 5$.

Mean chart
From Appendix B for $n = 5$, $A_2 = 0.58$ and $2/3\ A_2 = 0.39$

Mean control chart is set up with:

Upper action limit	$\overline{\overline{X}} + A_2\overline{R} = 3.50041$ *cm*
Upper warning limit	$\overline{\overline{X}} + 2/3\ A_2\overline{R} = 3.50027$ *cm*
Mean	$\overline{\overline{X}} = 3.5000$ *cm*
Lower warning limit	$\overline{\overline{X}} - 2/3\ A_2\overline{R} = 3.49973$ *cm*
Lower action limit	$\overline{\overline{X}} - A_2\overline{R} = 3.49959$ *cm.*

Range chart
From Appendix C $D'_{0.999} = 0.16$ $D'_{0.975} = 0.37$

$D'_{0.025} = 1.81$ $D'_{0.001} = 2.34$

Range control chart is set up with:

Upper action limit	$D'_{0.001}\overline{R} = 0.0016$ *cm*
Upper warning limit	$D'_{0.025}\overline{R} = 0.0013$ *cm*
Lower warning limit	$D'_{0.975}\overline{R} = 0.0003$ *cm*
Lower action limit	$D'_{0.999}\overline{R} = 0.0001$ *cm.*

(b) The process is correctly centred so:

From Appendix B $d_n = 2.326$

$\sigma = \overline{R}/d_n = 0.0007/2.326 = 0.0003$ *cm*

The process is in statistical control and capable. If mean and range charts are used for its control, significant changes should be detected by the first sample taken after the change. No further immediate action is suggested.

(c) The means and ranges of the results are given in the table above and are plotted on control charts in Figure 6.14.

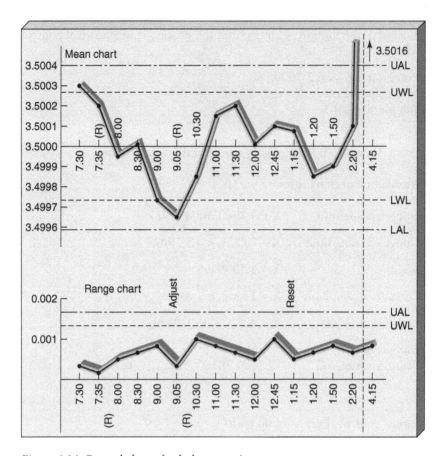

Figure 6.14 Control charts for lathe operation

Observations on the control charts

1   The 7.30 sample required a repeat sample to be taken to check the mean. The repeat sample at 7.35 showed that no adjustment was necessary.

2   The 9.00 sample mean was within the warning limits but was the fifth result in a downward trend. The operator correctly decided to take a repeat sample. The 9.05 mean result constituted a double warning since it remained in the downward trend and also fell in the warning zone. Adjustment of the mean was, therefore, justified.

3   The mean of the 13.15 sample was the fifth in a series above the mean and should have signalled the need for a repeat sample and *not* an adjustment. The adjustment, however, did not adversely affect control.

4   The whole of the batch completed at 14.35 was within specification and suitable for dispatch.

5   At 16.15 the machine was incorrectly reset.

General conclusions

There was a downward drift of the process mean during the manufacture of this batch. The drift was limited to the early period and appears to have stopped following the adjustment at 9.05. The special cause should be investigated.

The range remained in control throughout the whole period when it averaged 0.0007 cm, as in the original process capability study.

The operator's actions were correct on all but one occasion (the reset at 13.15); a good operator who may need a little more training, guidance or experience.

2 Control of dissolved iron in a dyestuff

Mean and range charts are to be used to maintain control on dissolved iron content of a dyestuff formulation in parts per million (ppm). After 25 subgroups of 5 measurements have been obtained.

$$\sum_{i=1}^{i=25} \overline{X}_i = 390 \text{ and } \sum_{i=1}^{i=25} R_i = 84$$

where

$\overline{X}_i$ = mean of ith subgroup

R_i = range of ith subgroup.

(a) Design the appropriate control charts.
(b) The specification on the process requires that no more than 18 ppm dissolved iron be present in the formulation. Assuming a normal distribution and that the process continues to be in statistical control with no change in average or dispersion, what proportion of the individual measurements may be expected to exceed this specification? (See also Chapter 9, Discussion question 5 and Chapter 10, Worked example 2.)

Solution

(a) *Control charts*

$$\text{Grand Mean,} \quad \overline{\overline{X}} = \frac{\sum \overline{X}_i}{k} = \frac{390}{25} = 15.6 \text{ ppm}$$

$$k = \text{No. of samples} = 25$$

$$\text{Mean Range,} \quad \overline{R} = \frac{\sum R_i}{k} = \frac{84}{25} = 3.36 \text{ ppm}$$

$$\sigma = \frac{\overline{R}}{d_n} = \frac{3.36}{2.326} = 1.445 \text{ ppm}$$

(d_n from Appendix B = 2.326, n = 5)

$$SE = \frac{\sigma}{\sqrt{n}} = \frac{1.445}{\sqrt{5}} = 0.646 \text{ ppm.}$$

Mean chart

$$\begin{aligned}
\text{Action Lines} &= \overline{\overline{X}} \pm \left(3 \times SE\right) \\
&= 15.6 \pm \left(3 \times 0.646\right) \\
&= 13.7 \; and \; 17.5 \; ppm \\
\text{Warning Lines} &= 15.6 \pm \left(2 \times 0.646\right) \\
&= 14.3 \; and \; 16.9 \; ppm.
\end{aligned}$$

Range chart

$$\text{Upper Action Line} = D'_{0.001} \overline{R} = 2.34 \times 3.36 = 7.9 \; ppm$$

$$\text{Upper Warning Line} = D'_{0.025} \overline{R} = 1.81 \times 3.36 = 6.1 \; ppm \,.$$

Alternative calculations of Mean Chart Control Lines

$$\begin{aligned}
\text{Action Lines} &= \overline{\overline{X}} \pm A_2 \overline{R} \\
&= 15.6 \pm \left(0.58 \times 3.36\right) \\
\text{Warning Lines} &= \overline{\overline{X}} \pm 2/3 A_2 \overline{R} \\
&= 15.6 \pm \left(0.39 \times 3.36\right)
\end{aligned}$$

A_2 and 2/3 A_2 from Appendix B.

(b) *Specification*

$$\begin{aligned}
Z_u &= \frac{U - \overline{X}}{\sigma} \\
&= \frac{18.0 - 15.6}{1.445} = 1.66.
\end{aligned}$$

From normal tables (Appendix A), proportion outside upper tolerance = 0.0485 or 4.85 per cent.

3 Pin manufacture

Samples are being taken from a pin manufacturing process every 15–20 minutes. The production rate is 350–400 per hour, and the specification limits on length are 0.820 and 0.840 cm. After 20 samples of 5 pins, the following information is available:

Sum of the sample means, $\quad \sum\limits_{i=1}^{i=20} \overline{X}_i = 16.88\,\mathrm{cm}$,

Sum of the sample ranges, $\quad \sum\limits_{i=1}^{i=20} R_i = 0.14\,\mathrm{cm}$,

where $\overline{X}$ and R_i are the mean and range of the *i*th sample, respectively:

(a) Set up mean and range charts to control the lengths of pins produced in the future.
(b) On the assumption that the pin lengths are normally distributed, what percentage of the pins would you estimate to have lengths outside the specification limits when the process is under control at the levels indicated by the data given?
(c) What would happen to the percentage defective pins if the process average should change to 0.837 cm?
(d) What is the probability that you could observe the change in (c) on your control chart on the first sample following the change?

(See also Chapter 10, Worked example 3)

Solution

(a) $\sum\limits_{i=1}^{i=20} \overline{X}_i = 16.88\,\mathrm{cm}$, $k =$ No. of samples $= 20$

Grand Mean, $\quad \overline{\overline{X}} = \sum \overline{X}_i / k = \dfrac{16.88}{20} = 0.844\,\mathrm{cm}$,

Mean Range, $\quad \overline{R} = \sum R_i / k = \dfrac{0.14}{20} = 0.007\,\mathrm{cm}$.

Mean chart

Action Lines at $\overline{\overline{X}} \pm A_2 \overline{R} = 0.834 \pm (0.594 \times 0.007)$

Upper Action Line $= 0.838\mathrm{cm}$

Lower Action Line $= 0.830\mathrm{cm}$.

Warning Lines at $\overline{\overline{X}} \pm 2/3\, A_2 \overline{R} = 0.834 \pm (0.377 \times 0.007)$

Upper Warning Line $= 0.837\mathrm{cm}$

Lower Warning Line $= 0.831\mathrm{cm}$.

The A_2 and 2/3 constants are obtained from Appendix B.

Range chart

Upper Action Line at $D'_{0.001} \overline{R} = 2.34 \times 0.007 = 0.0164$ *cm*

Upper Warning Line at $D'_{0.025} \overline{R} = 1.81 \times 0.007 = 0.0127\,cm$.

The D' constants are obtained from Appendix C.

(b) $\sigma = \dfrac{\overline{R}}{d_n} = \dfrac{0.007}{2.326} = 0.003\,\text{cm}.$

Upper tolerance

$$Zu = \frac{(U - \overline{X})}{\sigma} = \frac{(0.84 - 0.834)}{0.003} = 2.$$

Therefore percentage outside upper tolerance = 2.275 per cent (from Appendix A).

Lower tolerance

$$Z_l = \frac{(\overline{X} - L)}{\sigma} = \frac{0.834 - 0.82}{0.003} = 4.67.$$

Therefore percentage outside lower tolerance = 0

  Total outside both tolerances = 2.275 per cent

$$Zu = \frac{0.84 - 0.837}{0.003} = 1.$$

Therefore percentage outside upper tolerance will increase to 15.87 per cent (from Appendix A).

$$SE = \sigma/\sqrt{n} = \frac{0.003}{\sqrt{5}} = 0.0013.$$

Upper Warning Line (UWL)

  As $\mu = UWL$, the probability of sample point being outside UWL = 0.5 (50 per cent).

  Upper Action Line (UAL)

$$Z_{UAL} = \frac{0.838 - 0.837}{0.0013} = 0.769.$$

Therefore from tables, probability of sample point being outside UAL = 0.2206.

  Thus, the probability of observing the change to $\mu = 0.837\,cm$ on the first sample after the change is:

  0.50 – outside warning line (50 per cent or 1 in 2)

  0.2206 – outside action line (22.1 per cent or *ca*. 1 in 4.5).

4 Bale weight

(a) Using the bale weight data below, calculate the control limits for the mean and range charts to be used with these data.
(b) Using these control limits, plot the mean and range values onto the charts.
(c) Comment on the results obtained.

Bale weight data record (kg)

Sample number	Time	1	2	3	4	Mean X	Range R
1	10.18	34.07	33.99	33.99	34.12	34.04	0.13
2	10.03	33.98	34.08	34.10	33.99	34.04	0.12
3	10.06	34.19	34.21	34.00	34.00	34.15	0.21
4	10.09	33.79	34.01	33.77	33.82	33.85	0.24
5	10.12	33.92	33.98	33.70	33.74	33.84	0.28
6	10.15	34.01	33.98	34.20	34.13	34.08	0.22
7	10.18	34.07	34.30	33.80	34.10	34.07	0.50
8	10.21	33.87	33.96	34.04	34.05	33.98	0.18
9	10.24	34.02	33.92	34.05	34.18	34.04	0.26
10	10.27	33.67	33.96	34.04	34.31	34.00	0.64
11	10.30	34.09	33.96	33.93	34.11	34.02	0.18
12	10.33	34.31	34.23	34.18	34.21	34.23	0.13
13	10.36	34.01	34.09	33.91	34.12	34.03	0.21
14	10.39	33.76	33.98	34.06	33.89	33.92	0.30
15	10.42	33.91	33.90	34.10	34.03	33.99	0.20
16	10.45	33.85	34.00	33.90	33.85	33.90	0.15
17	10.48	33.94	33.76	33.82	33.87	33.85	0.18
18	10.51	33.69	34.01	33.71	33.84	33.81	0.32
19	10.54	34.07	34.11	34.06	34.08	34.08	0.05
20	10.57	34.14	34.15	33.99	34.07	34.09	0.16
					TOTAL	680.00	4.66

Solution

(a) $\overline{\overline{X}}$ = Grand (Process) Mean = $\dfrac{\text{Total of the means } (\overline{X})}{\text{Number of samples}}$

$= \dfrac{680.00}{20} = 34.00\,\text{kg}.$

$\overline{R}$ = Mean Range $= \dfrac{\text{Total of the ranges } (R)}{\text{Number of samples}}$

$= \dfrac{4.66}{20} = 0.233\,\text{kg}.$

$$\sigma = \overline{R} / d_n$$

for sample size $n = 4$, $d_n = 2.059$

$$\sigma = 0.233 / 2.059 = 0.113$$
$$\text{Standard Error} = \sigma / \sqrt{n} = 0.113 / \sqrt{4} = 0.057 .$$

Mean chart

$$\text{Action Lines} = \overline{\overline{X}} \pm 3\sigma / \sqrt{n}$$
$$= 34.00 \pm 3 \times 0.057$$
$$= 34.00 \pm 0.17$$

Upper Action Line = 34.17kg

Lower Action Line = 33.83kg.

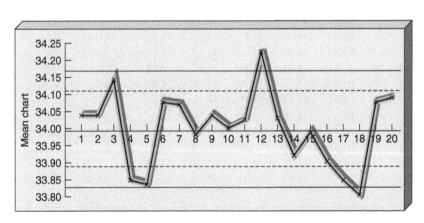

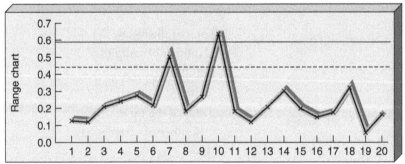

Figure 6.15 Bale weight data (kg)

Warning Lines $= \overline{\overline{X}} \pm 2\sigma / \sqrt{n}$

$= 34.00 \pm 2 \times 0.057$

$= 34.00 \pm 0.11$

Upper Warning Line $= 34.11$kg

Lower Warning Line $= 33.89$kg.

The mean of the chart is set by the specification or target mean.

Range chart

Action Line $= 2.57\overline{R} = 2.57 \times 0.233 = 0.599kg$

Warning Line $= 1.93\overline{R} = 1.93 \times 0.233 = 0.450kg$.

(b) The data are plotted in Figure 6.15.
(c) Comments on the mean and range charts.
The table below shows the actions that could have been taken had the charts been available during the production period.

Sample number	Chart	Observation	Interpretation
3	Mean	Upper warning	Acceptable on its own, resample required.
4	Mean	Lower warning	Two warnings must be in the same warning zone to be an action, resampling required. Note: Range chart has not picked up any problem.
5	Mean	Second lower warning; out of control	ACTION – increase weight setting on press by approximately 0.15kg.
7	Range	Warning	Acceptable on its own, resample required.
8	Range	No warning or action	No action required – sample 7 was a statistical event.
10	Range	ACTION – out of control	Actions could involve obtaining additional information but some possible actions could be:
			(a) check crumb size and flow rate
			(b) clean bale press
			(c) clean fabric bale cleaner
			Note: Mean chart indicates no problem, the mean value = target mean. (This emphasizes the need to plot and check both charts.)

(continued)

(continued)

Sample number	Chart	Observation	Interpretation
12	Mean	Upper action; out of control	Decrease weight setting on press by approximately 0.23kg.
17	Mean	Lower warning	Acceptable on its own, a possible downward trend is appearing, resample required.
18	Mean	Second lower warning/action; out of control	ACTION – Increase weight setting on press by 0.17kg.

7 Other types of control charts for variables

Objectives

- To understand how different types of data, including infrequent data, can be analysed using SPC techniques.
- To describe in detail charts for individuals (run charts) with moving range charts.
- To examine other types of control systems, including zone control and pre-control.
- To introduce alternative charts for central tendency: median, mid-range and multi-vari charts; and spread: standard deviation.
- To describe the setting up and use of moving mean, moving range and exponentially weighted moving average charts for infrequent data.
- To outline some techniques for short run SPC and provide reference for further study.

7.1 Life beyond the mean and range chart

Statistical process control is based on a number of basic principles that apply to all processes, including batch and continuous processes of the type commonly found in the manufacture of bulk chemicals, pharmaceutical products, speciality chemicals, processed foods and metals. The principles apply also to all processes in service and public sectors and commercial activities, including forecasting, claim processing and many financial transactions. One of these principles is that within any process variability is inevitable. As seen in earlier chapters variations are due to two types of causes; common (random) or special (assignable) causes. Common causes cannot easily be identified individually but these set the limits of the 'precision' of a process, whilst special causes reflect specific changes that either occur or are introduced.

If it is known that the difference between an individual observed result and a 'target' or average value is simply a part of the inherent process variation, there is no readily available means for correcting, adjusting or taking action on it. If the observed difference is known to be due to a special cause then a search for and a possible correction of this cause is sensible. Adjustments by instruments, computers, operators, instructions, etc. are often special causes of increased variation.

In many industrial and commercial situations, data are available on a large scale (dimensions of thousands of mechanical components, weights of millions of tablets, time, forecast/actual sales, etc.) and there is no doubt about the applicability of conventional SPC techniques here. The use of control charts is often thought, however, not to apply to situations in which a new item of data is available either in isolation or infrequently – one at a time, such as in batch processes where an analysis of the final product may reveal for the first time the characteristics of what has been manufactured or in continuous processes (including non-manufacturing) when data are available only on a one result per period basis. This is not the case.

Numerous papers have been published on the applications and modifications of various types of control charts. It is not possible to refer here to all the various innovations that have filled volumes of journals and, in this chapter, we shall not delve into the many refinements and modifications of control charts, but concentrate on some of the most important and useful applications.

The control charts for variables, first formulated by Shewhart, make use of the arithmetic mean and the range of samples to determine whether a process is in a state of statistical control. Several control chart techniques exist that make use of other measures.

Use of control charts

As we have seen in earlier chapters, control charts are used to investigate the *variability* of a process and this is essential when assessing the *capability* of a process. Data are often plotted on a control chart in the hope that this may help to find the causes of problems. Charts are also used to monitor or 'control' process performance.

In assessing past variability and/or capability, and in problem solving, all the data are to hand before plotting begins. This post-mortem analysis use of charting is very powerful. In monitoring performance, however, the data are plotted point by point as it becomes available in a real time analysis.

When using control charts it is helpful to distinguish between different types of processes:

1 Processes that give data that fall into natural subgroups. Here conventional mean and range charts are used for process monitoring, as described in Chapters 4–6.
2 Processes that give one-at-a-time data. Here an individuals chart or a moving mean chart with a (moving) range chart is better for process monitoring.

In after-the-fact or post-mortem analysis, of course, conventional mean and range charts may be used with *any* process.

Situations in which data are available infrequently or 'one at a time' include:

- measured quality of high value items, such as batches of chemical, turbine blades, large or complex castings. Because the value of each item is much greater than the cost of inspection, every 'item' is inspected;
- *Financial Times* all share index (daily);
- weekly sales or forecasts for a particular product;
- monthly, lost time accidents;
- quarterly, rate of return on capital employed.

Other data occur in a form that allows natural grouping, such as manufacture of low value items such as nails, plastic plugs, metal discs, and other 'widgets'. Because the value of each item is even less than the cost of inspection, only a small percentage are inspected, e.g. five items every 20 minutes.

When plotting naturally grouped data it is unwise to mix data from different groups, and in some situations it may be possible to group the data in several ways. For example, there may be three shifts, four teams and two machines.

7.2 Charts for individuals or run charts

The simplest variable chart that may be plotted is one for individual measurements. The individuals or run chart is often used with one-at-a-time data and the individual values, not means of samples, are plotted. The centreline (CL) is usually placed at:

- the centre of the specification; or
- the mean of past performance; or
- some other, suitable – perhaps target – value.

The action lines (UAL and LAL) or control limits (UCL and LCL) are placed three standard deviations from the centreline. Warning lines (upper and lower: UWL and LWL) may be placed at two standard deviations from the centreline.

Figure 7.1 shows measurements of batch moisture content from a process making a herbicide product. The specification tolerances in this case are 6.40 ± 0.015 per cent and these may be shown on the chart. When using the conventional sample mean chart the tolerances are not included, since the distribution of the means is much narrower than that of the process population, and confusion may be created if the tolerances are shown. The inclusion of the specification tolerances on the individuals chart may be sensible, but it may lead to over-control of the process as points are plotted near to the specification lines and adjustments are made.

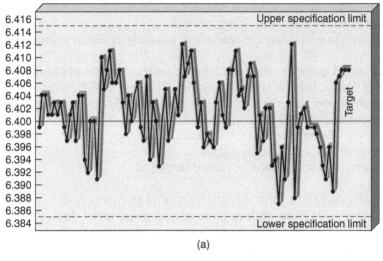

(a)

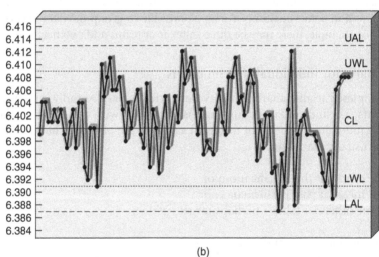

(b)

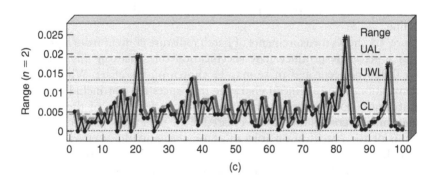

(c)

Figure 7.1 (a) Run chart for batch moisture content, (b) individuals control chart for batch moisture content, (c) moving range chart for batch moisture content ($n = 2$)

Setting up the individuals or run chart

The rules for the setting up and interpretation of individual or i-charts are similar to those for conventional mean and range charts. Measurements are taken from the process over a period of expected stability. The mean $(\overline{X})$ of the measurements is calculated together with the range or moving range between adjacent observations $(n = 2)$, and the mean range, $\overline{R}$. The control chart limits are found in the usual way.

In the example given, the centreline was placed at 6.40 per cent, which corresponds with the centre of the specification.

The standard deviation was calculated from previous data, when the process appeared to be in control. The mean range $(\overline{R}, n = 2)$ was 0.0047

$$\sigma = \overline{R} / d_n = 0.0047 / 1.128 = 0.0042 \text{ per cent}$$

i-Chart

Action Lines at $\overline{X} \pm 3\sigma$ or $\overline{X} \pm 3\overline{R} / d_n$ = 6.4126 and 6.3874

Warning Lines at $\overline{X} \pm 2\sigma$ or $\overline{X} \pm 2\overline{R} / d_n$ = 6.4084 and 6.3916

Central-line $\overline{X}$ which also corresponds with the target value = 6.40.

Moving range chart

Action Lines at $D'_{0.001}$ $\overline{R}$ = 0.0194

Warning Lines at $D'_{0.025}$ $\overline{R}$ = 0.0132.

The run chart with control limits for the herbicide data is shown in Figure 7.1b.

When plotting the individual results on the i-chart, the rules for out-of-control situations are:

- any points outside the 3σ limits;
- two out of three successive points outside the 2σ limits;
- eight points in a run on one side of the mean.

Owing to the relative insensitivity of i-charts, horizontal lines at $\pm 1\sigma$ either side of the mean are usually drawn, and action taken if four out of five points plot outside these limits.

How good is the individuals chart?

The individuals chart:

- is very simple;
- will indicate changes in the mean level (accuracy or centring);

- with careful attention, will even indicate changes in variability (precision or spread);
- is *not* so good at detecting small changes in process centring – a mean chart is much better at detecting quickly small changes in centring.

Charting with individual item values is always better than nothing. It is less satisfactory, however, than the charting of means and ranges, both because of its relative insensitivity to changes in process average and the lack of clear distinction between changes in accuracy and in precision. Whilst in general the chart for individual measurements is less sensitive than other types of control chart in detecting changes, it is often used with one-at-a-time data, and is far superior to a table of results for understanding variation. An improvement is the combined individual-moving range chart, which shows changes in the 'setting' or accuracy *and* spread of the process (Figure 7.1b and c).

The zone control chart and pre-control

The so-called 'zone control chart' is simply an adaptation of the individuals chart, or the mean chart. In addition to the action and warning lines, two lines are placed at one standard error from the mean.

Each point is given a score of 1, 2, 4 or 8, depending on which band it falls into. It is concluded that the process has changed if the cumulative score exceeds 7. The cumulative score is reset to zero whenever the plot crosses the centreline. An example of the zone control chart is given in Figure 7.2.

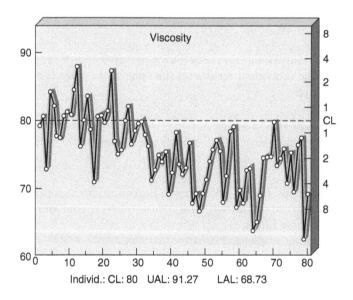

Figure 7.2 The zone control chart

In his book *World Class Quality*, Keki Bhote argued in favour of use of *pre-control* over conventional SPC control charts. The technique was developed many years ago and is very simple to introduce and operate. The technique was based on the product or service specification and its principles are shown in Figure 7.3.

The steps to set up are as follows:

1 Divide the specification width by four.
2 Set the boundaries of the middle half of the specification – the *green zone* or target area – as the upper and lower pre-control lines (UPCL and LPCL).
3 Designate the two areas between the pre-control lines and the specification limits as the *yellow zone*, and the two areas beyond the specification limits as *red zones*.

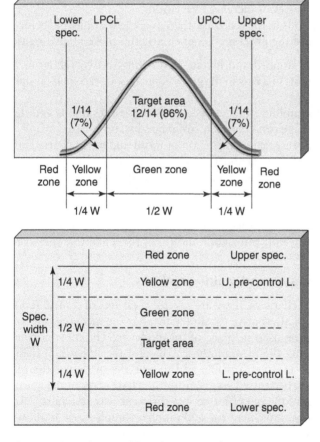

Figure 7.3 Basic principles of pre-control

The use and rules of pre-control are as follows:

4   Take an initial sample of five consecutive units or measurements from the process. If all five fall within the green zone, conclude that the process is in control and full production/operation can commence.[1]

If *one* or more of the five results is outside the green zone, the process is not in control, and an assignable cause investigation should be launched, as usual.

5   Once production/operation begins, take two consecutive units from the process periodically:
   - if both are in the green zone, *or* if one is in the green zone and the other in a yellow zone, continue operations;
   - if both units fall in the same yellow zone, adjust the process setting;
   - if the units fall in different yellow zones, stop the process and investigate the causes of increased variation;
   - if any unit falls in the red zone, there is a known out-of-specification problem and the process is stopped and the cause(s) investigated.

6   If the process is stopped and investigated owing to two yellow or a red result, the five units in a row in the green zone must be repeated on start up.

The frequency of sampling (time between consecutive results) is determined by dividing the average time between stoppages by six.

   In their excellent statistical comparison of mean and range charts with the method of pre-control, Barnett and Tong (1994) pointed out that pre-control is very simple and versatile and useful in a variety of applications. They showed, however, that conventional mean $(\overline{X})$ and range (R) charts are:

   - superior in picking up process changes – they are more sensitive;
   - more valuable in supporting continuous improvement than pre-control.

7.3 Median, mid-range and multi-vari charts

As we saw in earlier chapters, there are several measures of central tendency of variables data. An alternative to sample mean is the median, and control charts for this may be used in place of mean charts. The most convenient method for producing the *median chart* is to plot the individual item values for each sample in a vertical line and to ring the median – the middle item value. This has been used to generate the chart shown in Figure 7.4, which is derived from the data plotted in a different way in Figure 7.1. The method is only really convenient for odd number sample sizes. It allows the tolerances to be shown on the chart, provided the process data are normally distributed.

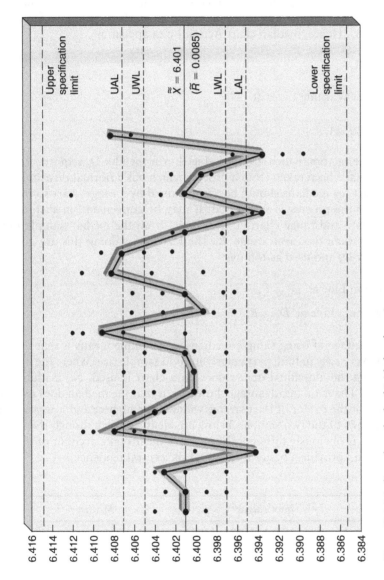

Figure 7.4 Median chart for herbicide batch moisture content

The control chart limits for this type of chart can be calculated from the median of sample ranges, which provides the measure of spread of the process. Grand or Process Median $(\widetilde{\widetilde{X}})$ – the median of the sample medians – and the Median Range $(\widetilde{R})$ – the median of the sample ranges –for the herbicide batch data previously plotted in Figure 7.1 are 6.401 per cent and 0.0085 per cent, respectively. The control limits for the median chart are calculated in a similar way to those for the mean chart, using the factors A_4 and $2/3\ A_4$. Hence, median chart Action Lines appear at

$$\widetilde{\widetilde{X}} \pm A_4\widetilde{R},$$

and the Warning Lines at

$$\widetilde{\widetilde{X}} \pm 2\,/\,3\,A_4\widetilde{R}.$$

Use of the factors, which are reproduced in Appendix D, requires that the samples have been taken from a process which has a normal distribution.

A chart for medians should be accompanied by a range chart so that the spread of the process is monitored. It may be convenient, in such a case, to calculate and range chart control limits from the median sample range $\widetilde{R}$ rather than the mean range $\bar{R}$. The factors for doing this are given in Appendix D, and used as follows:

Action Line at $D^m_{0.001}\ \widetilde{R}$,

Warning Line at $D^m_{0.025}\ \widetilde{R}$, .

The advantage of using sample medians over sample means is that the former are very easy to find, particularly for odd sample sizes where the method of circling the individual item values on a chart is used. No arithmetic is involved. The main disadvantage, however, is that the median does not take account of the extent of the extreme values – the highest and lowest. Thus, the medians of the two samples below are identical, even though the spread of results is obviously different. The sample means take account of this difference and provide a better measure of the central tendency.

Sample No.	Item values	Median	Mean
1	134, 134, 135, 139, 143	135	137
2	120, 123, 135, 136, 136	135	130

This failure of the median to give weight to the extreme values can be an advantage in situations where 'outliers' – item measurements with unusually high or low values – are to be treated with suspicion.

A technique similar to the median chart is the *chart for mid-range*. The middle of the range of a sample may be determined by calculating the average of the highest and lowest values. The mid-range ($\tilde{M}$) of the sample of 5, 553, 555, 561, 554, 551, is:

Highest   Lowest

$$\frac{561 + 551}{2} = 556.$$

The central-line on the mid-range control chart is the median of the sample mid-ranges $\tilde{M}_R$. The estimate of process spread is again given by the median of sample ranges and the control chart limits are calculated in a similar fashion to those for the median chart.

Hence,

Action Lines at $\tilde{M}_R \pm A_4 \tilde{R}$,

Warning Lines at $\tilde{M}_R \pm 2/3 A_4 \tilde{R}$.

Certain quality characteristics exhibit variation that derives from more than one source. For example, if cylindrical rods are being formed, their diameters may vary from piece to piece and along the length of each rod, due to taper. Alternatively, the variation in diameters may be due in part to the ovality within each rod. Such multiple variation may be represented on the *multi-vari* chart.

In the multi-vari chart, the specification tolerances are used as control limits. Sample sizes of three or five are commonly used and the results are plotted in the form of vertical lines joining the highest and lowest values in the sample, thereby representing the sample range. An example of such a chart used in the control of a heat treatment process is shown in Figure 7.5a. The longer the lines, the more variation exists within the sample. The chart shows dramatically the effect of an adjustment, or elimination or reduction of one major cause of variation.

The technique may be used to show within piece or batch, piece to piece, or batch to batch variation. Detection of trends or drift is also possible. Figure 7.5b illustrates all these applications in the measurement of piston diameters. The first part of the chart shows that the variation within each piston is very similar and relatively high. The middle section shows piece to piece variation to be high but a relatively small variation within each piston. The last section of the chart is clearly showing a trend of increasing diameter, with little variation within each piece.

One application of the multi-vari chart in the mechanical engineering, automotive and process industries is for trouble-shooting of variation caused by the position of equipment or tooling used in the production of similar parts, for example a multi-spindle automatic lathe, parts fitted to the same mandrel, multi-impression moulds or dies, parts held in string-milling fixtures. Use of

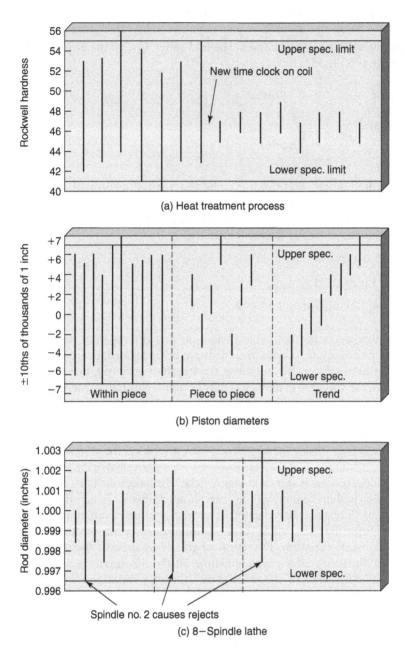

Figure 7.5 Multi-vari charts

multi-vari charts for parts produced from particular, identifiable spindles or positions can lead to the detection of the cause of faulty components and parts. Figure 7.5c shows how this can be applied to the control of ovality on an eight-spindle automatic lathe.

7.4 Moving mean, moving range and exponentially weighted moving average (EWMA) charts

As we have seen in Chapter 6, assessing changes in the average value and the scatter of grouped results – reflections of the centring of the process and the spread – is often used to understand process variation due to common causes and detect special causes. This applies to all processes, including batch, continuous and commercial.

When only one result is available at the conclusion of a batch process or when an isolated estimate is obtained of an important measure on an infrequent basis, however, one cannot simply ignore the result until more data are available with which to form a group. Equally it is impractical to contemplate taking, say, four samples instead of one and repeating the analysis several times in order to form a group – the costs of doing this would be prohibitive in many cases, and statistically this would be different to the grouping of less frequently available data.

An important technique for handling data that are difficult or time-consuming to obtain and, therefore, not available in sufficient numbers to enable the use of conventional mean and range charts is the moving mean and moving range chart. In the chemical industry, for example, the nature of certain production processes and/or analytical methods entails long time intervals between consecutive results. We have already seen in this chapter that plotting of individual results offers one method of control, but this may be relatively insensitive to changes in process average and changes in the spread of the process can be difficult to detect. On the other hand, waiting for several results in order to plot conventional mean and range charts may allow many tonnes of material to be produced outside specification before one point can be plotted.

In a polymerization process, one of the important process control measures is the unreacted monomer. Individual results are usually obtained once every 24 hours, often with a delay for analysis of the samples. Typical data from such a process appear in Table 7.1.

If the individual or run chart of these data (Figure 7.6) was being used alone for control during this period, the conclusions may include:

April 16 – warning and perhaps a repeat sample

April 18 – action signal – do something

April 23 – action signal – do something

April 29 – warning and perhaps a repeat sample.

From about 30 April a gradual decline in the values is being observed.

Table 7.1 Data on per cent of unreacted monomer at an intermediate stage in a polymerization process

Date	Daily value	Date	Daily value
April 1	0.29	25	0.16
2	0.18	26	0.22
3	0.16	27	0.23
		28	0.18
4	0.24	29	0.33
5	0.21	30	0.21
6	0.22	May 1	0.19
7	0.18		
8	0.22	2	0.21
9	0.15	3	0.19
10	0.19	4	0.15
		5	0.18
11	0.21	6	0.25
12	0.19	7	0.19
13	0.22	8	0.15
14	0.20		
15	0.25	9	0.23
16	0.31	10	0.16
17	0.21	11	0.13
		12	0.17
18	0.05	13	0.18
19	0.23	14	0.17
20	0.23	15	0.22
21	0.25		
22	0.16	16	0.15
23	0.35	17	0.14
24	0.26		

When using the individuals chart in this way, there is a danger that decisions may be based on the last result obtained. But it is not realistic to wait for another three days, or to wait for a repeat of the analysis three times and then group data in order to make a valid decision, based on the examination of a mean and range chart.

The alternative of *moving mean* and *moving range* charts uses the data differently and is generally preferred for the following reasons:

- By grouping data together, we will not be reacting to individual results and over-control is less likely.
- In using the moving mean and range technique we shall be making more meaningful use of the latest piece of data – two plots, one each on two different charts telling us different things, will be made from each individual result.
- There will be a calming effect on the process.

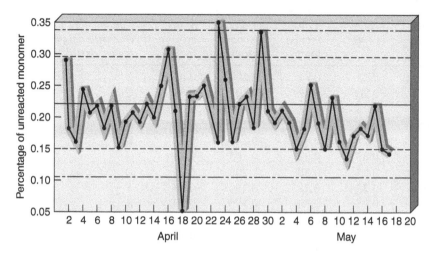

Figure 7.6 Daily values of unreacted monomer

The calculation of the moving means and moving ranges ($n = 4$) for the polymerization data is shown in Table 7.2. For each successive group of four, the earliest result is discarded and replaced by the latest. In this way it is possible to obtain and plot a 'mean' and 'range' every time an individual result is obtained – in this case every 24 hours. These have been plotted on charts in Figure 7.7.

Table 7.2 Moving means and moving ranges for data in unreacted monomer (Table 7.1)

Date	Daily value	4-day moving total	4-day moving mean	4-day moving range	Combination for conventional mean and range control charts
April 1	0.29				
2	0.18				
3	0.16				
4	0.24	0.87	0.218	0.13	A
5	0.21	0.79	0.198	0.08	B
6	0.22	0.83	0.208	0.08	C
7	0.18	0.85	0.213	0.06	D
8	0.22	0.83	0.208	0.04	A
9	0.15	0.77	0.193	0.07	B
10	0.19	0.74	0.185	0.07	C
11	0.21	0.77	0.193	0.07	D
12	0.19	0.74	0.185	0.06	A

(continued)

Table 7.2 (continued)

Date	Daily value	4-day moving total	4-day moving mean	4-day moving range	Combination for conventional mean and range control charts
13	0.22	0.81	0.203	0.03	B
14	0.20	0.82	0.205	0.03	C
15	0.25	0.86	0.215	0.06	D
16	0.31	0.98	0.245	0.11	A
17	0.21	0.97	0.243	0.11	B
18	0.05	0.82	0.205	0.26	C
19	0.23	0.80	0.200	0.26	D
20	0.23	0.72	0.180	0.18	A
21	0.25	0.76	0.190	0.20	B
22	0.16	0.87	0.218	0.09	C
23	0.35	0.99	0.248	0.19	D
24	0.26	1.02	0.255	0.19	A
25	0.16	0.93	0.233	0.19	B
26	0.22	0.99	0.248	0.19	C
27	0.23	0.87	0.218	0.10	D
28	0.18	0.79	0.198	0.07	A
29	0.33	0.96	0.240	0.15	B
30	0.21	0.95	0.238	0.15	C
May 1	0.19	0.91	0.228	0.15	D
2	0.21	0.94	0.235	0.14	A
3	0.19	0.80	0.200	0.02	B
4	0.15	0.74	0.185	0.06	C
5	0.18	0.73	0.183	0.06	D
6	0.25	0.77	0.193	0.10	A
7	0.19	0.77	0.193	0.10	B
8	0.15	0.77	0.193	0.10	C
9	0.23	0.82	0.205	0.10	D
10	0.16	0.73	0.183	0.08	A
11	0.13	0.67	0.168	0.10	B
12	0.17	0.69	0.173	0.10	C
13	0.18	0.64	0.160	0.05	D
14	0.17	0.65	0.163	0.05	A
15	0.22	0.74	0.185	0.05	B
16	0.15	0.72	0.180	0.07	C
17	0.14	0.68	0.170	0.08	D

The purist statistician would require that these points be plotted at the midpoint, thus the moving mean for the first four results should be placed on the chart at 2 April. In practice, however, the point is usually plotted at the last result time, in the case 4 April. In this way the moving average and moving range charts indicate the current situation, rather than being behind time.

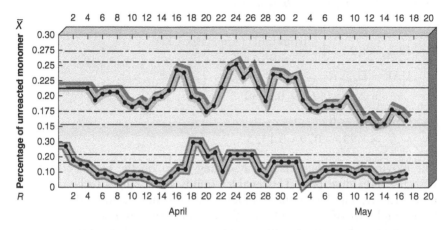

Figure 7.7 Four-day moving mean and moving range charts (unreacted monomer)

An earlier stage in controlling the polymerization process would have been to analyse the data available from an earlier period, say during February and March, to find the process mean and the mean range, and to establish the mean and range chart limits for the moving mean and range charts. The process was found to be in statistical control during February and March and capable of meeting the requirements of producing a product with less than 0.35 per cent monomer impurity. These observations had a process mean of 0.22 per cent and, with groups of $n = 4$, a mean range of 0.079 per cent. So the control chart limits, which are the same for both conventional and moving mean and range charts, would have been calculated before starting to plot the moving mean and range data onto charts. The calculations are shown below:

Moving mean and mean chart limits

$$n = 4$$
$$\overline{\overline{X}} = 0.22 \quad \text{from the results}$$
$$\overline{R} = 0.079 \quad \text{for February/March}$$

$$A_2 = 0.73 \quad \text{from table}$$
$$2/3\ A_2 = 0.49 \quad \text{(Appendix B)}$$

$$\text{UAL} = \overline{\overline{X}} + A_2 \overline{R}$$
$$= 0.22 + (0.73 \times 0.079) = 0.2777$$

$$\text{UWL} = \overline{\overline{X}} + 2/3\ A_2 \overline{R}$$
$$= 0.22 + (0.49 \times 0.079) = 0.2587$$

$$\text{LWL} = \overline{\overline{X}} - 2/3\ A_2 \overline{R}$$
$$= 0.22 - (0.49 \times 0.079) = 0.1813$$

$$\text{LAL} = \overline{\overline{X}} - A_2 \overline{R}$$
$$= 0.22 - (0.73 \times 0.079) = 0.1623$$

Moving range and range chart limits

$$D'_{0.001} = 2.57$$
$$D'_{0.025} = 1.93$$
from table (Appendix C)

$$\text{UAL} = D'_{0.001}\overline{R}$$
$$= 2.57 \times 0.079 = 0.2030$$

$$\text{UWL} = D'_{0.025}\overline{R}$$
$$= 1.93 \times 0.079 = 0.1525$$

The moving mean chart has a smoothing effect on the results compared with the individual plot. This enables trends and changes to be observed more readily. The larger the sample size the greater the smoothing effect. So a sample size of six would smooth even more the curves of Figure 7.7. A disadvantage of increasing sample size, however, is the lag in following any trend – the greater the size of the grouping, the greater the lag. This is shown quite clearly in Figure 7.8 in which sales data have been plotted using moving means of three and nine individual results. With such data the technique may be used as an effective forecasting method.

In the polymerization example one new piece of data becomes available each day and, if moving mean and moving range charts were being used, the result would be reviewed day by day. An examination of Figure 7.7 shows that:

- there was no abnormal behaviour of either the mean or the range on 16 April;
- the abnormality on 18 April was *not* caused by a change in the mean of the process, but an increase in the spread of the data, which shows as an action signal on the moving range chart. The result of nearly zero for the unreacted monomer (18th) is unlikely because it implies almost total polymerization. The resulting investigation revealed that the plant chemist had picked up the bottle containing the previous day's sample from which the unreacted monomers had already been extracted during analysis – so when he erroneously repeated the analysis the result was unusually low. This type of error is a human one – the process mean had not changed and the charts showed this;
- the plots for 19 April again show an action on the range chart. This is because the new mean and range plots are not independent of the previous ones. In reality, once a special cause has been identified, the individual 'outlier' result could be eliminated from the series. If this had been done the plot corresponding to the result from 19 April would not show an action on the moving range chart. The warning signals on 20 and 21 April are also due to the same isolated low result which is not removed from the series until 22 April.

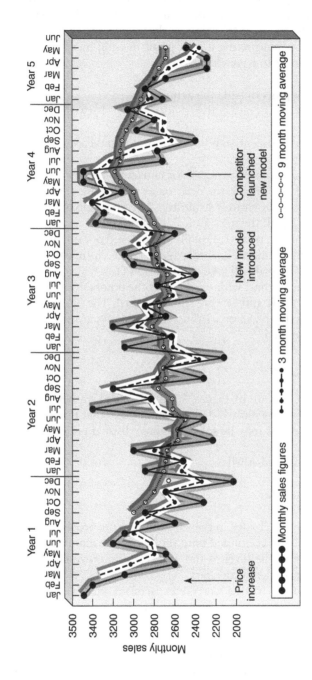

Figure 7.8 Sales figures and moving average charts

Supplementary rules for moving mean and moving range charts

The fact that the points on a moving mean and moving range chart are not independent affects the way in which the data are handled and the charts interpreted. Each value influences *four* (*n*) points on the four-point moving mean chart.

The rules for interpreting a four-point moving mean chart are that the process is assumed to have changed if:

1   ONE point plots outside the action lines.
2   THREE (*n* – 1) consecutive points appear between the warning and action lines.
3   TEN (2.5*n*) consecutive points plot on the same side of the centreline.

If the same data had been grouped for conventional mean and range charts, with a sample size of *n* = 4, the decision as to the date of starting the grouping would have been entirely arbitrary. The first sample group might have been 1, 2, 3, 4 April; the next 5, 6, 7, 8 April and so on; this is identified in Table 7.2 as combination A. Equally, 2, 3, 4, 5 April might have been combined; this is combination B. Similarly, 3, 4, 5, 6 April leads to combination C; and 4, 5, 6, 7 April will give combination D.

A moving mean chart with *n* = 4 is as if the points from four conventional mean charts A, B, C and D were superimposed. The plotted points on such charts are exactly the same as those on the moving mean and range plot previously examined.

The process overall

If the complete picture of Figure 7.7 is examined, rather than considering the values as they are plotted daily, it can be seen that the moving mean and moving range charts may be split into three distinct periods:

•   beginning to mid-April;
•   mid-April to early May;
•   early to mid-May.

Clearly, a dramatic change in the variability of the process took place in the middle of April and continued until the end of the month. This is shown by the general rise in the level of the values in the range chart and the more erratic plotting on the mean chart.

An investigation to discover the cause(s) of such a change is required. In this particular example, it was found to be due to a change in supplier of feedstock material, following a shut-down for maintenance work at the usual supplier's plant. When that supplier came back on stream in early May, not only did the variation in the impurity, unreacted monomer, return to normal, but its average level fell until on 13 May an action signal was given. Presumably this would have led to an investigation into the reasons for the low result, in order that this desirable situation might be repeated and maintained. This type of 'map-reading' of control charts, integrated into a good management system, is an indispensable part of SPC.

Moving mean and range charts are particularly suited to industrial processes in which results become available infrequently. This is often a consequence of either lengthy, difficult, costly or destructive analysis in continuous processes or product analyses in batch manufacture. The rules for moving mean and range charts are the same as for mean and range charts except that there is a need to understand and allow for non-independent results.

Exponentially weighted moving average

In mean and range control charts, the decision signal obtained depends largely on the last point plotted. In the use of moving mean charts some authors have questioned the appropriateness of giving equal importance to the most recent observation. The exponentially weighted moving average (EWMA) chart is a type of moving mean chart in which an 'exponentially weighted mean' is calculated each time a new result becomes available:

New weighted mean = ($a \times$ new result) + (($1 - a$) × previous mean),

where a is the 'smoothing constant'. It has a value between 0 and 1; many people use $a = 0.2$. Hence, new weighted mean = ($0.2 \times$ new result) + ($0.8 \times$ previous mean).

In the viscosity data plotted in Figure 7.9 the starting mean was 80.00. The results of the first few calculations are shown in Table 7.3.

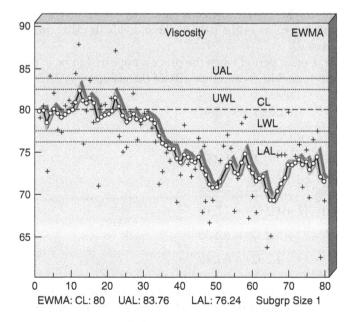

Figure 7.9 An EWMA chart

Table 7.3 Calculation of EWMA

Batch no.	Viscosity	Moving mean
–	–	80.00
1	79.1	79.82
2	80.5	79.96
3	72.7	78.50
4	84.1	79.62
5	82.0	80.10
6	77.6	79.60
7	77.4	79.16
8	80.5	79.43
•	•	•
•	•	•
•	•	•

When viscosity of batch 1 becomes available,

New weighted mean (1) = $(0.2 \times 79.1) + (0.8 \times 80.0) = 79.82$

When viscosity of batch 2 becomes available,

New weighted mean (2) = $(0.2 \times 80.5) + (0.8 \times 79.82) = 79.96$

Setting up the EWMA chart: the centreline was placed at the previous process mean (80.0 cSt.) as in the case of the individuals chart and in the moving mean chart.

Previous data, from a period when the process appeared to be in control, was grouped into four. The mean range ($\bar{R}$) of the groups was 7.733 cSt.

$$\sigma = \bar{R}/d_n = 7.733/2.059 = 3.756$$
$$SE = \sigma/\sqrt{[a/(2-a)]}$$
$$= 3.756\sqrt{[0.2/(2-0.2)]} = 1.252$$

LAL $= 80.0 - (3 \times 1.252) = 76.24$

LWL $= 80.0 - (2 \times 1.252) = 77.50$

UWL $= 80.0 + (2 \times 1.252) = 82.50$

UAL $= 80.0 + (3 \times 1.252) = 83.76.$

The choice of *a* has to be left to the judgement of the quality control specialist, the smaller the value of *a*, the greater the influence of the historical data.

Further terms can be added to the EWMA equation which are sometimes called the 'proportional', 'integral' and 'differential' terms in the process control engineer's basic proportional, integral, differential – or 'PID' – control equation (see Hunter, 1986).

The EWMA has been used by some organizations, particularly in the process industries, as the basis of new 'control/performance chart' systems. Great care must be taken when using these systems since they do not show changes in variability very well, and the basis for weighting data is often either questionable or arbitrary.

7.5 Control charts for standard deviation (σ)

Range charts are commonly used to control the precision or spread of processes. Ideally, a chart for standard deviation (σ) should be used but, because of the difficulties associated with calculations and understanding standard deviation, sample range is often substituted.

Significant advances in computing technology have led to the availability of cheap computers/calculators/software with which calculation of standard is quite simple. Using such technology, experiments in Japan have shown that the time required to calculate sample range is greater than that for σ, and the number of miscalculations is greater when using the former statistic. The conclusions of this work were that mean and standard deviation charts provide a simpler and better method of process control for variables than mean and range charts, when using modern computing technology.

The standard deviation chart is very similar to the range chart (see Chapter 6). The estimated standard deviation (s_i) for each sample being calculated, plotted and compared to predetermined limits.

Those using calculators for this computation must use the s or σ_{n-1} key and not the σ_n key. As we have seen in Chapter 5, the sample standard deviation calculated using the 'n' formula will tend to under-estimate the standard deviation of the whole process, and it is the value of $s(n-1)$ that is plotted on a standard deviation chart. The bias in the sample standard deviation is allowed for in the factors used to find the control chart limits.

Statistical theory allows the calculation of a series of constants (C_n) which enables the estimation of the process standard deviation (σ) from the average of the sample standard deviation ($\bar{s}$). The latter is the simple arithmetic mean of the sample standard deviations and provides the central-line on the standard deviation control chart.

The relationship between σ and $\bar{s}$ is given by the simple ratio:

$$\sigma = \bar{s} \; C_n,$$

where σ = estimated process standard deviation;

C_n = a constant, dependent on sample size. Values for C_n appear in Appendix E.

The control limits on the standard deviation chart, like those on the range chart, are asymmetrical, in this case about the average of the sample standard deviation ($\bar{s}$). The table in Appendix E provides four constants $B'_{.001}$, $B'_{.025}$, $B'_{.975}$ and $B'_{.999}$ which may be used to calculate the control limits for a standard deviation chart from $\bar{s}$. The table also gives the constants $B_{.001}$, $B_{.025}$, $B_{.975}$ and $B_{.999}$ which are used to find the warning and action lines from the estimated process standard deviation, σ. The control chart limits for the control chart are calculated as follows:

Upper Action Line at $B'_{.001} \bar{s}$ or $B_{.001} \sigma$

Upper Warning Line at $B'_{.025} \bar{s}$ or $B_{.025} \sigma$

Lower Warning Line at $B'_{.975} \bar{s}$ or $B_{.975} \sigma$

Lower Action Line at $B'_{.999} \bar{s}$ or $B_{.999} \sigma$.

Table 7.4 100 steel rod lengths as 25 samples of size 4

Sample number	Sample rod lengths				Sample mean	Sample range	Standard deviation
	(i)	(ii)	(iii)	(iv)	(mm)	(mm)	(mm)
1	144	146	154	146	147.50	10	4.43
2	151	150	134	153	147.00	19	8.76
3	145	139	143	152	144.75	13	5.44
4	154	146	152	148	150.00	8	3.65
5	157	153	155	157	155.50	4	1.91
6	157	150	145	147	149.75	12	5.25
7	149	144	137	155	146.25	18	7.63
8	141	147	149	155	148.00	14	5.77
9	158	150	149	156	153.25	9	4.43
10	145	148	152	154	149.75	9	4.03
11	151	150	154	153	152.00	4	1.83
12	155	145	152	148	150.00	10	4.40
13	152	146	152	142	148.00	10	4.90
14	144	160	150	149	150.75	16	6.70
15	150	146	148	157	150.25	11	4.79
16	147	144	148	149	147.00	5	2.16
17	155	150	153	148	151.50	7	3.11
18	157	148	149	153	151.75	9	4.11
19	153	155	149	151	152.00	6	2.58
20	155	142	150	150	149.25	13	5.38
21	146	156	148	160	152.50	14	6.61
22	152	147	158	154	152.75	11	4.57
23	143	156	151	151	150.25	13	5.38
24	151	152	157	149	152.25	8	3.40
25	154	140	157	151	150.50	17	7.42

An example should help to clarify the design and use of the sigma chart. Let us re-examine the steel rod cutting process which we met in Chapter 5, and for which we designed mean and range charts in Chapter 6. The data has been reproduced in Table 7.4 together with the standard deviation (s_i) for each sample of size four. The next step in the design of a sigma chart is the calculation of the average sample standard deviation (s). Hence:

$$\bar{s} = \frac{4.43 + 8.76 + 5.44 + \cdots + 7.42}{25}$$

$\bar{s} = 4.75$ mm.

The estimated process standard deviation (σ) may now be found. From Appendix E for a sample size $n = 4$, $C_n = 1.085$ and:

$\sigma = 4.75 \times 1.085 = 5.15$ mm.

This is very close to the value obtained from the mean range:

$\sigma = \bar{R} / dn = 10.8 / 2.059 = 5.25$ mm

The control limits may now be calculated using either σ and the B constants from Appendix E or $\bar{s}$ and the B′ constants:

Upper Action Line B′$_{.001}$ $\bar{s}$ = 2.522 × 4.75

or B$_{.001}$ σ = 2.324 × 5.15

= 11.97 mm

Upper Warning Line B′$_{.001}$ $\bar{s}$ = 1.911 × 4.75

or B$_{.001}$ σ = 1.761 × 5.15

= 9.09 mm

Lower Warning Line B′$_{.975}$ $\bar{s}$ = 0.291 × 4.75

or B$_{.975}$ σ = 0.2682 × 5.15

= 1.38 mm

Lower Action Line B′$_{.999}$ $\bar{s}$ = 0.098 × 4.75

or B$_{.999}$ σ = 0.090 × 5.15

= 0.46 mm.

Figure 7.10 shows control charts for sample standard deviation and range plotted using the data from Table 7.4. The range chart is, of course, exactly the same as that shown in Figure 6.8. The charts are very similar and either of them may be used to control the dispersion of the process, together with the mean chart to control process average.

If the standard deviation chart is to be used to control spread, then it may be more convenient to calculate the mean chart control limits from either the average sample standard deviation ($\bar{s}$) or the estimated process standard deviation (σ). The formula are:

Action Lines at $\overline{\overline{X}} \pm A_1\sigma$

or $\overline{\overline{X}} \pm A_3\bar{s}$.

Warning Lines at $\overline{\overline{X}} \pm 2/3A_1\sigma$

or $\overline{\overline{X}} \pm 2/3A_3\bar{s}$.

It may be recalled from Chapter 6 that the action lines on the mean chart are set at:

$$\overline{\overline{X}} \pm 3\sigma/\sqrt{n},$$

hence, the constant A_1 must have the value:

$$A_1 = 3/\sqrt{n},$$

which for a sample size of four:

$$A_1 = 3/\sqrt{4} = 1.5.$$

Similarly:

$$2/3\,A_1 = 2/\sqrt{n} \text{ and for } n = 4,$$
$$2/3\,A_1 = 2/\sqrt{4} = 1.0.$$

In the same way the values for the A_3 constants may be found from the fact that:

$$\sigma = \bar{s} \times C_n.$$

Hence, the action lines on the mean chart will be placed at:

$$\overline{\overline{X}} \pm 3\,\bar{s}\,C_n/\sqrt{n},$$

therefore, $A_3 = 3 \times C_n/\sqrt{n}$,

which for a sample size of four:

$$A_3 = 3 \times 1.085/\sqrt{4} = 1.628.$$

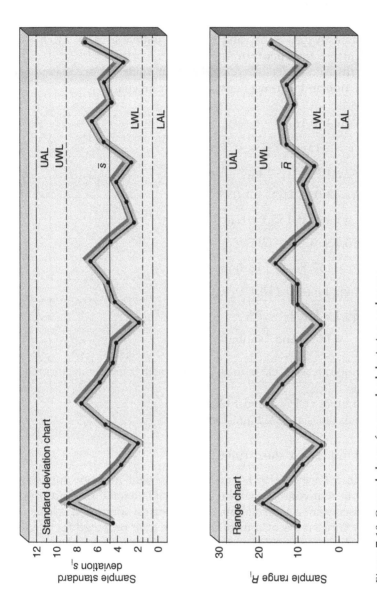

Figure 7.10 Control charts for standard deviation and range

Similarly:

$$2/3 \ A_3 = 2 \times C_n / \sqrt{n} \text{ and for } n = 4,$$
$$2/3 \ A_3 = 2 \times 1.085 / \sqrt{4} = 1.085.$$

The constants A_1, $2/3 \ A_1$, A_3, and $2/3 \ A_3$ for sample sizes $n = 2$ to $n = 12$ have been calculated and appear in Appendix B.

Using the data on lengths of steel rods in Table 7.4, we may now calculate the action and warning limits for the mean chart:

$$\overline{\overline{X}} = 150.1\text{mm}$$

$$\sigma = 5.15\text{mm} \quad \overline{s} = 4.75\text{mm}$$

$$A_1 = 1.5 \quad A_3 = 1.628$$

$$2/3 \ A_1 = 1.0 \quad 2/3 \ A_3 = 1.085$$

Action Lines at $150.1 \pm (1.5 \times 5.15)$

or $150.1 \pm (1.63 \times 4.75)$

$$= 157.8 \text{ and } 142.4\text{mm}.$$

Warning Lines at $150.1 \pm (1.0 \times 5.15)$

or $150.1 \pm (1.09 \times 4.75)$

$$= 155.3 \text{ and } 145.0\text{mm}.$$

These values are very close to those obtained from the mean range $\overline{R}$ in Chapter 6:

Action Lines at 158.2 and 142.0 mm.
Warning Lines at 155.2 and 145.0 mm.

7.6 Techniques for short run SPC

In Donald Wheeler's (1991) small but excellent book on this subject he pointed out that control charts may be easily adapted to short production runs to discover new information, rather than just confirming what is already known. Various types of control chart have been proposed for tackling this problem. The most usable are discussed in the next two subsections.

Difference charts

A very simple method of dealing with mixed data resulting from short runs of different product types is to subtract a 'target' value for each product from the results obtained. The differences are plotted on a chart which allows the underlying process variation to be observed.

The subtracted value is specific to each product and may be a target value or the historic grand mean. The centreline (CL) must clearly be zero.

The outer control limits for difference charts (also known as 'X-nominal' and 'X-target' charts) are calculated as follows:

$$\text{UCL/LCL} = 0.00 \pm 2.66 \; m\overline{R}.$$

The mean moving range, $m\overline{R}$, is best obtained from the moving ranges ($n = 2$) from the X-nominal values.

A moving range chart should be used with a *difference chart*, the centre-line of which is the mean moving range:

$$\text{CL}_R = m\overline{R}.$$

The upper control limit for this moving range chart will be:

$$\text{UCLR} = 3.268 \; m\overline{R}.$$

These charts will make sense, of course, only if the variation in the different products is of the same order. Difference charts may also be used with subgrouped data.

Z charts

The Z chart, like the difference chart, allows different target value products to be plotted on one chart. In addition it also allows products with different levels of dispersion or variation to be included. In this case, a target or nominal value for each product is required, plus a value for the products' standard deviations. The latter may be obtained from the product control charts.

The observed value (x) for each product is used to calculate a Z value by subtracting the target or nominal value (t) and dividing the difference by the standard deviation value (σ) for that product:

$$Z = \frac{x - t}{\sigma}.$$

The central-line for this chart will be zero and the outer limits placed at ± 3.0.

A variation on the Z chart is the Z* chart in which the difference between the observed value and the target or nominal value is divided by the mean range ($\overline{R}$):

$$Z^* = \frac{x - t}{\overline{R}}.$$

The centreline for this chart will again be zero and the outer control limits at ±2.66. Yet a further variation on this theme is the chart used with subgroup means.

7.7 Summarizing control charts for variables

There are many types of control chart and many types of processes. Charts are needed that will detect changes quickly, and are easily understood, so that they will help to control and improve the process.

With naturally grouped data conventional mean and range charts should be used. With one-at-a-time data use an individuals chart, moving mean and moving range charts or alternatively an EWMA chart should be used.

When choosing a control chart the following should be considered:

- Who will set up the chart?
- Who will plot the chart?
- Who will take what action and when?

A chart should always be chosen that the user can understand and that will detect changes quickly.

Chapter highlights

- SPC is based on basic principles that apply to all types of processes, including those in which isolated or infrequent data are available, as well as continuous processes – only the time scales differ. Control charts are used to investigate the variability of processes, help find the causes of changes, and monitor performance.
- Individual or run charts are often used for one-at-a-time data. Individual charts and range charts based on a sample of two are simple to use, but their interpretation must be carefully managed. They are not so good at detecting small changes in process mean.
- The zone control chart is an adaptation of the individuals or mean chart, on which zones with scores are set at one, two and three standard deviations from the mean. Keki Bhote's pre-control method uses similar principles, based on the product specification. Both methods are simple to use but inferior to the mean chart in detecting changes and supporting continuous improvement.
- The median and the mid-range may be used as measures of central tendency, and control charts using these measures are in use. The methods of setting up such control charts are similar to those for mean charts. In the multi-vari chart, the specification tolerances are used as control limits and the sample data are shown as vertical lines joining the highest and lowest values.

- When new data are available only infrequently they may be grouped into moving means and moving ranges. The method of setting up moving mean and moving range charts is similar to that for $\overline{X}$ and R charts. The interpretation of moving mean and moving range charts requires careful management as the plotted values do not represent independent data.
- Under some circumstances, the latest data point may require weighting to give a lower importance to older data and then use can be made of an exponentially weighted moving average (EWMA) chart.
- The standard deviation is an alternative measure of the spread of sample data. Whilst the range is often more convenient and more understandable, simple computers/calculators have made the use of standard deviation charts more accessible. Above sample sizes of 12, the range ceases to be a good measure of spread and standard deviations must be used.
- Standard deviation charts may be derived from both estimated standard deviations for samples and sample ranges. Standard deviation charts and range charts, when compared, show little difference in controlling variability.
- Techniques described in Donald Wheeler's book are available for short production runs. These include difference charts, which are based on differences from target or nominal values, and various forms of Z charts, based on differences and product standard deviations.
- When considering the many different types of control charts and processes, charts should be selected for their ease of detecting change, ease of understanding and ability to improve processes. With naturally grouped or past data, conventional mean and range charts should be used. For one-at-a-time data, individual (or run) charts, moving mean/moving range charts and EWMA charts may be more appropriate.

Note

1 Bhote claimed this demonstrates a minimum process capability of *Cpk* 1.33 – see Chapter 10.

References and further reading

Barnett, N. and Tong, P.F. (1994) 'A Comparison of Mean and Range Charts with Pre-Control Having Particular Reference to Short-Run Production', *Quality and Reliability Engineering International*, Vol. 10, No. 6, November/December, pp. 477–86.

Bhote, K.R. (1991) (Original 1925) *World Class Quality – Using Design of Experiments to Make it Happen*, American Management Association, New York, USA.

Hunter, J.S. (1986) 'The Exponentially Weighted Moving Average', *Journal of Quality Technology*, Vol. 18, pp. 203–210.

Wheeler, D.J. (1991) *Short Run SPC*, SPC Press, Knoxville TN, USA.

Wheeler, D.J. (2004) *Advanced Topics in SPC*, SPC Press, Knoxville TN, USA.

Discussion questions

1   Comment on the statement, 'a moving mean chart and a conventional mean chart would be used with different types of processes'.
2   The data in the next table shows the levels of contaminant in a chemical product:

   (a) Plot a histogram.
   (b) Plot an individuals or run chart.
   (c) Plot moving mean and moving range charts for grouped sample size $n = 4$.
       Interpret the results of these plots.

Levels of contamination in a chemical product

Sample	Result (ppm)	Sample	Result (ppm)
1	404.9	41	409.6
2	402.3	42	409.6
3	402.3	43	409.7
4	403.2	44	409.9
5	406.2	45	409.9
6	406.2	46	410.8
7	402.2	47	410.8
8	401.5	48	406.1
9	401.8	49	401.3
10	402.6	50	401.3
11	402.6	51	404.5
12	414.2	52	404.5
13	416.5	53	404.9
14	418.5	54	405.3
15	422.7	55	405.3
16	422.7	56	415.0
17	404.8	57	415.0
18	401.2	58	407.3
19	404.8	59	399.5
20	412.0	60	399.5
21	412.0	61	405.4
22	405.9	62	405.4
23	404.7	63	397.9
24	403.3	64	390.4
25	400.3	65	390.4
26	400.3	66	395.5
27	400.5	67	395.5
28	400.5	68	395.5

29	400.5	69	398.5
30	402.3	70	400.0
31	404.1	71	400.2
32	404.1	72	401.5
33	403.4	73	401.5
34	403.4	74	401.3
35	402.3	75	401.2
36	401.1	76	401.3
37	401.1	77	401.9
38	406.0	78	401.9
39	406.0	79	404.4
40	406.0	80	405.7

3   In a batch manufacturing process the viscosity of the compound increases during the reaction cycle and determines the end-point of the reaction. Samples of the compound are taken throughout the whole period of the reaction and sent to the laboratory for viscosity assessment. The laboratory tests cannot be completed in less than three hours. The delay during testing is a major source of under-utilization of both equipment and operators. Records have been kept of the laboratory measurements of viscosity and the power taken by the stirrer in the reactor during several operating cycles. When plotted as two separate moving mean and moving range charts this reveals the following data:

Date and time		*Moving mean viscosity*	*Moving mean stirrer power*
07/04	07.30	1020	21
	09.30	2250	27
	11.30	3240	28
	13.30	4810	35
	Batch completed and discharged		
	18.00	1230	22
	21.00	2680	22
08/04	00.00	3710	28
	03.00	3980	33
	06.00	5980	36
	Batch completed and discharged		
	13.00	2240	22
	16.00	3320	30

(continued)

(continued)

Date and time		Moving mean viscosity	Moving mean stirrer power
	19.00	3800	35
	22.00	5040	31
	Batch completed and discharged		
09/04	04.00	1510	25
	07.00	2680	27
	10.00	3240	28
	13.00	4220	30
	16.00	5410	37
	Batch completed and discharged		
	23.00	1880	19
10/04	02.00	3410	24
	05.00	4190	26
	08.00	4990	32
Batch completed and discharged			

Standard error of the means – viscosity – 490

Standard error of the means – stirrer power – 90

Is there a significant correlation between these two measured parameters? If the specification for viscosity is 4500 to 6000, could the measure of stirrer power be used for effective control of the process?

4   The catalyst for a fluid-bed reactor is prepared in single batches and used one at a time without blending. Tetrahydrofuran (THF) is used as a catalyst precursor solvent. During the impregnation (SIMP) step the liquid precursor is precipitated into the pores of the solid silica support. The solid catalyst is then reduced in the reduction (RED) step using aluminium alkyls. The THF level is an important process parameter and is measured during the SIMP and RED stages.

The following data were collected on batches produced during implementation of a new catalyst family. These data include the THF level on each batch at the SIMP step and the THF level on the final reduced catalyst.

The specifications are:	USL	LSL
THF–SIMP	15.0	12.2
THF–RED	11.6	9.5

Batch	THF SIMP	THF RED	Batch	THF SIMP	THF RED
196	14.2	11.1	371	13.7	11.0
205	14.5	11.4	372	14.4	11.5
207	14.6	11.7	373	14.3	11.9
208	13.7	11.6	374	13.7	11.2
209	14.7	11.5	375	14.0	11.6
210	14.6	11.1	376	14.2	11.5
231	13.6	11.6	377	14.5	12.2
232	14.7	11.6	378	14.4	11.6
234	14.2	12.2	379	14.5	11.8
235	14.4	12.0	380	14.4	11.5
303	15.0	11.9	381	14.1	11.5
304	13.8	11.7	382	14.1	11.4
317	13.5	11.5	383	14.1	11.3
319	14.1	11.5	384	13.9	10.8
323	14.6	10.7	385	13.9	11.6
340	13.7	11.5	386	14.3	11.5
343	14.8	11.8	387	14.3	12.0
347	14.0	11.5	389	14.1	11.3
348	13.4	11.4	390	14.1	11.8
349	13.2	11.0	391	14.8	12.4
350	14.1	11.2	392	14.7	12.2
359	14.5	12.1	394	13.9	11.4
361	14.1	11.6	395	14.2	11.6
366	14.2	12.0	396	14.0	11.6
367	13.9	11.6	397	14.0	11.1
368	14.5	11.5	398	14.0	11.4
369	13.8	11.1	399	14.7	11.4
370	13.9	11.5	400	14.5	11.7

Carry out an analysis of this data for the THF levels at the SIMP step and the final RED catalyst, assuming that the data were being provided infrequently, as the batches were prepared.

Assume that readings from previous similar campaigns had given the following data:

THF–SIMP $\bar{\bar{X}}$ 14.00 σ 0.30

THF–RED $\bar{\bar{X}}$ 11.50 σ 0.30.

5 The weekly demand of a product (in tonnes) is given below. Use appropriate techniques to analyse the data, assuming that information is provided at the end of each week.

Week	Demand (Tn)	Week	Demand (Tn)
1	7	25	8
2	5	26	7.5
3	8.5	27	7
4	7	28	6.5
5	8.5	29	10.5
6	8	30	9.5
7	8.5	31	8
8	10.5	32	10
9	8.5	33	8
10	11	34	4.5
11	7.5	35	10.5
12	9	36	8.5
13	6.5	37	9
14	6.5	38	7
15	6.5	39	7.5
16	7	40	10.5

6 Middshire Water Company discharges effluent, from a sewage treatment works, into the River Midd. Each day a sample of discharge is taken and analysed to determine the ammonia content. Results from the daily samples, over a 40-day period, are given below:

Day	Ammonia (ppm)	Temperature (°C)	Operator
1	24.1	10	A
2	26.0	16	A
3	20.9	11	B
4	26.2	13	A
5	25.3	17	B
6	20.9	12	C
7	23.5	12	A
8	21.2	14	A
9	23.8	16	B
10	21.5	13	B
11	23.0	10	C
12	27.2	12	A
13	22.5	10	C
14	24.0	9	C
15	27.5	8	B

16	19.1	11	B
17	27.4	10	A
18	26.9	8	C
19	28.8	7	B
20	29.9	10	A
21	27.0	11	A
22	26.7	9	C
23	25.1	7	C
24	29.6	8	B
25	28.2	10	B
26	26.7	12	A
27	29.0	15	A
28	22.1	12	B
29	23.3	13	B
30	20.2	11	C
31	23.5	17	B
32	18.6	11	C
33	21.2	12	C
34	23.4	19	B
35	16.2	13	C
36	21.5	17	A
37	18.6	13	C
38	20.7	16	C
39	18.2	11	C
40	20.5	12	C

Use suitable techniques to detect and demonstrate changes in ammonia concentration?

(See also Chapter 9, Discussion question 7)

7   The National Rivers Authority (NRA) also monitor the discharge of effluent into the River Midd. The NRA can prosecute the water company if 'the ammonia content exceeds 30 ppm for more than 5 per cent of the time'.

The current policy of Middshire Water Company is to achieve a mean ammonia content of 25 ppm. They believe that this target is a reasonable compromise between risk of prosecution and excessive use of electricity to achieve an unnecessary low level.

(a) Comment on the suitability of 25 ppm as a target mean, in the light of the day-to-day variations in the data in question 6.

(b) What would be a suitable target mean if Middshire Water Company could be confident of getting the process in control by eliminating the kind of changes demonstrated by the data?

(c) Describe the types of control chart that could be used to monitor the ammonia content of the effluent and comment briefly on their relative merits.

8   (a) Discuss the use of control charts for range and standard deviation, explaining their differences and merits.
   (b) Using process capability studies, processes may be classified as being in statistical control and capable. Explain the basis and meaning of this classification. Suggest conditions under which control charts may be used, and how they may be adapted to make use of data which are available only infrequently.

Worked example

Evan and Hamble manufacture shampoo which sells as an own-label brand in the Askway chain of supermarkets. The shampoo is made in two stages: a batch mixing process is followed by a bottling process. Each batch of shampoo mix has a value of £10,000, only one batch is mixed per day, and this is sufficient to fill 50,000 bottles.

Askway specify that the active ingredient content should lie between 1.2 per cent and 1.4 per cent. After mixing, a sample is taken from the batch and analysed for active ingredient content. Askway also insist that the net content of each bottle should exceed 248 ml. This is monitored by taking five bottles every half-hour from the end of the bottling line and measuring the content.

(a) Describe how you would demonstrate to the customer, Askway, that the bottling process was stable.
(b) Describe how you would demonstrate to the customer that the bottling process was capable of meeting the specification.
(c) If you were asked to demonstrate the stability and capability of the mixing process how would your analysis differ from that described in parts (a) and (b).

Solution

(a) Using data comprising five bottle volumes taken every half-hour for, say, 40 hours:

   i   calculate mean and range of each group of five;
   ii   calculate overall mean ($\overline{\overline{X}}$) and mean range ($\overline{R}$);
   iii calculate $\sigma = \overline{R} / d_n$
   iv   calculate action and warning values for mean and range charts;
   v   plot means on mean chart and ranges on range chart;
   vi   assess stability of process from the two charts using action lines, warning lines and supplementary rules.

(b) Using the data from part (a):

    i   draw a histogram;

    ii   calculate the standard deviation of all 200 volumes;

    iii compare the standard deviations calculated in parts (a) and (b), explaining any discrepancies with reference to the charts;

    iv compare the capability of the process with the specification;

    v   Discuss the capability indices with the customer, making reference to the histogram and the charts. (See Chapter 10.)

(c) The data should be plotted as an individuals chart, then put into arbitrary groups of, say, 4. (Data from 80 consecutive batches would be desirable.) Mean and range charts should be plotted as in part (a). A histogram should be drawn as in part (b). The appropriate capability analysis could then be carried out.

8 Process control by attributes

Objectives

- To introduce the underlying concepts behind using attribute data.
- To distinguish between the various types of attribute data.
- To describe in detail the use of control charts for attributes: np-, p-, c- and u-charts.
- To examine the use of attribute data analysis methods in non-manufacturing situations.

8.1 Underlying concepts

The quality of many products and services is dependent upon characteristics that cannot be measured as variables. These are called attributes and may be counted, having been judged simply as either present or absent, conforming or non-conforming, acceptable or defective. Such properties as bubbles of air in a windscreen, the general appearance of a paint surface, accidents, the particles of contamination in a sample of polymer, errors in an invoice and the number of telephone calls are all attribute parameters. It is clearly not possible to use the methods of measurement and control designed for variables when addressing attributes data.

An advantage of attributes is that they are in general more quickly assessed so, often, variables are converted to attributes for assessment. But, as we shall see, attributes are not so sensitive a measure as variables and, therefore, detection of small changes is less reliable.

The statistical behaviour of attribute data is different to that of variable data and this must be taken into account when designing process control systems for attributes. To identify which type of data distribution we are dealing with, we must know something about the product or service form and the attribute under consideration. The following types of attribute lead to the use of different types of control chart, which are based on different statistical distributions:

1 *Conforming or non-conforming units*, each of which can be wholly described as failing or not failing, acceptable or defective, present or not present, etc., e.g. ball-bearings, invoices, workers, respectively.

2 *Conformities or non-conformities*, which may be used to describe a product or service, e.g. number of defects, errors, faults or positive values such as sales calls, truck deliveries, goals scored.

Hence, a defective is an item or 'unit' that contains one or more flaws, errors, faults or defects. A defect is an individual flaw, error or fault.

When we examine a fixed sample of the first type of attribute, for example 100 ball-bearings or invoices, we can state how many are defective or non-conforming. We shall then very quickly be able to work out how many are acceptable or conforming. So in this case, if two ball-bearings or invoices are classified as unacceptable or defective, 98 will be acceptable. This is different to the second type of attribute. If we examine a product such as a windscreen and find four defects – scratches or bubbles – we are not able to make any statements about how many scratches/bubbles are not present. This type of defect data is similar to the number of goals scored in a football match. We can only report the number of goals scored. We are unable to report how many were not.

The two types of attribute data lead to the use of two types of control chart:

1 Number of non-conforming units (or defectives) chart.
2 Number of non-conformities (or defects) chart.

These are each further split into two charts, one for the situation in which the sample size (number of units, or length or volume examined or inspected) is constant, and one for the samples of varying size. Hence, the collection of charts for attributes becomes:

1 (a) Number of non-conforming units (defectives) (np) chart – for constant sample size.
  (b) Proportion of non-conforming units (defectives) (p) chart – for samples of varying size.
2 (a) Number of non-conformities (defects) (c) chart – for samples of same size every time.
  (b) Number of non-conformities (defects) per unit (u) chart – for varying sample size.

The specification

Process control can be exercised using these simple charts on which the number or proportion of units, or the number of incidents or incidents per unit are plotted. Before commencing to do this, however, it is absolutely vital to clarify what constitutes a defective, non-conformance, defect or error, etc. No process control system can survive the heated arguments that will surround badly defined non-conformances. It is evident that in the study of attribute data, there will be several degrees of imperfection. The description of attributes, such as defects and errors, is a subject in its own right, but it is clear that a scratch on a paintwork or table top surface may range from a deep gouge to a slight mark, hardly visible to the naked eye; the consequences of accidents may range from death or severe injury to mere inconvenience. To ensure the smooth control of a process using attribute data, it is often necessary to provide representative samples, photographs or other objective evidence to support the decision maker. Ideally a sample of an acceptable product and one that is just not acceptable should be provided. These will allow the attention and effort to be concentrated on improving the process rather than debating the issues surrounding the severity of non-conformances.

Attribute process capability and its improvement

When a process has been shown to be in statistical control, the average level of events, errors, defects per unit or whatever will represent the capability of the process when compared with the specification. As with variables, to improve process capability requires a systematic investigation of the whole process system – not just a diagnostic examination of particular apparent causes of lack of control. This places demands on management to direct action towards improving such contributing factors as:

- operator performance, training and knowledge;
- equipment performance, reliability and maintenance;
- material suitability, conformance and grade;
- methods, procedures and their consistent usage.

A philosophy of never-ending improvement is always necessary to make inroads into process capability improvement, whether using variables or attribute data. It is often difficult, however, to make progress in process improvement programmes when only relatively insensitive attribute data are being used. One often finds that some form of alternative variable data are available or can be obtained with a little effort and expense. The extra cost associated with providing data in the form of measurements may well be trivial compared with the savings that can be derived by reducing process variability.

8.2 *np*-charts for number of defectives or non-conforming units

Consider a process that is producing ball-bearings, 10 per cent of which are defective: p, the proportion of defects, is 0.1. If we take a sample of one ball from the process, the chance or probability of finding a defective is 0.1 or p.

Similarly, the probability of finding a non-defective ball-bearing is 0.90 or $(1 - p)$. For convenience we will use the letter q instead of $(1 - p)$ and add these two probabilities together:

$$p + q = 0.1 + 0.9 = 1.0.$$

A total of unity means that we have present all the possibilities, since the sum of the probabilities of all the possible events must be one. This is clearly logical in the case of taking a sample of one ball-bearing for there are only two possibilities – finding a defective or finding a non-defective.

If we increase the sample size to two ball-bearings, the probability of finding two defectives in the sample becomes:

$$p \times p = 0.1 \times 0.1 = 0.01 = p^2.$$

This is one of the first laws of probability – the *multiplication law*. When two or more events are required to follow consecutively, the probability of them all happening is the product of their individual probabilities. In other words, for A *and* B to happen, multiply the individual probabilities p_A and p_B.

We may take our sample of two balls and find zero defectives. What is the probability of this occurrence?

$$q \times q = 0.9 \times 0.9 = 0.81 = q^2.$$

Let us add the probabilities of the events so far considered:

Two defectives – probability = 0.01 (p^2)

Zero defectives – probability = 0.81 (q^2)

Total = 0.82.

Since the total probability of all possible events must be one, it is quite obvious that we have not considered all the possibilities. There remains, of course, the chance of picking out one defective followed by one non-defective. The probability of this occurrence is:

$$p \times q = 0.1 \times 0.9 = 0.09 = pq.$$

However, the single defective may occur in the second ball-bearing:

$$q \times p = 0.9 \times 0.1 = 0.09 = qp.$$

This brings us to a second law of probability – the *addition law*. If an event may occur by a number of alternative ways, the probability of the event is the sum of the probabilities of the individual occurrences. That is, for A *or* B to happen, add the probabilities p_A and p_B. So the probability of finding one defective in a sample of size two from this process is:

$pq + qp = 0.09 + 0.09 = 0.18 = 2pq.$

Now, adding the probabilities:

Two defectives – probability = 0.01 (p^2)

One defective – probability = 0.18 $(2pq)$

No defectives – probability = 0.81 (q^2)

Total probability = 1.00.

So, when taking a sample of two from this process, we can calculate the probabilities of finding one, two or zero defectives in the sample. Those who are familiar with simple algebra will recognize that the expression:

$p^2 + 2pq + q^2 = 1,$

is an expansion of:

$(p + q)^2 = 1,$

and this is called the *binomial* expression. It may be written in a general way:

$(p + q)^n = 1,$

where n = sample size (number of units);

p = proportion of defectives or 'non-conforming units' in the population from which the sample is drawn;

q = proportion of non-defectives or 'conforming units' in the population = $(1 - p)$.

To reinforce our understanding of the binomial expression, look at what happens when we take a sample of size four:

$n = 4$

$(p + q)^4 = 1$

expands to:

$$p^4 \quad + \quad 4p^3q \quad + \quad 6p^2q^2 \quad + \quad 4pq^3 \quad + \quad q^4$$

Probability	Probability	Probability	Probability	Probability
of 4	of 3	of 2	of 1	of zero
defectives	defectives	defectives	defective	defectives

in the sample

The mathematician represents the probability of finding x defectives in a sample of size n when the proportion present is p as:

$$P(x) = \left(\frac{n}{x} \right) p^x (1-p)^{(n-x)},$$

where $\left(\dfrac{n}{x} \right) = \dfrac{n!}{(n-x)!x!}$

$n!$ is $1 \times 2 \times 3 \times 4 \times \ldots \times n$

$x!$ is $1 \times 2 \times 3 \times 4 \times \ldots \times x$

For example, the probability $P(2)$ of finding two defectives in a sample of size five taken from a process producing 10 per cent defectives ($p = 0.1$) may be calculated:

$n = 5$

$x = 2$

$p = 0.1$

$$P(2) = \frac{5!}{(5-2)!2!} 0.1^2 \times 0.9^3$$

$$= \frac{5 \times 4 \times 3 \times 2 \times 1}{(3 \times 2 \times 1) \times (2 \times 1)} \times 0.1 \times 0.1 \times 0.9 \times 0.9 \times 0.9$$

$$= 10 \times 0.01 \times 0.729 = 0.0729.$$

This means that, on average, about 7 out of 100 samples of 5 ball-bearings taken from the process will have two defectives in them. The *average* number of defectives present in a sample of 5 will be 0.5.

It may be possible at this stage for the reader to see how this may be useful in the design of process control charts for the number of defectives

Table 8.1 Number of defectives found in samples of 100 ballpoint pen cartridges

2	2	2	2	1
4	3	4	1	3
1	0	2	5	0
0	3	1	3	2
0	1	6	0	1
4	2	0	2	2
5	3	3	2	0
3	1	1	1	4
2	2	2	3	2
3	1	1	1	1

or classified units. If we can calculate the probability of exceeding a certain number of defectives in a sample, we shall be able to draw action and warning lines on charts, similar to those designed for variables in earlier chapters.

To use the probability theory we have considered so far we must know the proportion of defective units being produced by the process. This may be discovered by taking a reasonable number of samples – say 50 – over a typical period, and recording the number of defectives or non-conforming units in each. Table 8.1 lists the number of defectives found in 50 samples of size $n = 100$ taken every hour from a process producing ballpoint pen cartridges. These results may be grouped into the frequency distribution of Table 8.2 and shown as the histogram of Figure 8.1. This is clearly a different type of histogram from the symmetrical ones derived from variables data in earlier chapters.

The average number of defectives per sample may be calculated by adding the number of defectives and dividing the total by the number of samples:

$$\frac{\text{Total number of defectives}}{\text{Number of samples}} = \frac{100}{50} = 2 \text{ (average number of defectives per sample).}$$

Table 8.2 Frequency distribution of defectives in sample

Number of defectives in sample	Tally chart (samples with that number of defectives)	Frequency
0	⟦⟧⟦⟧ 11	7
1	⟦⟧⟦⟧ ⟦⟧⟦⟧ 111	13
2	⟦⟧⟦⟧ ⟦⟧⟦⟧ 1111	14
3	⟦⟧⟦⟧ 1111	9
4	1111	4
5	11	2
6	1	1

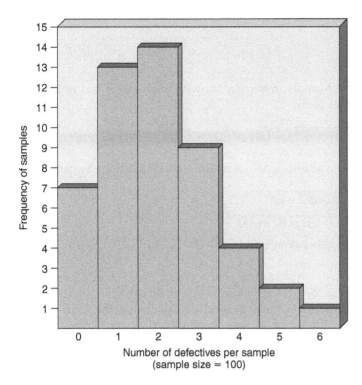

Figure 8.1 Histogram of results from Table 8.1

This value is $n\bar{p}$ – the sample size multiplied by the average proportion defective in the process.

Hence, $\bar{p}$ may be calculated:

$$\bar{p} = n\bar{p} \,/\, n = 2 \,/\, 100 = 0.02 \text{ or 2 per cent.}$$

The scatter of results in Table 8.1 is a reflection of sampling variation and not due to inherent variation within the process. Looking at Figure 8.1 we can see that at some point around 5 defectives per sample, results become less likely to occur and at around 7 they are very unlikely. As with mean and range charts, we can argue that if we find, say, 8 defectives in the sample, then there is a very small chance that the percentage of defectives being produced is still at 2 per cent, and it is likely that the percentage of defectives being produced has risen above 2 per cent.

We may use the binomial distribution to set action and warning lines for the so-called 'np- process control chart', sometimes known in the USA as a pn-chart. Attribute control chart practice in industry, however, is to set outer limits or action lines at three standard deviations (3σ) either side of the average number defective (or non-conforming units), and inner limits or warning lines at $\pm$ two standard deviations (2σ).

The standard deviation (σ) for a binomial distribution is given by the formula:

$$\sigma = \sqrt{n\bar{p}(1-\bar{p})}.$$

Use of this simple formula, requiring knowledge of only n and np, for the ballpoint cartridges gives:

$$\sigma = \sqrt{100 \times 0.02 \times 0.98} = 1.4.$$

Now, the upper action line (UAL) or control limit (UCL) may be calculated:

$$\begin{aligned}
\text{UWL} &= n\bar{p} + 3\sqrt{n\bar{p}(1-\bar{p})} \\
&= 2 + 3\sqrt{100 \times 0.02 \times 0.98} \\
&= 6.2, \text{ i.e. between 6 and 7.}
\end{aligned}$$

This result is the same as that obtained by setting the UAL at a probability of about 0.005 (1 in 200) using binomial probability tables.

This formula offers a simple method of calculating the UAL for the np-chart, and a similar method may be employed to calculate the upper warning line (UWL):

$$\begin{aligned}
\text{UWL} &= n\bar{p} + 2\sqrt{n\bar{p}(1-\bar{p})} \\
&= 2 + 2\sqrt{100 \times 0.02 \times 0.98} \\
&= 4.8, \text{ i.e. between 4 and 5.}
\end{aligned}$$

Again this gives the same result as that derived from using the binomial expression to set the warning line at about 0.05 probability (1 in 20).

It is not possible to find fractions of defectives in attribute sampling, so the presentation may be simplified by drawing the control lines between whole numbers. The sample plots then indicate clearly when the limits have been crossed. In our example, 4 defectives found in a sample indicates normal sampling variation, whilst 5 defectives gives a warning signal that another sample should be taken immediately because the process may have deteriorated. In control charts for attributes it is commonly found that only the upper limits are specified since we wish to detect an increase in defectives. Lower control lines may be useful, however, to indicate when a significant process improvement has occurred, or to indicate when suspicious results have been plotted. In the case under consideration, there are no lower action or warning lines, since it is expected that zero defectives will periodically be found in the samples of 100, when 2 per cent defectives are being generated by the process. This is shown by the negative values for ($np - 3\sigma$) and ($np - 2\sigma$).

As in the case of the mean and range charts, the attribute charts were invented by Shewhart and are sometimes called Shewhart charts. He recognized the need

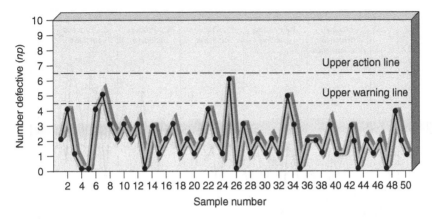

Figure 8.2 *np*-chart: number of defectives in samples of 100 ballpoint pen
  cartridges

for both the warning and the action limits. The use of warning limits is strongly
recommended since their use improves the sensitivity of the charts and tells the
'operator' what to do when results *approach* the action limits – take another
sample – but do not act until there is a clear signal to do so.

Figure 8.2 is an *np*-chart on which are plotted the data concerning the
ballpoint pen cartridges from Table 8.1. Since all the samples contain fewer
defectives than the action limit and only 3 out of 50 enter the warning
zone, and none of these are consecutive, the process is considered to be in
statistical control. We may, therefore, reasonably assume that the process is
producing a constant level of 2 per cent defective (that is the 'process capability') and the chart may be used to control the process. The method for
interpretation of control charts for attributes is similar to that described for
mean and range charts in earlier chapters.

Figure 8.3 shows the effect of increases in the proportion of defective
pen cartridges from 2 per cent through 3, 4, 5, 6 to 8 per cent in steps. For
each percentage defective, the run length to detection, that is the number of
samples which needed to be taken before the action line is crossed following
the increase in process defective, is given below:

Percentage process defective	*Run length to detection from Figure 8.3*
3	>10
4	9
5	4
6	3
8	1

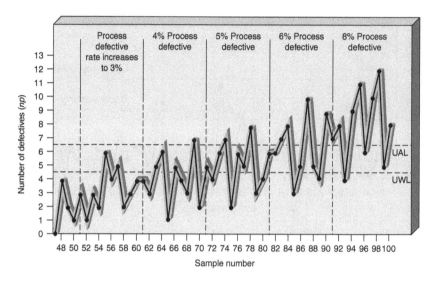

Figure 8.3 *np*-chart: defective rate of pen cartridges increasing

Clearly, this type of chart is not as sensitive as mean and range charts for detecting changes in process defective. For this reason, the action and warning lines on attribute control charts are set at the higher probabilities of approximately 1 in 200 (action) and approximately 1 in 20 (warning).

This lowering of the action and warning lines will obviously lead to the more rapid detection of a worsening process. It will also increase the number of incorrect action signals. Since inspection for attributes by, for example, using a go/no-go gauge is usually less costly than the measurement of variables, an increase in the amount of re-sampling may be tolerated.

If the probability of an event is – say – 0.25, on average it will occur every fourth time, as the average run length (ARL) is simply the reciprocal of the probability. Hence, in the pen cartridge case, if the proportion defective is 3 per cent ($p = 0.03$), and the action line is set between 6 and 7, the probability of finding 7 or more defectives may be calculated or derived from the binomial expansion as 0.0312 ($n = 100$). We can now work out the ARL to detection:

$$\text{ARL}(3\%) = 1/P(>7) = 1/0.0312 = 32.$$

For a process producing 5 per cent defectives, the ARL for the same sample size and control chart is:

$$\text{ARL}(5\%) = 1/P(>7) = 1/0.0234 = 4.$$

The ARL is quoted to the nearest integer.

The conclusion from the run length values is that, given time, the *np*-chart will detect a change in the proportion of defectives being produced. If the change is an increase of approximately 50 per cent, the *np*-chart will be very slow to detect it, on average. If the change is a decrease of 50 per cent, the chart will not detect it because, in the case of a process with 2 per cent defective, there are no lower limits. This is not true for all values of defective rate. Generally, *np*-charts are less sensitive to changes in the process than charts for variables.

8.3 *p*-charts for proportion defective or non-conforming units

In cases where it is not possible to maintain a constant sample size for attribute control, the *p*-chart, or proportion defective or non-conforming chart may be used. It is, of course, possible and quite acceptable to use the *p*-chart instead of the *np*-chart even when the sample size is constant. However, plotting directly the number of defectives in each sample onto an *np*-chart is simple and usually more convenient than having to calculate the proportion defective. The data required for the design of a *p*-chart are identical to those for an *np*-chart, both the sample size and the number of defectives need to be observed.

Table 8.3 Results from the issue of textile components in varying numbers

'Sample' number	Issue size	Number of rejects	Proportion defective
1	1135	10	0.009
2	1405	12	0.009
3	805	11	0.014
4	1240	16	0.013
5	1060	10	0.009
6	905	7	0.008
7	1345	22	0.016
8	980	10	0.010
9	1120	15	0.013
10	540	13	0.024
11	1130	16	0.014
12	990	9	0.009
13	1700	16	0.009
14	1275	14	0.011
15	1300	16	0.012
16	2360	12	0.005
17	1215	14	0.012
18	1250	5	0.004
19	1205	8	0.007

(continued)

Table 8.3 (continued)

'Sample' number	Issue size	Number of rejects	Proportion defective
20	950	9	0.009
21	405	9	0.022
22	1080	6	0.006
23	1475	10	0.007
24	1060	10	0.009

Table 8.3 shows the results from 24 deliveries of textile components. The batch (sample) size varies from 405 to 2860. For each delivery, the proportion defective has been calculated:

$$p_i = x_i / n_i,$$

where p_i = the proportion defective in delivery i;

x_i = the number of defectives in delivery i;

n_i = the size (number of items) of the ith delivery.

As with the np-chart, the first step in the design of a p-chart is the calculation of the average proportion defective ($\bar{p}$).

For the deliveries in question:

$$p = 280/27,930 = 0.010.$$

Control chart limits

If a constant 'sample' size is being inspected, the p-control chart limits would remain the same for each sample. When p-charts are being used with samples of varying sizes, the standard deviation and control limits change with n, and unique limits should be calculated for each sample size. However, for practical purposes, an average sample size (n) may be used to calculate action and warning lines. These have been found to be acceptable when the individual sample or lot sizes vary from n by no more than 25 per cent each way. For sample sizes outside this range, separate control limits must be calculated. There is no magic in this 25 per cent formula, it simply has been shown to work.

The next stage then in the calculation of control limits for the p-chart, with varying sample size, is to determine the average sample size (n) and the range 25 per cent either side.

Range of sample sizes with constant control chart limits equals:

$$n \pm 0.25\bar{n}.$$

For the deliveries under consideration:

$$\bar{n} = 27,930/24 = 1164.$$

Permitted range of sample size $= 1164 \pm (0.25 \times 1164)$

$$= 873 - 1455.$$

For sample sizes within this range, the control chart lines may be calculated using a value of σ given by:

$$\sigma = \frac{\sqrt{\bar{p}(1-\bar{p})}}{\sqrt{\bar{n}}} = \frac{\sqrt{0.010 \times 0.99}}{\sqrt{1164}} = 0.003.$$

Then, Action lines $= p \pm 3\sigma$

$$= 0.01 \pm 3 \times 0.003$$
$$= 0.019 \text{ and } 0.001.$$

Warning lines $= p \pm 2\sigma$

$$= 0.01 \pm 2 \times 0.003$$
$$= 0.016 \text{ and } 0.004.$$

Control lines for delivery numbers 3, 10, 13, 16 and 21 must be calculated individually as these fall outside the range 873–1455:

$$\text{Action lines} = \bar{p} \pm 3\sqrt{\bar{p}(1-\bar{p})}/\sqrt{n_i}.$$
$$\text{Warning lines} = \bar{p} \pm 2\sqrt{\bar{p}(1-\bar{p})}/\sqrt{n_i}.$$

Table 8.4 shows the detail of the calculations involved and the resulting action and warning lines. Figure 8.4 shows the p-chart plotted with the varying action and warning lines. It is evident that the design, calculation, plotting and interpretation of p-charts is more complex than that associated with np-charts.

The process involved in the delivery of the material is out of control. Clearly, the supplier has suffered some production problems during this period and some of the component deliveries are of doubtful quality. Complaints to the supplier after the delivery corresponding to sample 10 seemed to have a good effect until delivery 21 caused a warning signal. This type of control chart may improve substantially the dialogue and partnership between suppliers and customers.

Table 8.4 Calculation of p-chart lines for sample sizes outside the range 873–1455

General formulae:

$$\text{Action lines} = \bar{p} \pm 3 \sqrt{\bar{p}(1-\bar{p})}/\sqrt{n}$$

$$\text{Warning lines} = \bar{p} \pm 2 \sqrt{\bar{p}(1-\bar{p})}/\sqrt{n}$$

$$\bar{p} = 0.010$$

$$\text{and } \sqrt{\bar{p}(1-\bar{p})} = 0.0995$$

Sample number	Sample size	$\sqrt{\bar{p}(1-\bar{p})}/\sqrt{n}$	UAL	UWL	LWL	LAL
3	805	0.0035	0.021	0.017	0.003	neg. (i.e. 0)
10	540	0.0043	0.023	0.019	0.001	neg. (i.e. 0)
13	1700	0.0024	0.017	0.015	0.005	0.003
16	2360	0.0020	0.016	0.014	0.006	0.004
21	405	0.0049	0.025	0.020	neg. (i.e. 0)	neg. (i.e. 0)

Sample points falling below the lower action line also indicate a process that is out of control. Lower control lines are frequently omitted to avoid the need to explain to operating personnel why a very low proportion of defectives is classed as being out-of-control. When the p-chart is to be used by management, however, the lower lines are used to indicate when an investigation should be instigated to discover the cause of an unusually good performance. This may also indicate how it may be repeated. The lower control limits are given in Table 8.4. An examination of Figure 8.4 will show that none of the sample points fall below the lower action lines.

8.4 c-charts for number of defects/non-conformities

The control charts for attributes considered so far have applied to cases in which a random sample of definite size is selected and examined in some way. In the process control of attributes, there are situations where the number of events, defects, errors or non-conformities can be counted, but there is no information about the number of events, defects or errors which are *not* present. Hence, there is the important distinction between defectives and defects already given in Section 8.1. So far we have considered defectives where each item is classified either as conforming or non-conforming (a defective), which gives rise to the term *binomial* distribution. In the case of defects, such as holes in a fabric or fisheyes in plastic film, we know the number of defects present but we do not know the number of non-defects present. Other examples of these include the number of imperfections on a

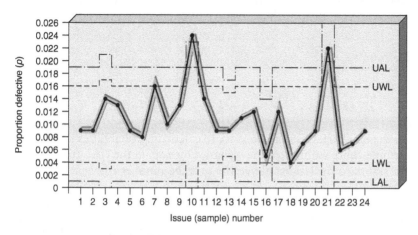

Figure 8.4 p-chart: for issued components

painted door, errors in a typed document, the number of faults in a length of woven carpet and the number of sales calls made. In these cases the binomial distribution does not apply.

This type of problem is described by the Poisson distribution, named after the Frenchman who first derived it in the early nineteenth century. Because there is no fixed sample size when counting the number of events, defects, etc., theoretically the number could tail off to infinity. Any distribution which does this must include something of the *exponential distribution* and the constant *e*. This contains the element of fading away to nothing since its value is derived from the formula:

$$e = \frac{1}{0!} + \frac{1}{1!} + \frac{1}{2!} + \frac{1}{3!} + \frac{1}{4!} + \frac{1}{5!} + \cdots + \frac{1}{\infty!}.$$

If the reader cares to work this out, the value *e* = 2.7183 is obtained.

The equation for the Poisson distribution includes the value of *e* and looks rather formidable at first. The probability of observing *x* defects in a given unit is given by the equation:

$$P(x) = e^{-\bar{c}}(\bar{c}^{x}/x!),$$

where *e* = exponential constant, 2.7183;

$\bar{c}$ = average number of defects per unit being produced by the process.

The reader who would like to see a simple derivation of this formula should refer to the excellent book *Facts from Figures* by Moroney (1983).

So the probability of finding three bubbles in a windscreen from a process which is producing them with an average of one bubble present is given by:

$$P(3) = e^{-1} \times \frac{1^3}{3 \times 2 \times 1}$$

$$= \frac{1}{2.7183} \times \frac{1}{6} = 0.0613.$$

As with the *np*-chart, it is not necessary to calculate probabilities in this way to determine control limits for the *c*-chart. Once again the UAL (UCL) is set at three standard deviations above the average number of events, defects, errors, etc.

Let us consider an example in which, as for *np*-charts, the sample is constant in number of units, or volume, or length, etc. In a polythene film process, the number of defects – fisheyes – on each identical length of film are being counted. Table 8.5 shows the number of fisheyes that have been found on inspecting 50 lengths, randomly selected, over a 24-hour period. The total number of defects is 159 and, therefore, the average number of defects $\bar{c}$ is calculated.

In this example,

$$\bar{c} = 159/50 = 3.2.$$

The standard deviation of a Poisson distribution is very simply the square root of the process average. Hence, in the case of defects,

$$\sigma = \sqrt{\bar{c}},$$

and for our polyethylene process

$$\sigma = \sqrt{3.2} = 1.79.$$

Table 8.5 Number of fisheyes in identical pieces of polythene film (10 m²)

4	2	6	3	6
2	4	1	4	3
1	3	5	5	1
3	0	2	1	3
2	6	3	2	2
4	2	4	0	4
1	4	3	4	2
5	1	5	3	1
3	3	4	2	5
7	5	2	8	3

The UAL (UCL) may now be calculated:

$$\text{UAL (UCL)} = \bar{c} + 3\sqrt{\bar{c}}$$
$$= 3.2 + 3\sqrt{3.2}$$
$$= 8.57, \text{ i.e. between 8 and 9.}$$

This sets the UAL at approximately 0.005 probability, using a Poisson distribution. In the same way, an UWL may be calculated:

$$\text{UWL} = \bar{c} + 2\sqrt{\bar{c}}$$
$$= 3.2 + 2\sqrt{3.2}$$
$$= 6.78, \text{ i.e. between 6 and 7.}$$

Figure 8.5, which is a plot of the 50 polythene film inspection results used to design the *c*-chart, shows that the process is in statistical control, with an average of 3.2 defects on each length. If this chart is now used to control the process, we may examine what happens over the next 25 lengths, taken over a period of 12 hours. Figure 8.6 is the *c*-chart plot of the results. The picture tells us that all was running normally until sample 9, which shows 8 defects on the unit being inspected, this signals a warning and another sample is taken immediately. Sample 10 shows that the process has drifted out of control and results in an investigation to find the assignable cause. In this case, the film extruder filter was suspected of being blocked and so it was cleaned. An immediate resample after restart of the process shows the process to be back in control. It continues to remain in that state for at least the next 14 samples.

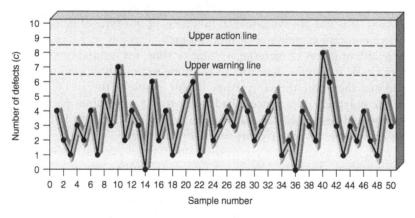

Figure 8.5 *c*-chart: polythene fisheyes – process in control

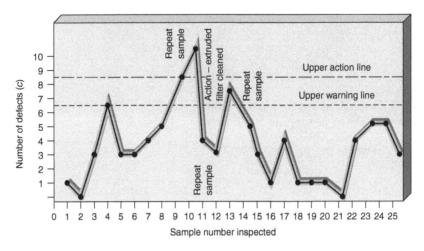

Figure 8.6 *c*-chart: polythene fisheyes

As with all types of control chart, an improvement in quality and productivity is often observed after the introduction of the *c*-chart. The confidence of having a good control system that derives as much from knowing when to leave the process alone as when to take action leads to more stable processes, less variation and fewer interruptions from unnecessary alterations.

8.5 *u*-charts for number of defects/non-conformities per unit

We saw in the previous section how the *c*-chart applies to the number of events, defects or errors in a constant size of sample, such as a table, a length of cloth, the hull of a boat, a specific volume, a windscreen, an invoice or a time period. It is not always possible, however, in this type of situation to maintain a constant sample size or unit of time.

The length of pieces of material, volume or time, for instance, may vary. At other times, it may be desirable to continue examination until a defect is found and then note the sample size. If, for example, the average value of *c* in the polythene film process had fallen to 0.5, the values plotted on the chart would be mostly 0 and 1, with an occasional 2. Control of such a process by a whole number *c*-chart would be nebulous.

The *u*-chart is suitable for controlling this type of process, as it measures the number of events defects, or non-conformities per unit or time period, and the 'sample' size can be allowed to vary. In the case of inspection of cloth or other surfaces, the area examined may be allowed to vary and the *u*-chart will show the number of defects per unit area, e.g. per square metre. The statistical theory behind the *u*-chart is very similar to that for the *c*-chart.

The design of the *u*-chart is similar to the design of the *p*-chart for proportion defective. The control lines will vary for each sample size, but for practical purposes may be kept constant if sample sizes remain with 25 per cent either side of the average sample size, $\bar{n}$.

As in the *p*-chart, it is necessary to calculate the process average defect rate. In this case we introduce the symbol *u*:

$$\bar{u} = \text{Process average defects per unit}$$

$$= \frac{\text{Total number of defects}}{\text{Total sample inspected}}$$

The defects found per unit (*u*) will follow a Poisson distribution, the standard deviation σ of which is the square root of the process average. Hence:

Action lines $= \bar{u} \pm 3\sqrt{\bar{u}}/\sqrt{\bar{n}}$.

Warning lines $= \bar{u} \pm 2\sqrt{\bar{u}}/\sqrt{\bar{n}}$.

A summary table

Table 8.6 shows a summary of all four attribute control charts in common use. Appendix J gives some approximations to assist in process control of attributes.

8.6 Attribute data in non-manufacturing

Activity sampling

Activity or work sampling is a simple technique based on the binomial theory. It is used to obtain a realistic picture of productive time, or time spent on particular activities, by both human and technological resources.

An exercise should begin with discussions with the staff involved, explaining to them the observation process, and the reasons for the study. This would be followed by an examination of the processes, establishing the activities to be identified. A preliminary study is normally carried out to confirm that the set of activities identified is complete, familiarize people with the method and reduce the intrusive nature of work measurement, and to generate some preliminary results in order to establish the number of observations required in the full study. The preliminary study would normally cover 50–100 observations, made at random points during a representative period of time, and may include the design of a check sheet on which to record the data. After the study it should be possible to determine the number of observations required in the full study using the formula:

$$N = \frac{4P(100 - P)}{L^2} \text{ (for 95 per cent confidence)}$$

Table 8.6 Attribute data: control charts

What is measured	Chart name	Attribute charted	Centreline	Warning lines	Action or control lines	Comments
Number of defectives in sample of constant size n	'np' chart or 'pn' chart	$n\bar{p}$ number of defectives in sample of size n	$n\bar{p}$	$n\bar{p} \pm 2\sqrt{n\bar{p}(1-\bar{p})}$	$n\bar{p} \pm 3\sqrt{n\bar{p}(1-\bar{p})}$	n = sample size p = proportion defective $\bar{p}$ = average of p
Proportion defective in a sample of variable size	'p' chart	p – the ratio of defectives to sample size	$\bar{p}$	$\bar{p} \pm 2\sqrt{\dfrac{\bar{p}(1-\bar{p})}{\bar{n}}}$ *	$\bar{p} \pm 3\sqrt{\dfrac{\bar{p}(1-\bar{p})}{\bar{n}}}$ *	$\bar{n}$ = average sample size $\bar{p}$ = average value of p
Number of defects/flaws in sample of constant size	'c' chart	c number of defects/flaws in sample of constant size	$\bar{c}$	$\bar{c} \pm 2\sqrt{\bar{c}}$	$\bar{c} \pm 3\sqrt{\bar{c}}$	$\bar{c}$ = average number of defects/flaws in sample of constant size
Average number of flaws/defects in sample of variable size	'u' chart	u – the ratio of defects to sample size	$\bar{u}$	$\bar{u} \pm 2\sqrt{\dfrac{\bar{u}}{\bar{n}}}$ *	$\bar{u} \pm 3\sqrt{\dfrac{\bar{u}}{\bar{n}}}$ *	u = defects/flaws per sample $\bar{u}$ = average value of u n = sample size $\bar{n}$ = average value of n

*Only valid when n is in zone $\bar{n}$ ±25 per cent.

where N = number of observations;

P = percentage occurrence of any one activity;

L = required precision in the estimate of P.

If the first study indicated that 45 per cent of the time is spent on productive work, and it is felt that an accuracy of 2 per cent is desirable for the full study (i.e. we want to be reasonably confident that the actual value lies between 43 and 47 per cent assuming the study confirms the value of 45 per cent), then the formula tells us we should make:

$$\frac{4 \times 45 \times (100 - 45)}{2 \times 2} = 2475 \text{ observations.}$$

If the work centre concerned has five operators, this implies 495 tours of the centre in the full study. It is now possible to plan the main study with 495 tours covering a representative period of time.

Having carried out the full study, it is possible to use the same formula, suitably arranged, to establish the actual accuracy in the percentage occurrence of each activity:

$$L = \sqrt{\frac{4P(100 - P)}{N}}.$$

The technique of activity sampling, although quite simple, is very powerful. It can be used in a variety of ways, in a variety of environments, both manufacturing and non-manufacturing. While it can be used to indicate areas which are worthy of further analysis, using for example process improvement techniques, it can also be used to establish time standards themselves.

Absenteeism

Figure 8.7 is a simple demonstration of how analysis of attribute data may be helpful in a non-manufacturing environment. A manager joined the Personnel Department of a gas supply company at the time shown by the plot for week 14 on the 'employees absent in 1 week chart'. She attended an (SPC) course two weeks later (week 16), but at this time control charts were not being used in the Personnel Department. She started plotting the absenteeism data from week 15 onwards. When she plotted the dreadful result for week 17, she decided to ask the SPC co-ordinator for his opinion of the action to be taken, and to set up a meeting to discuss the alarming increase in absenteeism. The SPC co-ordinator examined the history of absenteeism and established the average value as well as the warning and action lines,

Date	UAL 11.5			UWL 9.5			Mean 4.83			LWL 0.5			LAL			Specification									
Time/sample no.	1	2	3	4	5	6	7	8	9	10	11	12	13	14	15	16	17	18	19	20	21				
Total inspected. *n*		C		O		N		S		T		A		N		T									
Total absent. *np*	6	4	3	2	7	8	5	6	1	3	5	2	8	4	3	5	8	7	5	4	5				
	1	2	3	4	5	6	7	8	9	10	11	12	13	14	15	16	17	18	19	20	21	22	23	24	25

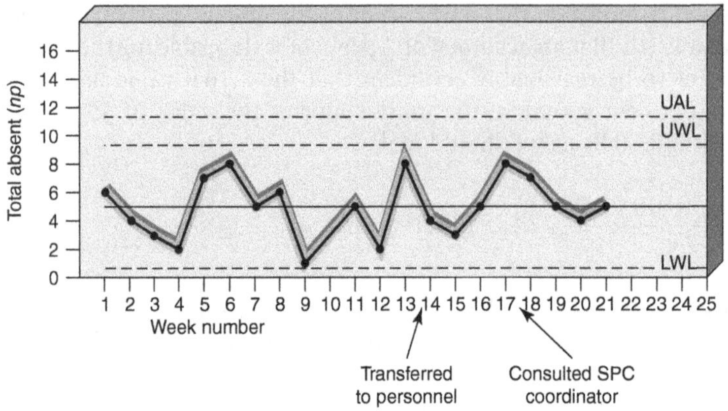

Figure 8.7 Attribute chart of number of employee-days absent each week

both of which he added to the plot. Based on this he persuaded her to take no action and to cancel the proposed meeting since there was no significant event to discuss.

Did the results settle down to a more acceptable level after this? No, the results continued to be randomly scattered about the average – there had been no special cause for the observation in week 17 and hence no requirement for a solution. In many organizations the meeting would not only have taken place, but the management would have congratulated themselves on their 'evident' success in reducing absenteeism. Over the whole period there were no significant changes in the 'process' and absenteeism was running at an average of approximately five per week, with random or common variation about that value. No assignable or special causes had occurred. If there was an item for the agenda of a meeting about absenteeism, it should have been to discuss the way in which the average could be reduced and the discussion would be helped by looking at the general causes which give rise to this average, rather than specific periods of apparently high absenteeism.

In both manufacturing and non-manufacturing, and when using both attributes and variables, the temptation to take action when a 'change' is assumed to have occurred is high, and reacting to changes that are not significant is a frequent cause of adding variation to otherwise stable processes. This is sometimes known as management interference; it may be recognized by the stable running of a process during the night shift, or at weekends, when the managers are at home!

Chapter highlights

- Attributes, things that are counted and are generally more quickly assessed than variables, are often used to determine quality. These require different control methods to those used for variables.
- Attributes may appear as numbers of non-conforming or defective units, or as numbers of non-conformities or defects. In the examination of samples of attribute data, control charts may be further categorized into those for constant sample size and those for varying sample size. Hence, there are charts for:

  number defective (non-conforming) np

  proportion defective (non-conforming) p

  number of defects (non-conformities) c

  number of defects (non-conformities) per unit u

- It is vital, as always, to define attribute specifications. The process capabilities may then be determined from the average level of defectives or defects measured. Improvements in the latter require investigation of the whole process system. Never-ending improvement applies equally well to attributes, and variables should be introduced where possible to assist this.
- Control charts for number (np) and proportion (p) defective are based on the binomial distribution. Control charts for number of defects (c) and number of defects per unit (u) are based on the Poisson distribution.
- A simplified method of calculating control chart limits for attributes is available, based on an estimation of the standard deviation σ.
- Np- and c-charts use constant sample sizes and, therefore, the control limits remain the same for each sample. For p- and u-charts, the sample size (n) varies and the control limits vary with n. In practice, an 'average sample size' ($\bar{n}$) may be used in most cases.
- The concepts of processes being in and out of statistical control applies to attributes. Attribute charts are not so sensitive as variable control charts for detecting changes in non-conforming processes. Attribute control chart performance may be measured, using the average run length (ARL) to detection.
- Attribute data is frequently found in non-manufacturing. Activity sampling is a technique based on the binomial theory and is used to obtain a realistic picture of time spent on particular activities. Attribute control charts may be useful in the analysis of absenteeism, invoice errors, etc.

References and further reading

Grant, E.L. and Leavenworth, R.S. (1996) *Statistical Quality Control*, 7th edn, McGraw-Hill, New York, USA.

Lockyer, K.G., Muhlemann, A.P. and Oakland, J.S. (1992) *Production and Operations Management*, 6th edn, Pitman, London, UK.

Montgomery, D. (2008) *Statistical Process Control: A Modern Introduction*, ASQ Press, Milwaukee WI, USA.

Moroney, M.J. (1983) *Facts from Figures*, Pelican (reprinted), London, UK.

Pyzdek, T. (1990) *Pyzdek's Guide to SPC, Vol. 1: Fundamentals*, ASQC Quality Press, Milwaukee WI, USA.

Shewhart, W.A. (1931) *Economic Control of Quality from the Viewpoint of Manufactured Product*, Van Nostrand (Republished in 1980 by ASQC Quality Press, Milwaukee WI, USA.)

Wheeler, D.J. and Chambers, D.S. (1992) *Understanding Statistical Process Control*, 2nd edn, SPC Press, Knoxville TN, USA.

Discussion questions

1 (a) Process control charts may be classified under two broad headings, 'variables' and 'attributes'. Compare these two categories and indicate when each one is most appropriate.

   (b) In the context of quality control explain what is meant by a *number of defectives* (*np-*) chart.

2 Explain the difference between:

    np-chart,
    p-chart,
    c-chart.

3 Write down the formulae for the probability of obtaining r defectives in a sample of size n drawn from a population proportion p defective based on:

  i the binomial distribution;
  ii the Poisson distribution.

4 A factory finds that on average 20 per cent of the bolts produced by a machine are defective. Determine the probability that out of 4 bolts chosen at random:

  (a) 1, (b) 0, (c) at most 2 bolts will be defective.

5 The following record shows the number of defective items found in a sample of 100 taken twice per day.

Sample number	Number of defectives	Sample number	Number of defectives
1	4	7	3
2	2	8	1
3	4	9	1
4	3	10	5
5	2	11	4
6	6	12	4

13	1	27	0
14	2	28	1
15	1	29	3
16	4	30	0
17	1	31	0
18	0	32	2
19	3	33	1
20	4	34	1
21	2	35	4
22	1	36	0
23	0	37	2
24	3	38	3
25	2	39	2
26	2	40	1

Set up a Shewhart *np*-chart, plot the above data and comment on the results. (See also Chapter 9, Discussion question 3.)

6 Twenty samples of 50 polyurethane foam products are selected. The sample results are:

Sample No.	1	2	3	4	5	6	7	8	9	10
Number defective	2	3	1	4	0	1	2	2	3	2
Sample No.	11	12	13	14	15	16	17	18	19	20
Number defective	2	2	3	4	5	1	0	0	1	2

Design an appropriate control chart.
Plot these values on the chart and interpret the results.

3 Given in the table below are the results from the inspection of filing cabinets for scratches and small indentations.

Cabinet No.	1	2	3	4	5	6	7	8	
Number of defects	1	0	3	6	3	3	4	5	
Cabinet No.	9	10	11	12	13	14	15	16	
Number of defects	10	8	4	3	7	5	3	1	
Cabinet No.	17	18	19	20	21	22	23	24	25
Number of defects	4	1	1	1	0	4	5	5	5

Set up a control chart to monitor the number of defects. What is the average run length to detection when 6 defects are present?

Plot the data on the chart and comment upon the process. (See also Chapter 9, Discussion question 2.)

7 A control chart for a new kind of plastic is to be initiated. Twenty-five samples of 100 plastic sheets from the assembly line were inspected for flaws during a pilot run. The results are given below. Set up an appropriate control chart.

Sample No.	1	2	3	4	5	6	7	8	
Number of flaws/sheet	2	3	0	2	4	2	8	4	
Sample No.	9	10	11	12	13	14	15	16	17
Number of flaws/sheet	5	8	3	5	2	3	1	2	3
Sample No.	18	19	20	21	22	23	24	25	
Number of flaws/sheet	4	1	0	3	2	4	2	1	

Worked examples

1 Injury data

In an effort to improve safety in their plant, a company decided to chart the number of injuries that required first aid, each month. Approximately the same amount of hours were worked each month. The table below contains the data collected over a 2-year period.

Year 1 Month	Number of injuries (c)	Year 2 Month	Number of injuries (c)
January	6	January	10
February	2	February	5
March	4	March	9
April	8	April	4
May	5	May	3
June	4	June	2
July	23	July	2
August	7	August	1
September	3	September	3
October	5	October	4
November	12	November	3
December	7	December	1

Use an appropriate charting method to analyse the data.

Solution

As the same number of hours were worked each month, a *c*-chart should be utilized:

$$\Sigma c = 133.$$

From these data, the average number of injuries per month ($\bar{c}$) may be calculated:

$$\bar{c} = \frac{\Sigma c}{k} = \frac{133}{24} = 5.44 \text{ (centreline).}$$

The control limits are as follows:

$$\text{UAL/LAL} = \bar{c} \pm 3\sqrt{\bar{c}} = 5.54 \pm 3\sqrt{5.54}$$

$$\text{UAL} = 12.6 \text{ injuries (there is no LAL)}$$

$$\text{UWL/LWL} = \bar{c} \pm 2\sqrt{\bar{c}} = 5.54 \pm 2\sqrt{5.54} = 10.25 \text{ and } 0.83.$$

Figure 8.8 shows the control chart. In July Year 1, the reporting of 23 injuries resulted in a point above the UCL. The assignable cause was a large amount of holiday leave taken during that month. Untrained people and excessive overtime were used to achieve the normal number of hours worked for a

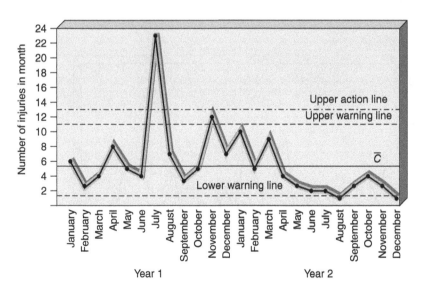

Figure 8.8 *c*-chart of injury data

month. There was also a run of nine points in a row below the centreline starting in April Year 2. This indicated that the average number of reported first aid cases per month had been reduced. This reduction was attributed to a switch from wire to plastic baskets for the carrying and storing of parts and tools which greatly reduced the number of injuries due to cuts. If this trend continues, the control limits should be recalculated when sufficient data were available.

2 Herbicide additions

The active ingredient in a herbicide product is added in two stages. At the first stage 160 litres of the active ingredient is added to 800 litres of the inert ingredient. To get a mix ratio of exactly 5 to 1 small quantities of either ingredient are then added. This can be very time consuming as sometimes a large number of additions are made in an attempt to get the ratio just right. The recently appointed Production Manager has introduced a new procedure for the first mixing stage. To test the effectiveness of this change he recorded the number of additions required for 30 consecutive batches, 15 with the old procedure and 15 with the new. Figure 8.9 is based on these data:

(a) What conclusions would you draw from the control chart in Figure 8.9, regarding the new procedure?
(b) Explain how the position of the control and warning lines were calculated for Figure 8.9.

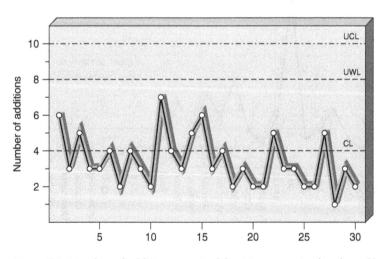

Figure 8.9 Number of additions required for 30 consecutive batches of herbicide

Solution

(a) This is a *c*-chart, based on the Poisson distribution. The centreline is drawn at 4, which is the mean for the first 15 points. UAL is at $4 + 3\sqrt{4}$. No lower action line has been drawn. $(4 - 3\sqrt{4}-$, would be negative; a Poisson with $\bar{c} = 4$ would be rather skewed.) Thirteen of the last 15 points are at or below the centreline. This is strong evidence of a decrease but might not be noticed by someone using rigid rules. A cusum chart may be useful here (see Chapter 9, Worked example 4).

(b) Based on the Poisson distribution:

$$\text{UAL} = \bar{c} + 3\sqrt{\bar{c}} = 4 + 3\sqrt{4} = 10.$$

$$\text{UWL} = \bar{c} + 2\sqrt{\bar{c}} = 4 + 2\sqrt{4} = 8.$$

9 Cumulative sum (cusum) charts

Objectives

- To introduce the technique of cusum charts for detecting change.
- To show how cusum charts should be used and interpreted.
- To demonstrate the use of cusum charts in product screening and selection.
- To cover briefly the decision procedures for use with cusum charts, including V-masks.

9.1 Introduction to cusum charts

In Chapters 5–8 we have considered Shewhart control charts for variables and attributes, named after the man who first described them in the 1920s. The basic rules for the operation of these charts predominantly concern the interpretation of each sample plot. Investigative and possibly corrective action is taken if an individual sample point falls outside the action lines, or if two consecutive plots appear in the warning zone – between warning and action lines. A repeat sample is usually taken immediately after a point is plotted in the warning zone. Guidelines have been set down in Chapter 6 for the detection of trends and runs above and below the average value but, essentially, process control by Shewhart charts considers each point as it is plotted. There are alternative control charts that consider more than one sample result.

The moving average and moving range charts described in Chapter 7 take into account part of the previous data, but a technique that uses all the information available is the Cumulative Sum or CUSUM method. This type of chart was developed in Britain in the 1950s and is one of the most powerful management tools available for the detection of trends and slight changes in data.

The advantage of plotting the cusum chart in highlighting small but persistent changes may be seen by an examination of some simple accident data. Table 9.1 shows the number of minor accidents per month in a large organization. Looking at the figures alone will not give the reader any clear picture of the safety performance of the business. Figure 9.1 is a c-chart on which the results have been plotted. The control limits have been calculated using the method given in Chapter 8.

Table 9.1 Number of minor accidents per month in a large organization

Month	Number of accidents	Month	Number of accidents	Month	Number of accidents	Month	Number of accidents
1	1	11	3	21	2	31	1
2	4	12	4	22	1	32	4
3	3	13	2	23	2	33	1
4	5	14	3	24	3	34	3
5	4	15	7	25	1	35	1
6	3	16	3	26	2	36	5
7	6	17	5	27	6	37	5
8	3	18	1	28	0	38	2
9	2	19	3	29	5	39	3
10	5	20	3	30	2	40	4

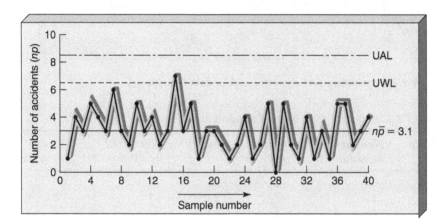

Figure 9.1 The *c*-chart of minor accidents per month

The average number of accidents per month is approximately three. The 'process' is obviously in statistical control since none of the sample points lie outside the action line and only one of the 40 results is in the warning zone. It is difficult to see from this chart any significant changes, but careful examination will reveal that the level of minor accidents is higher between months 2 and 17 than that between months 18 and 40. However, we are still looking at individual data points on the chart.

In Figure 9.2 the same data are plotted as cumulative sums on a 'cusum' chart. The calculations necessary to achieve this are extremely simple and are shown in Table 9.2. The average number of defectives, 3, has been subtracted from each sample result and the residues cumulated to give the cusum 'Score', Sr, for each sample. Values of Sr are plotted on the chart. The

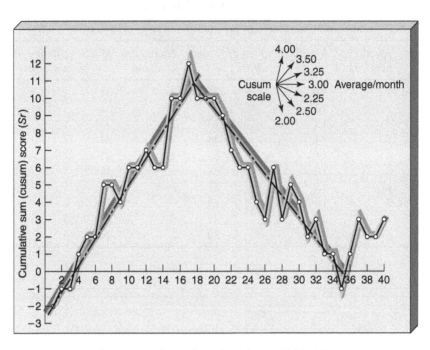

Figure 9.2 Cumulative sum chart of accident data in Table 9.1

Table 9.2 Cumulative sum values of accident data from Table 9.1 ($\bar{c} = 3$)

Month	Number of accidents − $\bar{c}$	Cusum score, Sr	Month	Number of accidents − $\bar{c}$	Cusum score, Sr
1	−2	−2	21	−1	9
2	1	−1	22	−2	7
3	0	−1	23	−1	6
4	2	1	24	0	6
5	1	2	25	−2	4
6	0	2	26	−1	3
7	3	5	27	3	6
8	0	5	28	−3	3
9	−1	4	29	2	5
10	2	6	30	−1	4
11	0	6	31	−2	2
12	1	7	32	1	3
13	−1	6	33	−2	1
14	0	6	34	0	1
15	4	10	35	−2	−1
16	0	10	36	2	1
17	2	12	37	2	3
18	−2	10	38	−1	2
19	0	10	39	0	2
20	0	10	40	1	3

difference in accident levels is shown dramatically. It is clear, for example, that from the beginning of the chart up to and including month 17, the level of minor accidents is on average higher than 3, since the cusum plot has a positive slope. Between months 18 and 35 the average accident level has fallen and the cusum slope becomes negative. Is there an increase in minor accidents commencing again over the last 5 months? Recalculation of the average number of accidents per month over the two main ranges gives:

Months (inclusive)	Total number of accidents	Average number of accidents per month
1–17	63	3.7
18–35	41	2.3

This confirms that the signal from the cusum chart was valid. The task now begins of diagnosing the special cause of this change. It may be, for example, that the persistent change in accident level is associated with a change in operating procedures or systems. Other factors, such as a change in materials used may be responsible. Only careful investigation will confirm or reject these suggestions. The main point is that the change was identified because the cusum chart takes account of past data.

Cusum charts are useful for the detection of short- and long-term changes and trends. Their interpretation requires care because it is not the actual cusum score which signifies the change, but the overall slope of the graph. For this reason the method is often more suitable as a management technique than for use on the shop floor. Production operatives, for example, will require careful training and supervision if cusum charts are to replace conventional mean and range charts or attribute charts at the point of manufacture.

The method of cumulating differences and plotting them has great application in many fields of management, and they provide powerful monitors in such areas as:

forecasting	– actual versus forecasted sales
absenteeism production levels }	– detection of slight changes
plant breakdowns	– maintenance performance

and many others in which data must be used to detect changes.

9.2 Interpretation of simple cusum charts

The interpretation of cusum charts is concerned with the assessment of gradients or slopes of graphs. Careful design of the charts is, therefore, necessary so that the appropriate sensitivity to change is obtained.

The calculation of the cusum score, Sr, is very simple and may be represented by the formula:

$$Sr = \sum_{i=1}^{r}(x_i - t),$$

where Sr = cusum score of the rth sample;
$\quad\quad x_i$ = result from the individual sample i (x_i may be a sample mean, $\bar{x}_i$);
$\quad\quad t$ = the target value.

The choice of the value of t is dependent upon the application of the technique. In the accident example we considered earlier, t, was given the value of the average number of accidents per month over 40 months. In a forecasting application, t may be the forecast for any particular period. In the manufacture of tablets, t may be the target weight or the centre of a specification tolerance band. It is clear that the choice of the t value is crucial to the resulting cusum graph. If the graph is always showing a positive slope, the data are constantly above the target or reference quantity. A high target will result in a continuously negative or downward slope. The rules for interpretation of cusum plots may be summarized.

- the cusum slope is *upwards*, the observations are *above* target;
- the cusum slope is *downwards*, the observations are *below* target;
- the cusum slope is *horizontal*, the observations are *on* target;
- the cusum slope *changes*, the observations are *changing* level;
- the absolute value of the cusum score has little meaning.

Setting the scales

As we are interested in the slope of a cusum plot the control chart design must be primarily concerned with the choice of its vertical and horizontal scales. This matter is particularly important for variables if the cusum chart is to be used in place of Shewhart charts for sample-to-sample process control at the point of operation.

In the design of conventional mean and range charts for variables data, we set control limits at certain distances from the process average. These corresponded to multiples of the standard error of the means, SE ($\sigma/\sqrt{n}$). Hence, the warning lines were set 2SE from the process average and the action lines at 3SE (Chapter 6). We shall use this convention in the design of cusum charts for variables, not in the setting of control limits, but in the calculation of vertical and horizontal scales.

When we examine a cusum chart, we would wish that a major change – such as a change of 2SE in sample mean – shows clearly, yet not so obtusely that the cusum graph is oscillating wildly following normal variation. This requirement may be met by arranging the scales such that a shift in sample

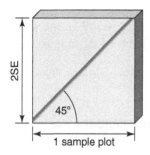

Figure 9.3 Slope of cusum chart for a change of 2SE in sample mean

mean of 2SE is represented on the chart by *ca* 45° slope. This is shown in Figure 9.3. It requires that the distance along the horizontal axis which represents one sample plot is approximately the same as that along the vertical axis representing 2SE. An example should clarify the explanation.

In Chapter 6, we examined a process manufacturing steel rods. Data on rod lengths taken from 25 samples of size four had the following characteristics:

Grand or Process Mean Length, $\bar{\bar{X}}$ =150.1mm

Mean Sample Range, $\bar{R}$ =10.8mm.

We may use our simple formula from Chapter 6 to provide an estimate of the process standard deviation, σ:

$$\sigma = \bar{R} / d_n,$$

where d_n is Hartley's Constant = 2.059 for sample size $n = 4$.

Hence, $\sigma = 10.8/2.059 = 5.25$ mm.

This value may in turn be used to calculate the standard error of the means:

$$SE = \sigma / \sqrt{n}$$
$$SE = 5.25 \sqrt{4} = 2.625$$

and

$$2SE = 2 \times 2.625 = 5.25 \text{ mm.}$$

We are now in a position to set the vertical and horizontal scales for the cusum chart. Assume that we wish to plot a sample result every 1 cm along the horizontal scale (abscissa) – the distance between each sample plot is 1 cm.

To obtain a cusum slope of *ca* 45° for a change of 2SE in sample mean, 1 cm on the vertical axis (ordinate) should correspond to the value of 2SE or thereabouts. In the steel rod process, 2SE = 5.25 mm. No one would be happy plotting a graph which required a scale 1 cm = 5.25 mm, so it is necessary to round up or down. Which shall it be?

Guidance is provided on this matter by the scale ratio test. The value of the scale ratio is calculated as follows:

$$\text{Scale ratio} = \frac{\text{Linear distance between plots along abscissa}}{\text{Linear distance representing 2SE along ordinate}}.$$

The value of the scale ratio should lie between 0.8 and 1.5. In our example if we round the ordinate scale to 1 cm = 4 mm, the following scale ratio will result:

Linear distance between plots along abscissa = 1 cm

Linear distance representing 2SE (5.25 mm) = 1.3125 cm

and scale ratio = 1 cm/1.3125 cm   = 0.76.

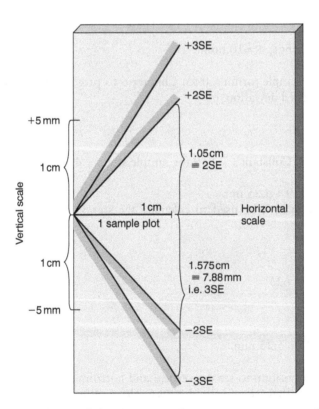

Figure 9.4 Scale key for cusum plot

This is outside the required range and the chose scales are unsuitable. Conversely, if we decide to set the ordinate scale at 1 cm = 5 mm, the scale ratio becomes 1 cm/1.05 cm = 0.95, and the scales chosen are acceptable. Having designed the cusum chart for variables, it is usual to provide a key showing the slope that corresponds to changes of two and three SE (Figure 9.4). A similar key may be used with simple cusum charts for attributes. This is shown in Figure 9.2.

We may now use the cusum chart to analyse data. Table 9.3 shows the sample means from 30 groups of four steel rods, which were used in plotting the mean chart of Figure 9.5a (from Chapter 5). The process average

Table 9.3 Cusum values of sample means (*n* = 4) for steel rod cutting process

Sample number	Sample mean, x̄ (mm)	(x̄ − t) mm (t = 150.1mm)	Sr
1	148.50	−1.60	−1.60
2	151.50	1.40	−0.20
3	152.50	2.40	2.20
4	146.00	−4.10	−1.90
5	147.75	−2.35	−4.25
6	151.75	1.65	−2.60
7	151.75	1.65	−0.95
8	149.50	−0.60	−1.55
9	154.75	4.65	3.10
10	153.00	2.90	6.00
11	155.00	4.90	10.90
12	159.00	8.90	19.80
13	150.00	−0.10	19.70
14	154.25	4.15	23.85
15	151.00	0.90	24.75
16	150.25	0.15	24.90
17	153.75	3.65	28.55
18	154.00	3.90	32.45
19	157.75	7.65	40.10
20	163.00	12.90	53.00
21	137.50	−12.60	40.40
22	147.50	−2.60	37.80
23	147.50	−2.60	35.20
24	152.50	2.40	37.60
25	155.50	5.40	43.00
26	159.00	8.90	51.90
27	144.50	−5.60	46.30
28	153.75	3.65	49.95
29	155.00	4.90	54.85
30	158.50	8.40	63.25

of 150.1 mm has been subtracted from each value and the cusum values calculated. The latter have been plotted on the previously designed chart to give Figure 9.5b.

If the reader compares this chart with the corresponding mean chart certain features will become apparent. First, an examination of sample plots 11 and 12 on both charts will demonstrate that the mean chart more

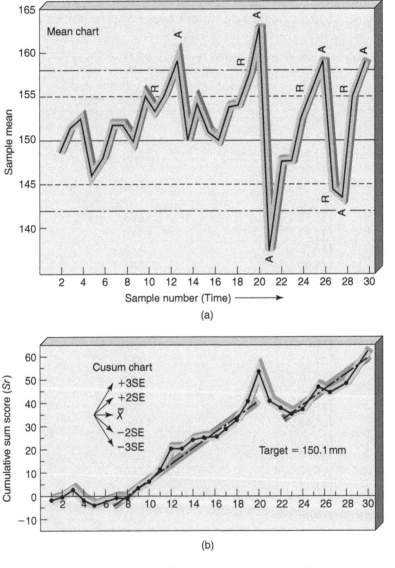

Figure 9.5 Shewhart and cusum charts for means of steel rods

readily identifies large changes in the process mean. This is by virtue of the sharp 'peak' on the chart and the presence of action and warning limits. The cusum chart depends on comparison of the gradients of the cusum plot and the key. Secondly, the zero slope or horizontal line on the cusum chart between samples 12 and 13 shows what happens when the process is perfectly in control. The actual cusum score of sample 13 is still high at 19.80, even though the sample mean (150.00 mm) is almost the same as the reference value (150.1 mm).

The care necessary when interpreting cusum charts is shown again by sample plot 21. On the mean chart there is a clear indication that the process has been over-corrected and that the length of rods are too short. On the cusum plot the negative slope between plots 20 and 21 indicates the same effects, but it must be understood by all who use the chart that the rod length should be increased, even though the cusum score remains high at over 40 mm. The power of the cusum chart is its ability to detect persistent changes in the process mean and this is shown by the two parallel trend lines drawn on Figure 9.5b. More objective methods of detecting significant changes, using the cusum chart, are introduced in Section 9.4.

9.3 Product screening and pre-selection

Cusum charts can be used in categorizing process output. This may be for the purposes of selection for different processes or assembly operations, or for despatch to different customers with slightly varying requirements. To perform the screening or selection, the cusum chart is divided into different sections of average process mean by virtue of changes in the slope of the cusum plot. Consider, for example, the cusum chart for rod lengths in Figure 9.5. The first eight samples may be considered to represent a stable period of production and the average process mean over that period is easily calculated:

$$\sum_{i=1}^{8} \bar{x}_i/8 = t + (S_8 - S_0)/8$$
$$= 150.1 + (-1.55 - 0)/8 = 149.91.$$

The first major change in the process occurs at sample 9 when the cusum chart begins to show a positive slope. This continues until sample 12. Hence, the average process mean may be calculated over that period:

$$\sum_{i=9}^{12} \bar{x}_i/4 = t + (S_{12} - S_8)/4$$
$$= 150.1 + (19.8 - (-1.55))/4 = 155.44.$$

In this way the average process mean may be calculated from the cusum score values for each period of significant change.

For samples 13–16, the average process mean is:

$$\sum_{i=13}^{16} \bar{x}_i/4 = t + (S_{16} - S_{12})/4$$

$$= 150.1 + (24.9 - 19.8)/4 = 151.38.$$

For samples 17–20:

$$\sum_{i=17}^{20} \bar{x}_i/4 = t + (S_{20} - S_{16})/4$$

$$= 150.1 + (53.0 - 24.9)/4 = 157.13.$$

For samples 21–23:

$$\sum_{i=21}^{23} \bar{x}_i/3 = t + (S_{23} - S_{20})/3$$

$$= 150.1 + (35.2 - 53.0)/3 = 144.17.$$

For samples 24–30:

$$\sum_{i=24}^{30} \bar{x}_i/7 = t + (S_{30} - S_{23})/7$$

$$= 150.1 + (63.25 - 35.2)/7 = 154.11.$$

This information may be represented on a Manhattan diagram, named after its appearance. Such a graph has been drawn for the above data in Figure 9.6. It shows clearly the variation in average process mean over the time-scale of the chart.

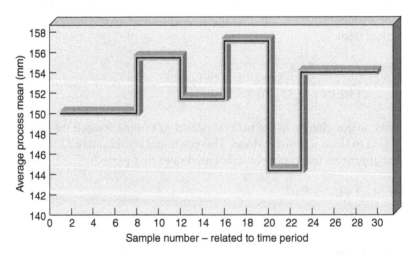

Figure 9.6 Manhattan diagram – average process mean with time

9.4 Cusum decision procedures

Cusum charts are used to detect when changes have occurred. The extreme sensitivity of cusum charts, which was shown in the previous sections, needs to be controlled if unnecessary adjustments to the process and/or stoppages are to be avoided. The largely subjective approaches examined so far are not very satisfactory. It is desirable to use objective decision rules, similar to the control limits on Shewhart charts, to indicate when significant changes have occurred. Several methods are available, but two in particular have practical application in industrial situations, and these are described here. They are:

(i) V-masks,
(ii) Decision intervals.

The methods are theoretically equivalent, but the mechanics are different. These need to be explained.

V-masks

In 1959 G.A. Barnard described a V-shaped mask that could be superimposed on the cusum plot. This is usually drawn on a transparent overlay or by a computer and is as shown in Figure 9.7. The mask is placed over the chart so that the line AO is parallel with the horizontal axis, the vertex O points forwards, and the point A lies on top of the last sample plot. A significant change in the process is indicated by part of the cusum plot being covered by either limb of the V-mask, as in Figure 9.7. This should be followed by a search for assignable causes. If all the points previously plotted fall within the V-shape, the process is assumed to be in a state of statistical control.

The design of the V-mask obviously depends upon the choice of the lead distance d (measured in number of sample plots) and the angle θ. This may be made empirically by drawing a number of masks and testing out each one on past data. Since the original work on V-masks, many quantitative methods of design have been developed.

The construction of the mask is usually based on the standard error of the plotted variable, its distribution and the average number of samples up to the point at which a signal occurs, i.e. the average run length (ARL) properties. The essential features of a V-mask, shown in Figure 9.8, are:

- a point A, which is placed over any point of interest on the chart (this is often the most recently plotted point);
- the vertical half distances, AB and AC – the decision intervals, often ±5SE;
- the sloping decision lines BD and CE – an out of control signal is indicated if the cusum graph crosses or touches either of these lines;

- the horizontal line AF, which may be useful for alignment on the chart – this line represents the zero slope of the cusum when the process is running at its target level;
- AF is often set at 10 sample points and DF and EF at ±10SE.

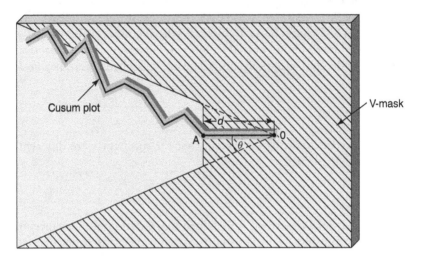

Figure 9.7 V-mask for cusum chart

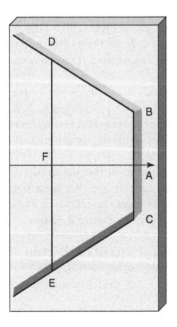

Figure 9.8 V-mask features

The geometry of the truncated V-mask shown in Figure 9.8 is the version recommended for general use and has been chosen to give properties broadly similar to the traditional Shewhart charts with control limits.

Decision intervals

Procedures exist for detecting changes in one direction only. The amount of change in that direction is compared with a predetermined amount – the decision interval h, and corrective action is taken when that value is exceeded. The modern decision interval procedures may be used as one- or two-sided methods. An example will illustrate the basic concepts.

Suppose that we are manufacturing pistons, with a target diameter (t) of 10.0 mm and we wish to detect when the process mean diameter decreases – the tolerance is 9.6 mm. The process standard deviation is 0.1 mm. We set a reference value, k, at a point half-way between the target and the so-called Reject Quality Level (RQL), the point beyond which an unacceptable proportion of reject material will be produced. With a normally distributed variable, the RQL may be estimated from the specification tolerance (T) and the process standard deviation (σ). If, for example, it is agreed that no more than one piston in 1000 should be manufactured outside the tolerance, then the RQL will be approximately 3σ inside the specification limit. So for the piston example with the lower tolerance T_L:

$$RQL_L = T_L + 3\sigma$$
$$= 9.6 + 0.3 = 9.9\,\text{mm}.$$

and the reference value is:

$$k_L = (t + RQL_L)/2$$
$$= (10.0 + 9.9)/2 = 9.95\,\text{mm}.$$

For a process having an upper tolerance limit:

$$RQL_U = T_U - 3\sigma$$

and

$$k_U = (RQL_U + t)/2.$$

Alternatively, the RQL may be set nearer to the tolerance value to allow a higher proportion of defective materials. For example, the RQL_L set at $T_L + 2\sigma$ will allow *ca.* 2.5 per cent of the products to fall below the lower specification limit. For the purposes of this example, we shall set the RQL_L at 9.9 mm and k_L at 9.95 mm.

Cusum values are calculated as before, but subtracting k_L instead of t from the individual results:

$$Sr = \sum_{i=1}^{r} (x_i - k_L).$$

This time the plot of Sr against r will be expected to show a rising trend if the target value is obtained, since the subtracting k_L will always lead to a positive result. For this reason, the cusum chart is plotted in a different way. As soon as the cusum rises above zero, a new series is started, only negative values and the first positive cusums being used. The chart may have the appearance of Figure 9.9. When the cusum drops below the decision interval, $-h$, a shift of the process mean to a value below k_L is indicated. This procedure calls attention to those downward shifts in the process average that are considered to be of importance.

The one-sided procedure may, of course, be used to detect shifts in the positive direction by the appropriate selection of k. In this case k will be higher than the target value and the decision to investigate the process will be made when Sr has a positive value that rises above the interval h.

It is possible to run two one-sided schemes concurrently to detect both increases and decreases in results. This requires the use of two reference values k_L and k_U, which are respectively half-way between the target value and the lower and upper tolerance levels, and two decision intervals $-h$ and h. This gives rise to the so-called two-sided decision procedure.

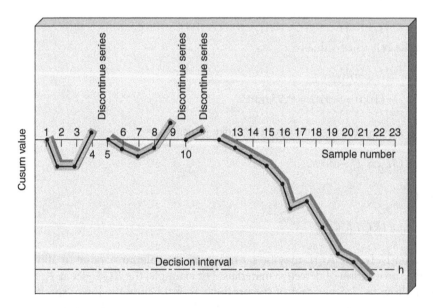

Figure 9.9 Decision interval one-sided procedure

Two-sided decision intervals and V-masks

When two one-sided schemes are run with upper and lower reference values, k_U and k_L, the overall procedure is equivalent to using a V-shaped mask. If the distance between two plots on the horizontal scale is equal to the distance on the vertical scale representing a change of v, then the two-sided decision interval scheme is the same as the V-mask scheme if:

$$k_U - t = t - k_L = v - \tan\theta$$

and

$$h = -h = dv\tan\theta = d|t - k|.$$

A demonstration of this equivalence is given by K.W. Kemp in *Applied Statistics* (1962, p. 20).

Most software packages for statistical process control (SPC), such as Minitab, will perform all these decision intervals and V-masks with cusum charts.

Chapter highlights

- Shewhart charts allow a decision to be made after each plot. Whilst rules for trends and runs exist for use with such charts, cumulating process data can give longer-term information. The cusum technique is a method of analysis in which data is cumulated to give information about longer-term trends.
- Cusum charts are obtained by determining the difference between the values of individual observations and a 'target' value, and cumulating these differences to give a cusum score which is then plotted.
- When a line drawn through a cusum plot is horizontal, it indicates that the observations were scattered around the target value; when the slope of the cusum is positive the observed values are above the target value; when the slope of the cusum plot is negative the observed values lie below the target value; when the slope of the cusum plot changes the observed values are changing.
- The cusum technique can be used for attributes and variables by pre-determining the scale for plotting the cusum scores, choosing the target value and setting up a key of slopes corresponding to predetermined changes.
- The behaviour of a process can be comprehensively described by using the Shewhart and cusum charts in combination. The Shewhart charts are best used at the point of control, whilst the cusum chart is preferred for a later review of data.

- Shewhart charts are more sensitive to rapid changes within a process, whilst the cusum is more sensitive to the detection of small sustained changes.
- Various decision procedures for the interpretation of cusum plots are possible including the use of V-masks.
- The construction of the V-mask is usually based on the standard error of the plotted variable, its distribution and the ARL properties. The most widely used V-mask has decision lines: ±5SE at sample zero ± 10SE at sample 10.

References and further readings

Barnard, G.A. (1959) 'Decision Interval V-masks for Use in Cumulative Sum Charts', *Applied Statistics*, Vol. 1, p. 132.

Kemp, K.W. (1962) 'The Use of Cumulative Sums for Sampling Inspection Schemes', *Applied Statistics*, Vol. 11, p. 20.

Discussion questions

1 (a) Explain the principles of Shewhart control charts for sample mean and sample range, and cumulative sum control charts for sample mean and sample range. Compare the performance of these charts.

   (b) A chocolate manufacturer takes a sample of six boxes at the end of each hour in order to verify the weight of the chocolates contained within each box. The individual chocolates are also examined visually during the check-weighing and the various types of major and minor faults are counted.

   The manufacturer equates 1 major fault to 4 minor faults and accepts a maximum equivalent to 2 minor physical faults/chocolate, in any box. Each box contains 24 chocolates.
   Discuss how the cusum chart techniques can be used to monitor the physical defects. Illustrate how the chart would be set up and used.

2 In Chapter 8, Discussion question 7, the table gives the results from the inspection of cabinets for scratches and small indentations.
   Plot the data on a suitably designed cusum chart and comment on the results.

3 In Chapter 8, Discussion question 5, the table records the number of defective items found in a sample of 100 taken twice per day.
   Set up and plot a cusum chart. Interpret your findings. (Assume a target value of 2 defectives.)

4 In Chapter 6, Discussion question 10, the table gives the average width (mm) for each of 20 samples of five panels. Also given is the range (mm) of each sample.

Design cumulative sum (cusum) charts to control the process. Explain the differences between these charts and Shewhart charts for means and ranges.

5 Shewhart charts are to be used to maintain control on dissolved iron content of a dyestuff formulation in parts per million (ppm). After 25 subgroups of 5 measurements have been obtained,

$$\sum_{i=1}^{i=25} \bar{x}_i = 390 \text{ and } \sum_{i=1}^{i=25} \bar{R}_i = 84,$$

where $\bar{x}_i$ = mean of ith subgroup;

$\bar{R}_i$ = range of ith subgroup;

Design appropriate cusum charts for control of the process mean and sample range and describe how the charts might be used in continuous production for product screening.
(see also Chapter 6, Worked example 2)

6 The following data were obtained when measurements were made on the diameter of steel balls for use in bearings. The mean and range values of sixteen samples of size 5 are given in the table:

Sample number	Mean dia. (0.001 mm)	Sample range (mm)	Sample number	Mean dia. (0.001 mm)	Sample range (mm)
1	250.2	0.005	9	250.4	0.004
2	251.3	0.005	10	250.0	0.004
3	250.4	0.005	11	249.4	0.0045
4	250.2	0.003	12	249.8	0.0035
5	250.7	0.004	13	249.3	0.0045
6	248.9	0.004	14	249.1	0.0035
7	250.2	0.005	15	251.0	0.004
8	249.1	0.004	16	250.6	0.0045

Design a mean cusum chart for the process and plot the results on the chart.

Interpret the cusum chart and explain briefly how it may be used to categorize production in pre-selection for an operation in the assembly of the bearings.

7 In Chapter 7, Discussion question 6, are given the results of daily samples of discharge analysed to determine the ammonia content.

(a) Examine the data using a cusum plot of the ammonia data. What conclusions do you draw concerning the ammonia content of the effluent during the 40-day period?

(b) What other techniques could you use to detect and demonstrate changes in ammonia concentration. Comment on the relative merits of these techniques compared to the cusum plot.

(c) Comment on the assertion that 'the cusum chart could detect changes inaccuracy but could not detect changes in precision'.

8   Small plastic bottles are made from preforms supplied by Britanic Polymers. It is possible that the variability in the bottles is due in part to the variation in the preforms. Thirty preforms are sampled from the extruder at Britanic Polymers, one preform every 5 minutes for two and a half hours. The weights of the preforms are (g):

32.9	33.7	33.4	33.4	33.6	32.8	33.3	33.1	32.9	33.0
33.2	32.8	32.9	33.3	33.1	33.0	33.7	33.4	33.5	33.6
33.2	33.8	33.5	33.9	33.7	33.4	33.5	33.6	33.2	33.6

(The data should be read from left to right along the top row, then the middle row, etc.)

Carry out a cusum analysis of the preform weights and comment on the stability of the process.

9   The data given below are taken from a process of acceptable mean value $\mu_0 = 8.0$ and unacceptable mean value $\mu_1 = 7.5$ and known standard deviation of 0.45.

Sample number	$\overline{X}$	Sample number	$\overline{X}$
1	8.04	11	8.11
2	7.84	12	7.80
3	8.46	13	7.86
4	7.73	14	7.23
5	8.44	15	7.33
6	7.50	16	7.30
7	8.28	17	7.67
8	7.62	18	6.90
9	8.33	19	7.38
10	7.60	20	7.44

Plot the data on a cumulative sum chart, using any suitable type of chart with the appropriate correction values and decision procedures.

What are the ARLs at μ_0 and μ_1 for your chosen decision procedure?

10 A cusum scheme is to be installed to monitor gas consumption in a chemical plant where a heat treatment is an integral part of the process. The engineers know from intensive studies that when the system is operating as it was designed the average amount of gas required in a period of 8 hours would be 250 therms, with a standard deviation of 25 therms.

The following table shows the gas consumption and shift length for 20 shifts.

Shift number	Hours operation (H)	Gas consumption (G)
1	8	256
2	4	119
3	8	278
4	4	122
5	6	215
6	6	270
7	8	262
8	8	216
9	3	103
10	8	206
11	3	83
12	8	214
13	3	95
14	8	234
15	8	266
16	4	150
17	8	284
18	3	118
19	8	298
20	4	138

Standardize the gas consumption to an 8-hour shift length, i.e. standardized gas consumption X is given by:

$$X = \left(\frac{G}{H}\right) \times 8.$$

Using a reference value of 250 hours construct a cumulative sum chart based on X. Apply a selected V-mask after each point is plotted.

When you identify a significant change, state when the change occurred, and start the cusum chart again with the same reference value of 250 therms assuming that appropriate corrective action has been taken.

Worked examples

1 Three packaging processes

Figure 9.10 shows a certain output response from three parallel packaging processes operating at the same time. From this chart all three processes seem to be subjected to periodic swings and the responses appear to become closer together with time. The cusum charts shown in Figure 9.11 confirm the periodic swings and show that they have the same time period, so some external factor is probably affecting all three processes. The cusum charts also show that process 3 was the nearest to target – this can also be seen on the individuals chart but less obviously. In addition, process 4 was initially above target and process 5 even more so. Again, once this is pointed out, it can also be seen in Figure 9.10. After an initial separation of the cusum plots they remain parallel and some distance apart. By referring to the individuals plot we see that this distance was close to zero. Reading the two charts together gives a very complete picture of the behaviour of the processes.

2 Profits on sales

A company in the financial sector had been keeping track of the sales and the percentage of the turnover as profit. The sales for the last 25 months had remained relatively constant due to the large percentage of agency business. During the previous few months profits as a percentage of turnover had been below average and the information Table 9.4 had been collected.

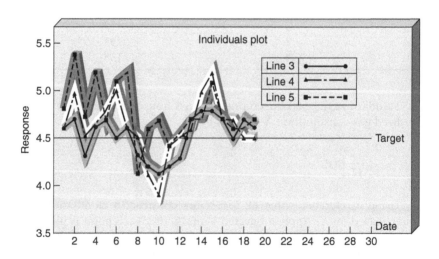

Figure 9.10 Packaging processes output response

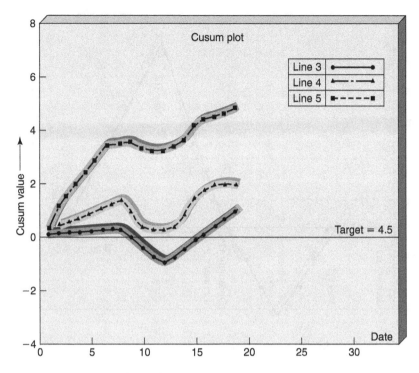

Figure 9.11 Cusum plot of data in Figure 9.10

Table 9.4 Profit, as per cent of turnover, for each 25 months

Year 1		Year 2	
Month	*Profit (%)*	*Month*	*Profit (%)*
January	7.8	January	9.2
February	8.4	February	9.6
March	7.9	March	9.0
April	7.6	April	9.9
May	8.2	May	9.4
June	7.0	June	8.0
July	6.9	July	6.9
August	7.2	August	7.0
September	8.0	September	7.3
October	8.8	October	6.7
November	8.8	November	6.9
December	8.7	December	7.2
		January Year 3	7.6

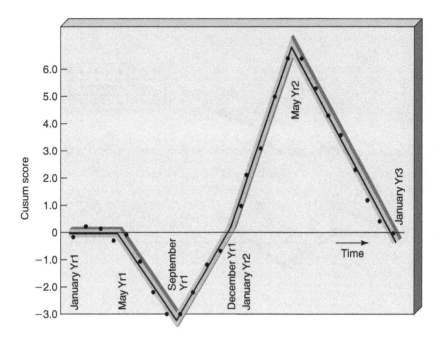

Figure 9.12 Cusum chart of data on profits

After receiving SPC training, the company accountant decided to analyse the data using a cusum chart. He calculated the average profit over the period to be 8.0 per cent and subtracted this value from each month's profit figure. He then cumulated the differences and plotted them as in Figure 9.12.

The dramatic changes that took place in approximately May and September in Year 1, and in May in Year 2 were investigated and found to be associated with the following assignable causes:

May   Year 1 Introduction of 'efficiency' bonus payment scheme.
September   Year 1 Introduction of quality improvement teams.
May   Year 2 Revision of efficiency bonus payment scheme.

The motivational (or otherwise) impact of managerial decisions and actions often manifests itself in business performance results in this way. The cusum technique is useful in highlighting the change points so that possible causes may be investigated.

3 Forecasting income

The three divisions of an electronics company were required to forecast sales income on an annual basis and update the forecasts each month. These forecasts were critical to staffing and prioritizing resources in the organization.

Table 9.5 Three month income forecast (unit × 1000) and actual (unit × 1000)

Month	Division A		Division B		Division C	
	Forecast	*Actual*	*Forecast*	*Actual*	*Forecast*	*Actual*
1	200	210	250	240	350	330
2	220	205	300	300	420	430
3	230	215	130	120	310	300
4	190	200	210	200	340	345
5	200	200	220	215	320	345
6	210	200	210	190	240	245
7	210	205	230	215	200	210
8	190	200	240	215	300	320
9	210	220	160	150	310	330
10	200	195	340	355	320	340
11	180	185	250	245	320	350
12	180	200	340	320	400	385
13	180	240	220	215	400	405
14	220	225	230	235	410	405
15	220	215	320	310	430	440
16	220	220	320	315	330	320
17	210	200	230	215	310	315
18	190	195	160	145	240	240
19	190	185	240	230	210	205
20	200	205	130	120	330	320

Forecasts were normally made 1 year in advance. The 1 month forecast was thought to be reasonably reliable. If the 3 months forecast had been reliable, the material scheduling could have been done more efficiently. Table 9.5 shows the 3 months forecasts made by the three divisions for 20 consecutive months. The actual income for each month is also shown. Examine the data using the appropriate techniques.

Solution

The cusum chart was used to examine the data, the actual sales being subtracted from the forecast and the differences cumulated. The resulting cusum graphs are shown in Figure 9.13. Clearly there is a vast difference in forecasting performance of the three divisions. Overall, division B is over-forecasting resulting in a constantly rising cusum. A and C were generally under-forecasting during months 7–12 but, during the latter months of the period, their forecasting improved resulting in a stable, almost horizontal line cusum plot. Periods of improved performance such as this may be useful in identifying the causes of the earlier

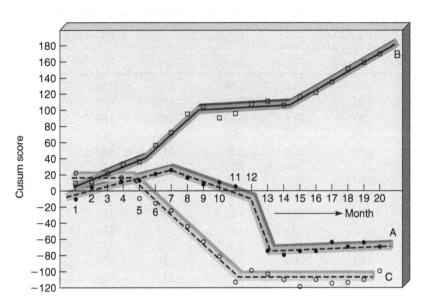

Figure 9.13 Cusum charts of forecast versus actual sales for three divisions

over-forecasting and the generally poor performance of division B's forecasting system. The points of change in slope may also be useful indicators of assignable causes, if the management system can provide the necessary information.

Other techniques useful in forecasting include the moving mean and moving range charts and exponential smoothing (see Chapter 7).

4 Herbicide ingredient (see also Chapter 8, Worked example 2)

The active ingredient in a herbicide is added in two stages. At the first stage 160 litres of the active ingredient is added to 800 litres of the inert ingredient. To get a mix ratio of exactly 5 to 1 small quantities of either ingredient are then added. This can be very time-consuming as sometimes a large number of additions are made in an attempt to get the ratio just right. The recently appointed Production Manager has introduced a new procedure for the first mixing stage. To test the effectiveness of this change he recorded the number of additions required for 30 consecutive batches, 15 with the old procedure and 15 with the new. Figure 9.14 is a cusum chart based on these data.

What conclusions would you draw from the cusum chart in Figure 9.14?

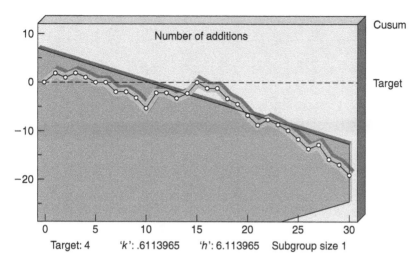

Figure 9.14 Herbicide additions

Solution

The cusum in Figure 9.14 uses a target of 4 and shows a change of slope at batch 15. The V-mask indicates that the means from batch 15 are significantly different from the target of 4. Thus the early batches (1–15) have a horizontal plot. The V-mask shows that the later batches are significantly lower on average and the new procedure appears to give a lower number of additions.

Part IV
Process capability

10 Process capability for variables and its measurement

Objectives

- To introduce the idea of measuring process capability.
- To describe process capability indices and show how they are calculated.
- To give guidance on the interpretation of capability indices.
- To illustrate the use of process capability analysis in a service environment.

10.1 Will it meet the requirements?

In managing variables the usual aim is not to achieve exactly the same length for every steel rod, the same diameter for every piston, the same weight for every tablet, sales figures exactly as forecast, but to reduce the variation of products and process parameters around a target value. No adjustment of a process is called for as long as there has been no identified change in its accuracy or precision. This means that, in controlling a process, it is necessary to establish first that it is in statistical control, and then to compare its centring and spread with the specified target value and specification tolerance.

We have seen in previous chapters that, if a process is not in statistical control, special causes of variation may be identified with the aid of control charts. Only when all the special causes have been accounted for, or eliminated, can process capability be sensibly assessed. The variation due to common causes may then be examined and the 'natural specification' compared with any imposed specification or tolerance zone.

The relationship between process variability and tolerances may be formalized by consideration of the standard deviation, σ, of the process. In order to manufacture within the specification, the distance between the upper specification limit (USL) or upper tolerance ($+T$) and lower specification limit (LSL) or lower tolerance ($-T$), i.e. (USL–LSL) or $2T$ must be equal to or greater than the width of the base of the process bell, i.e. 6σ. This is shown in Figure 10.1. The relationship between (USL–LSL) or $2T$ *and* 6σ gives rise to three levels of precision of the process (Figure 10.2).

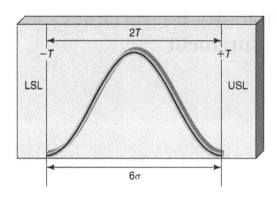

Figure 10.1 Process capability

- High Relative Precision, where the tolerance band is very much greater than 6σ ($2T \gg 6\sigma$) (Figure 10.2a);
- Medium Relative Precision, where the tolerance band is just greater than 6σ ($2T > 6\sigma$) (Figure 10.2b);
- Low Relative Precision, where the tolerance band is less than 6σ ($2T < 6\sigma$) (Figure 10.2c).

For example, if the specification for the lengths of the steel rods discussed in Chapters 5 and 6 had been set at 150 ± 10 mm and on three different machines the processes were found to be in statistical control, centred correctly but with different standard deviations of 2, 3 and 4mm, we could represent the results in Figure 10.2. Figure 10.2a shows that when the standard deviation (σ) is 2 mm, the bell value of 6σ is 12 mm, and the total process variation is far less than the tolerance band of 20 mm. Indeed there is room for the process to 'wander' a little and, provided that any change in the centring or spread of the process is detected early, the tolerance limits will not be crossed. With a standard deviation of 3 mm (Figure 10.2b) the room for movement before the tolerance limits are threatened is reduced, and with a standard deviation of 4 mm (Figure 10.2c) the production of material outside the specification is inevitable.

10.2 Process capability indices

A process capability index is a measure relating the actual performance of a process to its specified performance, where processes are considered to be a combination of the plant or equipment, the method, the people, the materials and the environment. The absolute minimum requirement is that three process standard deviations each side of the process mean are contained within the specification limits. This means that *ca.* 99.7 per cent of output will be within

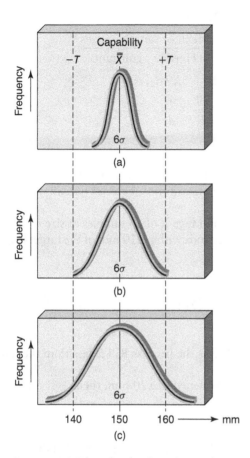

Figure 10.2 Three levels of precision of a process

the tolerances. A more stringent requirement is often stipulated to ensure that produce of the correct quality is consistently obtained over the long term.

When a process is under statistical control (i.e. only random or common causes of variation are present), a process capability index may be calculated. Process capability indices are simply a means of indicating the variability of a process relative to the product specification tolerance.

The situations represented in Figure 10.2 may be quantified by the calculation of several indices, as discussed in the following sections.

Relative Precision Index

This is the oldest index being based on a ratio of the mean range of samples with the tolerance band. In order to avoid the production of defective material, the specification width must be greater than the process variation, hence:

$$2T > 6\sigma$$

we know that $\sigma = \dfrac{\bar{R}}{d_n} = \dfrac{\text{Mean of sample ranges}}{\text{Hartley's constant}}$,

so: $2T > 6\bar{R}/d_n$,

therefore: $\dfrac{2T}{\bar{R}} > \dfrac{6}{d_n}$.

$2T/\bar{R}$ is known as the Relative Precision Index (RPI) and the value of $6/d_n$ is the minimum RPI to avoid the generation of material outside the specification limit.

In our steel rod example, the mean range $\bar{R}$ of 25 samples of size $n = 4$ was 10.8 mm. If we are asked to produce rods within ±10 mm of the target length:

RPI $= 2T/\bar{R} = 20/10.8 = 1.852$.

Minimum RPI $= \dfrac{6}{d_n} = \dfrac{6}{2.059} = 2.914$.

Clearly, reject material is inevitable as the process RPI is less than the minimum required.

If the specified tolerances were widened to ±20 mm, then:

RPI $= 2T/\bar{R} = 40/10.8 = 3.704$

and reject material can be avoided, if the centring and spread of the process are adequately controlled (Figure 10.3, the change from a to b). RPI provided a quick and simple way of quantifying process capability. It does not, of course, comment on the centring of a process as it deals only with relative spread or variation.

Cp index

In order to manufacture within a specification, the difference between the USL and the LSL must be less than the total process variation. So a comparison of 6σ with (USL–LSL) or $2T$ gives an obvious process capability index, known as the Cp of the process:

$$Cp = \dfrac{\text{USL - LSL}}{6\sigma} \quad \text{or} \quad \dfrac{2T}{6\sigma}.$$

Clearly, any value of Cp below 1 means that the process variation is greater than the specified tolerance band so the process is incapable. For

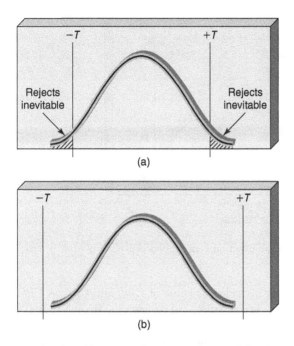

(a)

(b)

Figure 10.3 Changing relative process capability by widening the specification

increasing values of *Cp* the process becomes increasingly capable. The *Cp* index, like the RPI, makes no comment about the centring of the process, it is a simple comparison of total variation with tolerances.

Cpk index

It is possible to envisage a relatively wide tolerance band with a relatively small process variation, but in which a significant proportion of the process output lies outside the tolerance band (Figure 10.4). This does not invalidate the use of *Cp* as an index to measure the 'potential capability' of a process when centred, but suggests the need for another index that takes account of both the process variation and the centring. Such an index is the *Cpk*, which is widely accepted as a means of communicating process capability.

For upper and lower specification limits, there are two *Cpk* values, Cpk_u and Cpk_l. These relate the difference between the process mean and the upper and the lower specification limits, respectively, to 3σ (half the total process variation) (Figure 10.5):

$$Cpk_u = \frac{\text{USL} - \bar{X}}{3\sigma}, \quad Cpk_l = \frac{\bar{X} - \text{LSL}}{3\sigma}.$$

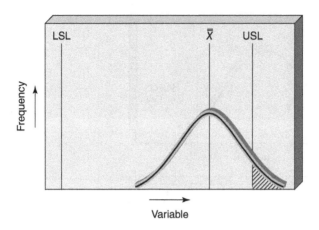

Figure 10.4 Process capability: non-centred process

The overall process *Cpk* is the lower value of Cpk_u and Cpk_l. A *Cpk* of 1 or less means that the process variation and its centring is such that at least one of the tolerance limits will be exceeded and the process is incapable. As in the case of *Cp*, increasing values of *Cpk* correspond to increasing capability. It may be possible to increase the *Cpk* value by centring the process so that its mean value and the mid-specification or target, coincide. A comparison of the *Cp* and the *Cpk* will show zero difference *if* the process is centred on the target value.

The *Cpk* can be used when there is only one specification limit, upper or lower – a one-sided specification. This occurs quite frequently and the *Cp* index cannot be used in this situation.

Examples should clarify the calculation of *Cp* and *Cpk* indices:

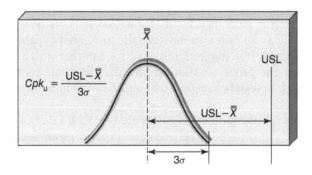

Figure 10.5 Process capability index Cpk_u

(i) In tablet manufacture, the process parameters from 20 samples of size $n = 4$ are:

Mean Range $(\bar{R}) = 91\,\text{mg}$, Process mean $(\bar{\bar{X}}) = 2500\,\text{mg}$,

Specified requirements USL $= 2650\,\text{mg}$, LSL $= 2350\,\text{mg}$,

$\sigma = \bar{R}/d_n = 91/2.059 = 44.2\,\text{mg}$,

$$Cp = \frac{\text{USL} - \text{LSL}}{6\sigma} \quad \text{or} \quad \frac{2T}{6\sigma} = \frac{2650 - 2350}{6 \times 44.2}$$

$$= \frac{300}{265.2} = 1.13,$$

$$Cpk = \text{lesser of } \frac{\text{USL} - \bar{\bar{X}}}{3\sigma} \quad \text{or} \quad \frac{\bar{\bar{X}} - \text{LSL}}{3\sigma}$$

$$= \frac{2650 - 2500}{3 \times 44.2} \quad \text{or} \quad \frac{2500 - 2350}{3 \times 44.2} = 1.13.$$

Conclusion: The process is centred $(Cp = Cpk)$ and of low capability since the indices are only just greater than 1.

(ii) If the process parameters from 20 samples of size $n = 4$ are:

Mean range $(\bar{R}) = 91\,\text{mg}$, Process mean $(\bar{\bar{X}}) = 2650\,\text{mg}$,

Specified requirements USL $= 2750\,\text{mg}$, LSL $= 2250\,\text{mg}$,

$\sigma = \bar{R}/d_n = 91/2.059 = 44.2\,\text{mg}$,

$$Cp = \frac{\text{USL} - \text{LSL}}{6\sigma} \quad \text{or} \quad \frac{2T}{6\sigma} = \frac{2750 - 2250}{6 \times 44.2} = \frac{500}{265.2} = 1.89,$$

$$Cpk = \text{lesser of } \frac{2750 - 2650}{3 \times 44.2} \quad \text{or} \quad \frac{2650 - 2250}{3 \times 44.2}$$

$$= \text{lesser of } 0.75 \text{ or } 3.02 = 0.75.$$

Conclusion: The Cp at 1.89 indicates a potential for higher capability than in example (i), but the low Cpk shows that this potential is not being realized because the process is not centred.

It is important to emphasize that in the calculation of all process capability indices, no matter how precise they may appear, the results are only ever approximations – we never actually *know* anything, progress lies in obtaining successively closer approximations to the truth. In the case of the process capability this is the case because:

- there is always some variation due to sampling;
- no process is ever fully in statistical control;
- no output exactly follows the normal distribution or indeed any other standard distribution.

Interpreting process capability indices without knowledge of the source of the data on which they are based can give rise to serious misinterpretation.

10.3 Interpreting capability indices

In the calculation of process capability indices so far, we have derived the standard deviation, σ, from the mean range $(\bar{R})$ and recognized that this estimates the short-term variations within the process. This short term is the period over which the process remains relatively stable, but we know that processes do not remain stable for all time and so we need to allow within the specified tolerance limits for:

- some movement of the mean,
- the detection of changes of the mean,
- possible changes in the scatter (range),
- the detection of changes in the scatter,
- the possible complications of non-normal distributions.

Taking these into account, the following values of the *Cpk* index represent the given level of confidence in the process capability:

- *Cpk* < 1 A situation in which the producer is not capable and there will inevitably be non-conforming output from the process (Figure 10.2c).
- *Cpk* = 1 A situation in which the producer is not really capable, since any change within the process will result in some undetected non-conforming output (Figure 10.2b).
- *Cpk* = 1.33 A still far from acceptable situation since non- conformance is not likely to be detected by the process control charts.
- *Cpk* = 1.5 Not yet satisfactory since non-conforming output will occur and the chances of detecting it are still not good enough.
- *Cpk* = 1.67 Promising, non-conforming output will occur but there is a very good chance that it will be detected.
- *Cpk* = 2 High level of confidence in the producer, provided that control charts are in regular use (Figure 10.2a).

10.4 The use of control chart and process capability data

The *Cpk* values so far calculated have been based on estimates of σ from $\bar{R}$, obtained over relatively short periods of data collection and should more properly be known as the $Cpk_{(potential)}$. Knowledge of the $Cpk_{(potential)}$ is available only to those who have direct access to the process and can assess the short-term variations that are typically measured during process capability studies.

An estimate of the standard deviation may be obtained from any set of data using calculations that can be performed by appropriate software, of course. For example, a customer can measure the variation within a delivered batch of material, or between batches of material supplied over time, and use the data to calculate the corresponding standard deviation. This will provide some knowledge of the process from which the examined product was obtained. The customer may also estimate the process mean values and, coupled with the specification, calculate a *Cpk* using the usual formula. This practice is recommended, provided that the results are interpreted correctly.

An example may help to illustrate the various types of *Cpks* that may be calculated. A pharmaceutical company carried out a process capability study on the weight of tablets produced and showed that the process was in statistical control with a process mean ($\bar{X}$) of 2504 mg and a mean range ($\bar{R}$) from samples of size $n = 4$ of 91 mg. The specification was USL = 2800 mg and LSL = 2200 mg.

Hence, $\sigma = \bar{R} / d_n = 91 / 2.059 = 44.2 \text{mg}$

and

$$Cpk_{(potential)} = (USL - \bar{X}) / 3\sigma = 296 / (3 \times 44.2) = 2.23.$$

The mean and range charts used to control the process on a particular day are shown in Figure 10.6. In a total of 23 samples, there were four warning signals and six action signals, from which it is clear that during this day the process was no longer in statistical control. The data from which this chart was plotted are given in Table 10.1. It is possible to use the tablet weights in Table 10.1 to compute the grand mean as 2513 mg and the standard deviation as 68 mg. Then:

$$Cpk = \frac{USL - \bar{\bar{X}}}{3\sigma} = \frac{2800 - 2513}{3 \times 68} = 1.41.$$

The standard deviation calculated by this method reflects various components, including the common-cause variations, all the assignable causes

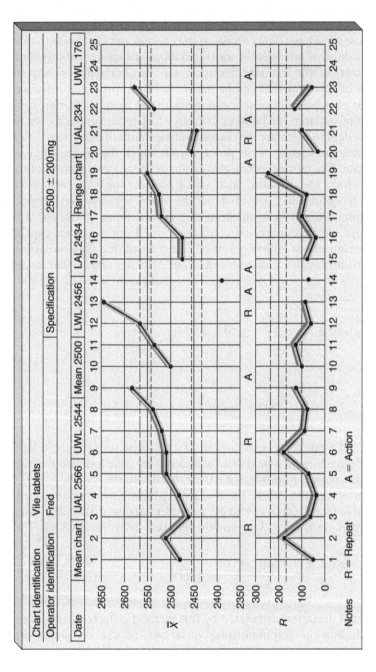

Figure 10.6 Mean and range control charts: tablet weights

apparent from the mean and range chart, and the limitations introduced by using a sample size of four. It clearly reflects more than the inherent random variations and so the *Cpk* resulting from its use is not the $Cpk_{(potential)}$, but the $Cpk_{(production)}$ – a capability index of the day's output and a useful way of monitoring, over a period, the actual performance of any process. The symbol *Ppk* is sometimes used to represent $Cpk_{(production)}$ which includes the common and special causes of variation and cannot be greater than the $Cpk_{(potential)}$. If it appears to be greater, it can only be that the process has improved. A record of the $Cpk_{(production)}$ reveals how the production performance varies and takes account of both the process centring and the spread.

The mean and range control charts could be used to classify the product and only products from 'good' periods could be dispatched. If 'bad' product is defined as that produced in periods prior to an action signal as well as any periods prior to warning signals which were followed by action signals, from the charts in Figure 10.6 this requires eliminating the product from the periods preceding samples 8, 9, 12, 13, 14, 19, 20, 21 and 23.

Excluding from Table 10.1 the weights corresponding to those periods, 56 tablet weights remain from which may be calculated the process mean at 2503 mg and the standard deviation at 49.4 mg. Then:

$$Cpk = (USL - \overline{X}) / 3\sigma = (2800 - 2503) / (3 \times 49.4) = 2.0.$$

This is the $Cpk_{(delivery)}$. If this selected output from the process were dispatched, the customer should find on sampling a similar process mean, standard deviation and $Cpk_{(delivery)}$ and should be reasonably content. It is not surprising that the *Cpk* should be increased by the elimination of the product known to have been produced during 'out-of-control' periods. The term $Csk_{(supplied)}$ is sometimes used to represent the $Cpk_{(delivery)}$.

Only the producer can know the $Cpk_{(potential)}$ and the method of product classification used. Not only the product, but the justification of its classification should be available to the customer. One way in which the latter may be achieved is by letting the customer have copies of the control charts and the justification of the $Cpk_{(potential)}$. Both of these requirements are becoming standard in those industries which understand and have assimilated the concepts of process capability and the use of control charts for variables.

There are two important points that should be emphasized:

• The use of control charts not only allows the process to be controlled, it also provides all the information required to complete product classification.
• The producer, through the data coming from the process capability study and the control charts, can judge the performance of a process – the process performance cannot be judged equally well from the product alone.

Table 10.1 Samples of tablet weights (*n* = 4) with means and ranges

Sample number	Weight in mg				Mean	Range
1	2501	2461	2512	2468	2485	51
2	2416	2602	2482	2526	2507	186
3	2487	2494	2428	2443	2463	66
4	2471	2462	2504	2499	2484	42
5	2510	2543	2464	2531	2512	79
6	2558	2412	2595	2482	2512	183
7	2518	2540	2555	2461	2519	94
8	2481	2540	2569	2571	2540	90
9	2504	2599	2634	2590	2582	130
10	2541	2463	2525	2559	2500	108
11	2556	2457	2554	2588	2539	131
12	2544	2598	2531	2586	2565	67
13	2591	2644	2666	2678	2645	87
14	2353	2373	2425	2410	2390	72
15	2460	2509	2433	2511	2478	78
16	2447	2490	2477	2498	2478	51
17	2523	2579	2488	2481	2518	98
18	2558	2472	2510	2540	2520	86
19	2579	2644	2394	2572	2547	250
20	2446	2438	2453	2475	2453	37
21	2402	2411	2470	2499	2446	97
22	2551	2454	2549	2584	2535	130
23	2590	2600	2574	2540	2576	60

If a customer knows that a supplier has a $Cpk_{(potential)}$ value of at least 2 and that the supplier uses control charts for both control and classification, then the customer can have confidence in the supplier's process and method of product classification. This is very different from an 'inspect and reject' approach to quality.

10.5 A service industry example: process capability analysis in a bank

A project team in a small bank was studying the productivity of the operations. Work during the implementation of statistical process control had identified variation in transaction (deposit/withdrawal) times as a potential area for improvement. The operators of the process agreed to collect data on transaction times in order to study the process.

Once an hour, each operator recorded in time the seconds required to complete the next seven transactions. After three days, the operators developed control charts for this data. All the operators calculated control limits for their own data. The totals of the $\bar{X}$ s and Rs for 24 sub-groups (3 days times 8 hours per day) for one operator were: $\Sigma\bar{X} = 5640$ seconds, $\Sigma R = 1900$ seconds. Control limits for this operator's $\bar{X}$ and R chart were calculated and the process was shown to be stable.

An 'efficiency standard' had been laid down that transactions should average 3 minutes (180 seconds), with a maximum of 5 minutes (300 seconds) for any one transaction. The process capability was calculated as follows:

$$\bar{\bar{X}} = \frac{\Sigma\bar{X}}{k} = \frac{5640}{24} = 235 \text{ seconds,}$$

$$\bar{R} = \frac{\Sigma R}{k} = \frac{1900}{24} = 79.2 \text{ seconds,}$$

$$\sigma = \bar{R}/d_n, \text{ for } n = 7, \ \sigma = 79.2/2.704 = 29.3 \text{ seconds,}$$

$$Cpk = \frac{USL - \bar{\bar{X}}}{3\sigma} = \frac{300 - 235}{3 \times 29.3} = 0.74.$$

i.e. not capable, and not centred on the target of 180 seconds.

As the process was not capable of meeting the requirements, management led an effort to improve transaction efficiency. This began with a flowcharting of the process (see Chapter 2). In addition, a brainstorming session involving the operators was used to generate the cause and effect diagram (see Chapter 11). A quality improvement team was formed, further data collected, and the 'vital' areas of incompletely understood procedures and operator training were tackled. This resulted over a period of 6 months, in a reduction in average transaction time to 190 seconds, with standard deviation of 15 seconds ($Cpk = 2.44$). (see also Chapter 11, Worked example 2.)

Chapter highlights

* Process capability is assessed by comparing the width of the specification tolerance band with the overall spread of the process. Processes may be classified as low, medium or high relative precision.
* Capability can be assessed by a comparison of the standard deviation (σ) and the width of the tolerance band. This gives a process capability index.
* The RPI is the relative precision index, the ratio of the tolerance band ($2T$) to the mean sample range ($\bar{R}$).
* The Cp index is the ratio of the tolerance band to six standard deviations (6σ). The Cpk index is the ratio of the band between the process mean and the closest tolerance limit, to three standard deviations (3σ).

- *Cp* measures the potential capability of the process, if centred; *Cpk* measures the capability of the process, including its centring. The *Cpk* index can be used for one-sided specifications.
- Values of the standard deviation, and hence the *Cp* and *Cpk*, depend on the origin of the data used, as well as the method of calculation. Unless the origin of the data and method is known the interpretation of the indices will be confused.
- If the data used is from a process which is in statistical control, the *Cpk* calculation from $\bar{R}$ is the $Cpk_{(potential)}$ of the process.
- The $Cpk_{(potential)}$ measures the confidence one may have in the control of the process, and classification of the output, so that the presence of non-conforming output is at an acceptable level.
- For all sample sizes a $Cpk_{(potential)}$ of 1 or less is unacceptable, since the generation of non-conforming output is inevitable.
- If the $Cpk_{(potential)}$ is between 1 and 2, the control of the process and the elimination of non-conforming output will be uncertain.
- A *Cpk* value of 2 gives high confidence in the producer, provided that control charts are in regular use.
- If the standard deviation is estimated from all the data collected during normal running of the process, it will give rise to a $Cpk_{(production)}$, which will be less than the $Cpk_{(potential)}$. The $Cpk_{(production)}$ is a useful index of the process performance during normal production.
- If the standard deviation is based on data taken from selected deliveries of an output it will result in a $Cpk_{(delivery)}$ which will also be less than the $Cpk_{(potential)}$, but may be greater than the $Cpk_{(production)}$, as the result of output selection. This can be a useful index of the delivery performance.
- A customer should seek from suppliers information concerning the potential of their processes, the methods of control and the methods of product classification used.
- The concept of process capability may be used in service environments and capability indices calculated.

References and further reading

Grant, E.L. and Leavenworth, R.S. (2000) *Statistical Quality Control*, 7th edn, McGraw-Hill, New York, USA.

Owen, M. (1993) *SPC and Business Improvement*, IFS Publications, Bedford, UK.

Porter, L.J. and Oakland, J.S. (1991) 'Process Capability Indices: An Overview of Theory and Practice', *Quality and Reliability Engineering International*, Vol. 7, pp. 437–449.

Pyzdek, T. (1990) *Pyzdek's Guide to SPC, Vol. 1: Fundamentals*, ASQC Quality Press, Milwaukee WI, USA.

Wheeler, D.J. (2001) *Process Evaluation Handbook*, SPC Press, Knoxville TN, USA.

Wheeler, D.J. (2010) *Understanding Statistical Process Control*, 3rd edn, SPC Press, Knoxville TN, USA.

Discussion questions

1 (a) Using process capability studies, processes may be classified as being in statistical control and capable. Explain the basis and meaning of this classification.

   (b) Define the process capability indices *Cp* and *Cpk* and describe how they may be used to monitor the capability of a process, its actual performance and its performance as perceived by a customer.

2 Using the data given in Discussion question No. 5 in Chapter 6, calculate the appropriate process capability indices and comment on the results.

3 From the results of your analysis of the data in Discussion question No. 6, Chapter 6, show quantitatively whether the process is capable of meeting the specification given.

4 Calculate *Cp* and *Cpk* process capability indices for the data given in Discussion question No. 8 in Chapter 6 and write a report to the Development Chemist.

5 Show the difference, if any, between Machine I and Machine II in Discussion question No. 9 in Chapter 6, by the calculation of appropriate process capability indices.

6 In Discussion question No. 10 in Chapter 6, the specification was given as 540 ± 5 mm, comment further on the capability of the panel making process using process capability indices to support your arguments.

Worked examples

1. Lathe operation

Using the data given in Worked example No. 1 (Lathe operation) in Chapter 6, answer question 1(b) with the aid of process capability indices.

Solution

$$\sigma = \bar{R}/d_n = 0.0007/2.326 = 0.0003 \text{cm}$$

$$Cp = Cpk = \frac{(USL - \bar{\bar{X}})}{3\sigma} = \frac{(\bar{\bar{X}} - LSL)}{3\sigma}$$

$$= \frac{0.002}{0.0009} = 2.22.$$

2. Control of dissolved iron in a dyestuff

Using the data given in Worked example No. 2 (Control of dissolved iron in a dyestuff) in Chapter 6, answer question 1(b) by calculating the *Cpk* value.

Solution

$$Cpk = \frac{\text{USL} - \overline{\overline{X}}}{3\sigma}$$

$$= \frac{18.0 - 15.6}{3 \times 1.445} = 0.55.$$

With such a low *Cpk* value, the process is not capable of achieving the required specification of 18 ppm. The *Cp* index is not appropriate here as there is a one-sided specification limit.

3. Pin manufacture

Using the data given in Worked example No. 3 (Pin manufacture) in Chapter 6, calculate *Cp* and *Cpk* values for the specification limits 0.820 cm and 0.840 cm, when the process is running with a mean of 0.834 cm.
   Solution

$$Cp = \frac{\text{USL} - \text{LSL}}{6\sigma} = \frac{0.84 - 0.82}{6 \times 0.003} = 1.11.$$

The process is potentially capable of just meeting the specification. Clearly the lower value of *Cpk* will be:

$$Cpk = \frac{\text{USL} - \overline{\overline{X}}}{3\sigma} = \frac{0.84 - 0.834}{3 \times 0.003} = 0.67.$$

The process is not centred and not capable of meeting the requirements.

Part V
Process improvement

11 Process problem solving and improvement

Objectives

- To introduce and provide a framework for process problem solving and improvement.
- To describe the major problem-solving tools.
- To illustrate the use of the tools with worked examples.
- To provide an understanding of how the techniques can be used together to aid process improvement.

11.1 Introduction

Process improvements are often achieved through specific opportunities, commonly called problems, being identified or recognized. A focus on improvement opportunities should lead to the creation of teams whose membership is determined by their work on and detailed knowledge of the process, and their ability to take improvement action. The teams must then be provided with good leadership and the right tools to tackle the job.

By using reliable methods, creating a favourable environment for team-based problem solving, and continuing to improve using systematic techniques, the never-ending improvement cycle of plan, do, check, act will be engaged. This approach demands the real time management of data, and actions on processes – inputs, controls and resources, not outputs. It will require a change in the language of many organizations from percentage defects, percentage 'prime' product and number of errors, to *process capability*. The climate must change from the traditional approach of 'If it meets the specification, there are no problems and no further improvements are necessary'. The driving force for this will be the need for better internal and external customer satisfaction levels, which will lead to the continuous improvement question, 'Could we do the job better?'

In Chapter 1 some basic tools and techniques were briefly introduced. Some of these are very useful in a problem identification and solving context, namely Pareto analysis, cause and effect analysis, scatter diagrams and stratification.

The effective use of these tools requires their application by the people who actually work on the processes. Their commitment to this will be possible only if they are assured that management cares about improving quality. Managers must show they are serious by establishing a systematic approach and providing the training and implementation support required.

The systematic approach mapped out in Figure 11.1 should lead to the use of factual information, collected and presented by means of proven techniques, to open a channel of communications not available to the many organizations that do not follow this or a similar approach to problem solving and improvement. Continuous improvements in the quality of products, services and processes can often be obtained without major capital investment, if an organization marshals its resources, through an understanding and breakdown of its processes in this way.

Organizations that embrace the concepts of total quality and operational excellence should recognize the value of problem-solving techniques in *all* areas, including sales, purchasing, invoicing, finance, distribution, training, etc., that are outside production or operations – the traditional area for SPC use. A Pareto analysis, a histogram, a flowchart or a control chart is a vehicle for communication. Data are data and, whether the numbers represent defects or invoice errors, the information relating to machine settings, process variables, prices, quantities, discounts, customers or supply points is irrelevant – the techniques can always be used to good effect.

Some of the most effective applications of SPC and problem-solving tools have emerged from organizations and departments that, when first introduced to the methods, could see little relevance to their own activities. Following appropriate training, however, they have learned how to, for example:

- *Pareto analyse* sales turnover by product and injury data;
- *brainstorm* and *cause and effect analyse* reasons for late payment and poor purchase invoice matching;
- *histogram* absenteeism and arrival times of trucks during the day;
- *control chart* the movement in currency and weekly demand of a product.

Distribution staff have used *p*-charts to monitor the proportion of deliveries that are late and Pareto analysis to look at complaints involving the distribution system. Computer and call-centre operators have used cause and effect analysis and histograms to represent errors and problems from their service. Moving average and cusum charts have immense potential for improving forecasting in all areas including marketing, demand, output, currency value and commodity prices.

Those organizations that have made most progress in implementing a company-wide approach to improvement have recognized at an early stage that SPC is for the whole organization. Restricting it to traditional

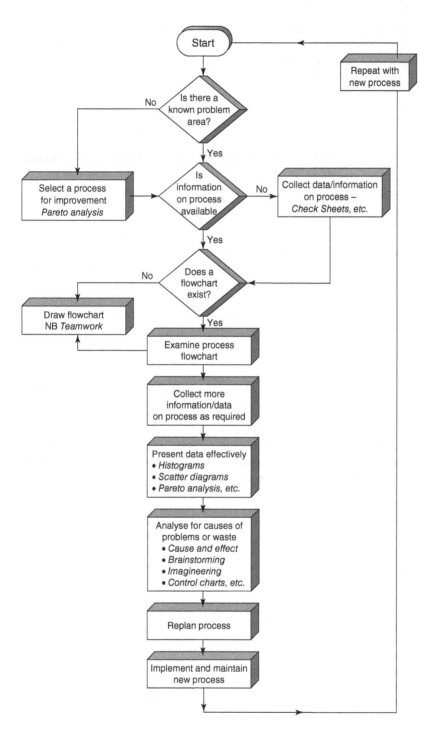

Figure 11.1 Strategy for continuous process improvement

manufacturing or operational activities means that a window of opportunity has been closed. Applying the methods and techniques outside
manufacturing will make it easier, not harder, to gain maximum benefit
from an SPC programme.

Sales and marketing is one area that often resists training in SPC on the
basis that it is difficult to apply. Personnel in this vital function need to be
educated in SPC methods for two reasons:

(i) They need to understand the way the manufacturing and/or service
    producing processes in their organizations work. This enables them to
    have more meaningful and involved dialogues with customers about the
    whole product/service system capability and control. It will also enable
    them to influence customers' thinking about specifications and create a
    competitive advantage from improving process capabilities.
(ii) They need to identify and improve the marketing processes and activities.
    A significant part of the sales and marketing effort is clearly associated
    with building relationships, which are best based on facts (data) and not
    opinions. There are also opportunities to use SPC techniques directly in
    such areas as forecasting, demand levels, market requirements, monitor-
    ing market penetration, marketing control and product development,
    all of which must be viewed as processes.

SPC has considerable applications for non-manufacturing organizations, in
both the public and the private sectors. Data and information on patients in
hospitals, students in universities and schools, people who pay (and do not
pay) tax, draw benefits, shop at Sainsbury or Walmart are available in abun-
dance. If it were to be used in a systematic way, and all operations treated
as processes, far better decisions could be made concerning the past, present
and future performances of these operations.

11.2 Pareto analysis

In many things we do in life we find that most of our problems arise from a
few of the sources. The Italian economist Vilfredo Pareto used this concept
when he approached the distribution of wealth in his country at the turn
of the century. He observed that 80–90 per cent of Italy's wealth lay in the
hands of 10–20 per cent of the population. A similar distribution has been
found empirically to be true in many other fields. For example, 80 per cent
of the defects will probably arise from 20 per cent of the causes; 80 per
cent of the complaints originate from 20 per cent of the customers. These
observations have become known as part of Pareto's Law or the 80/20 rule.

The technique of arranging data according to priority or importance and
tying it to a problem-solving framework is called Pareto analysis. This is a
formal procedure that is readily teachable, easily understood and very effective. Pareto diagrams or charts are used extensively by improvement teams

all over the world; indeed the technique has become fundamental to their operation for identifying the really important problems and establishing priorities for action.

Pareto analysis procedures

There are always many aspects of business operations that require improvement: the number of errors, process capability, rework, sales, etc. Each problem comprises many smaller problems and it is often difficult to know which ones to tackle to be most effective. For example, Table 11.1 gives some data on the reasons for batches of a dyestuff product being scrapped or reworked. A definite procedure is needed to transform this data to form a basis for action.

It is quite obvious that two types of Pareto analysis are possible here to identify the areas that should receive priority attention. One is based on the frequency of each cause of scrap/rework and the other is based on cost. It is reasonable to assume that both types of analysis will be required. The identification of the most frequently occurring reason should enable the total number of batches scrapped or requiring rework to be reduced. This may be necessary to improve plant operator morale which may be adversely affected by a high proportion of output being rejected. Analysis using cost as the basis will be necessary to derive the greatest financial benefit from the effort exerted. We shall use a generalizable stepwise procedure to perform both of these analyses.

Table 11.1 List of reasons for scrap/rework

SCRIPTAGREEN – A Plant B		Batches scrapped/reworked Period 05–07 incl.		
Batch No.	Reason for scrap/rework	Labour cost (£)	Material cost (£)	Plant cost (£)
05–005	Moisture content high	500	50	100
05–011	Excess insoluble matter	500	nil	125
05–018	Dyestuff contamination	4000	22000	14000
05–022	Excess insoluble matter	500	nil	125
05–029	Low melting point	1000	500	3500
05–035	Moisture content high	500	50	100
05–047	Conversion process failure	4000	22000	14000
05–058	Excess insoluble matter	500	nil	125
05–064	Excess insoluble matter	500	nil	125
05–066	Excess insoluble matter	500	nil	125
05–076	Low melting point	1000	500	3500
05–081	Moisture content high	500	50	100

(continued)

Table 11.1 (continued)

SCRIPTAGREEN – A Plant B		Batches scrapped/reworked Period 05–07 incl.		
Batch No.	Reason for scrap/rework	Labour cost (£)	Material cost (£)	Plant cost (£)
05–086	Moisture content high	500	50	100
05–104	High iron content	500	nil	2000
05–107	Excess insoluble matter	500	nil	125
05–111	Excess insoluble matter	500	nil	125
05–132	Moisture content high	500	50	100
05–140	Low melting point	1000	500	3500
05–150	Dyestuff contamination	4000	22000	14000
05–168	Excess insoluble matter	500	nil	125
05–170	Excess insoluble matter	500	nil	125
05–178	Moisture content high	500	50	100
05–179	Excess insoluble matter	500	nil	125
05–179	Excess insoluble matter	500	nil	125
05–189	Low melting point	1000	500	3500
05–192	Moisture content high	500	50	100
05–208	Moisture content high	500	50	100
06–001	Conversion process failure	4000	22000	14000
06–003	Excess insoluble matter	500	nil	125
06–015	Phenol content .1%	1500	1300	2000
06–024	Moisture content high	500	50	100
06–032	Unacceptable application	2000	4000	4000
06–041	Excess insoluble matter	500	nil	125
06–057	Moisture content high	500	50	100
06–061	Excess insoluble matter	500	nil	125
06–064	Low melting point	1000	500	3500
06–069	Moisture content high	500	50	100
06–071	Moisture content high	500	50	100
06–078	Excess insoluble matter	500	nil	125
06–082	Excess insoluble matter	500	nil	125
06–094	Low melting point	1000	500	3500
06–103	Low melting point	1000	500	3500
06–112	Excess insoluble matter	500	nil	125
06–126	Excess insoluble matter	500	nil	125
06–131	Moisture content high	500	50	100
06–147	Unacceptable absorption spectrum	500	50	400
06–150	Excess insoluble matter	500	nil	125
06–151	Moisture content high	500	50	100

06–161	Excess insoluble matter	500	nil	125
06–165	Moisture content high	500	50	100
06–172	Moisture content high	500	50	100
06–186	Excess insoluble matter	500	nil	125
06–198	Low melting point	1000	500	3500
06–202	Dyestuff contamination	4000	22000	14000
06–214	Excess insoluble matter	500	nil	125
07–010	Excess insoluble matter	500	nil	125
07–021	Conversion process failure	4000	22000	14000
07–033	Excess insoluble matter	500	nil	125
07–051	Excess insoluble matter	500	nil	125
07–057	Phenol content .1%	1500	1300	2000
07–068	Moisture content high	500	50	100
07–072	Dyestuff contamination	4000	22000	14000
07–077	Excess insoluble matter	500	nil	125
07–082	Moisture content high	500	50	100
07–087	Low melting point	1000	500	3500
07–097	Moisture content high	500	50	100
07–116	Excess insoluble matter	500	nil	125
07–117	Excess insoluble matter	500	nil	125
07–118	Excess insoluble matter	500	nil	125
07–121	Low melting point	1000	500	3500
07–131	High iron content	500	nil	2000
07–138	Excess insoluble matter	500	nil	125
07–153	Moisture content high	500	50	100
07–159	Low melting point	1000	500	3500
07–162	Excess insoluble matter	500	nil	125
07–168	Moisture content high	500	50	100
07–174	Excess insoluble matter	500	nil	125
07–178	Moisture content high	500	50	100
07–185	Unacceptable chromatogram	500	1750	2250
07–195	Excess insoluble matter	500	nil	125
07–197	Moisture content high	500	50	100

Step 1: List all the elements

This list should be exhaustive to preclude the inadvertent drawing of inappropriate conclusions. In this case the reasons may be listed as they occur in Table 11.1. They are moisture content high, excess insoluble matter, dyestuff contamination, low melting point, conversion process failure, high iron content, phenol content >1 per cent, unacceptable application, unacceptable absorption spectrum, unacceptable chromatogram.

Step 2: Measure the elements

It is essential to use the same unit of measure for each element. It may be in cash value, time, frequency, number or amount, depending on the element. In the scrap and rework case, the elements – reasons – may be measured in terms of frequency, labour cost, material cost, plant cost and total cost. We shall use the first and the last – frequency and total cost. The tally chart, frequency distribution and cost calculations are shown in Table 11.2.

Step 3: Rank the elements

This ordering takes place according to the measures and not the classification. This is the crucial difference between a Pareto distribution and the usual frequency distribution and is particularly important for numerically classified elements. For example, Figure 11.2 shows the comparison between the frequency and Pareto distributions from the same data on pin lengths.

Table 11.2 Frequency distribution and total cost of dyestuff batches scrapped/reworked

Reason for scrap/ rework	Tally	Frequency	Cost per batch (£)	Total cost (£)
Moisture content high	₩₩ ₩₩ ₩₩ ₩₩ |||	23	650	14950
Excess insoluble matter	₩₩ ₩₩ ₩₩ ₩₩ ₩₩ ||	32	625	20000
Dyestuff contamination	||||	4	40000	160000
Low melting point	₩₩ ₩₩		11	5000	55000
Conversion process failure	|||	3	40000	120000
High iron content	||	2	2500	5000
Phenol content >1%	||	2	4800	9600
Unacceptable application			1	10000	10000
Unacceptable absorption spectrum			1	950	950
Unacceptable chromatogram			1	4500	4500

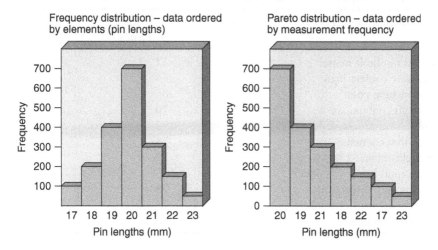

Figure 11.2 Comparison between frequency and Pareto distribution (pin lengths)

The two distributions are ordered in contrasting fashion with the frequency distribution structured by element value and the Pareto arranged by the measurement values on the element.

To return to the scrap and rework case, Table 11.3 shows the reasons ranked according to frequency of occurrence, whilst Table 11.4 has them in order of decreasing cost.

Step 4: Create cumulative distributions

The measures are cumulated from the highest ranked to the lowest, and each cumulative frequency shown as a percentage of the total. The elements are also cumulated and shown as a percentage of the total. Tables 11.3 and 11.4 show these calculations for the scrap and rework data – for frequency of occurrence and total cost, respectively. The important thing to remember about the cumulative element distribution is that the gaps between each element should be equal. If they are not, then an error has been made in the calculations or reasoning. The most common mistake is to confuse the frequency of measure with elements.

Step 5: Draw the Pareto curve

The cumulative percentage distributions are plotted as linear graphs. The cumulative percentage measure is plotted on the vertical axis against the cumulative percentage element along the horizontal axis. Figures 11.3 and 11.4 are the respective Pareto curves for frequency and total cost of reasons

Table 11.3 Scrap/rework – Pareto analysis of frequency of reasons

Reason for scrap/rework	Frequency	Cumulative frequency	Percentage of total
Excess insoluble matter	32	32	40.00
Moisture content high	23	55	68.75
Low melting point	11	66	82.50
Dyestuff contamination	4	70	87.50
Conversion process failure	3	73	91.25
High iron content	2	75	93.75
Phenol content >1%	2	77	96.25
Unacceptable:			
Absorption spectrum	1	78	97.50
Application	1	79	98.75
Chromatogram	1	80	100.00

Table 11.4 Scrap/rework – Pareto analysis of total costs

Reason for scrap/rework	Total cost	Cumulative cost	Cumulative percentage of grand total
Dyestuff contamination	160000	160000	40.0
Conversion process failure	120000	280000	70.0
Low melting point	55000	335000	83.75
Excess insoluble matter	20000	355000	88.75
Moisture content high	14950	369950	92.5
Unacceptable application	10000	379950	95.0
Phenol content >1%	9600	389550	97.4
High iron content	5000	395550	98.65
Unacceptable chromatogram	4500	399050	99.75
Unacceptable absorption spectrum	950	400000	100.0

for the scrapped/reworked batches of dyestuff product. Of course, there are many software programs, such as Minitab which will perform these analyses.

Step 6: Interpret the Pareto curves

The aim of Pareto analysis in problem solving is to highlight the elements that should be examined first. A useful first step is to draw a vertical line from the 20 to 30 per cent area of the horizontal axis. This has been done in both Figures 11.3 and 11.4 and shows that:

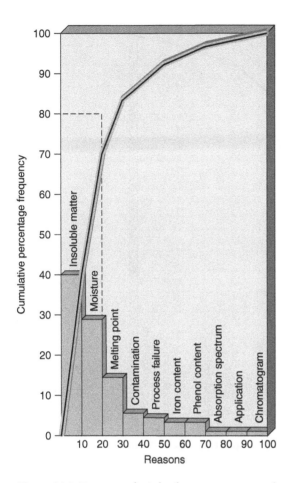

Figure 11.3 Pareto analysis by frequency: reasons for scrap/rework

1 30 per cent of the reasons are responsible for 82.5 per cent of all the batches being scrapped or requiring rework. The reasons are:

- excess insoluble matter (40 per cent),
- moisture content high (28.75 per cent),
- low melting point (13.75 per cent).

2 30 per cent of the reasons for scrapped or reworked batches cause 83.75 per cent of the total cost. The reasons are:

- dyestuff contamination (40 per cent),
- conversion process failure (30 per cent),
- low melting point (13.75 per cent).

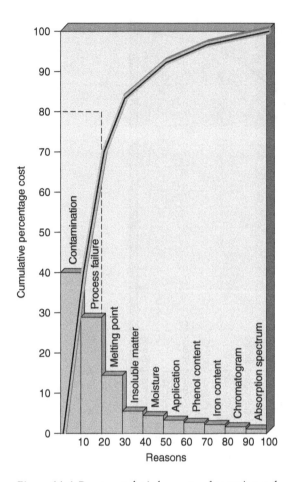

Figure 11.4 Pareto analysis by costs of scrap/rework

These are often called the 'A' items or the 'vital few' that have been high-lighted for special attention. It is quite clear that, if the objective is to reduce costs, then contamination must be tackled as a priority. Even though this has occurred only four times in 80 batches, the costs of scrapping the whole batch are relatively very large. Similarly, concentration on the problem of excess insoluble matter will have the biggest effect on reducing the number of batches that require to be reworked.

It is conventional to further arbitrarily divide the remaining 70–80 per cent of elements into two classifications – the B elements and the C elements, the so-called 'trivial many'. This may be done by drawing a vertical line from the 50–60 per cent mark on the horizontal axis. In this case only 5 per cent of the costs come from the 50 per cent of the 'C' reasons. This type of classification of elements gives rise to the alternative name for this technique – ABC analysis.

Procedural note

ABC or Pareto analysis is a powerful 'narrowing down' tool but it is based on empirical rules that have no mathematical foundation. It should always be remembered, when using the concept, that it is not rigorous and that elements or reasons for problems need not stand in line until higher ranked ones have been tackled. In the scrap and rework case, for example, if the problem of phenol content >1 per cent can be removed by easily replacing a filter costing a small amount, then let it be done straight away. The aim of the Pareto technique is simply to ensure that the maximum reward is returned for the effort expelled, but it is not a requirement of the systematic approach that 'small', easily solved problems must be made to wait until the larger ones have been resolved.

11.3 Cause and effect analysis

In any study of a problem, the *effect* – such as a particular defect or a certain process failure – is usually known. Cause and effect analysis may be used to elicit all possible contributing factors, or *causes* of the effect. This technique comprises usage of cause and effect diagrams and brainstorming.

The cause and effect diagram is often mentioned in passing as, 'one of the techniques used by quality circles'. Whilst this statement is true, it is also needlessly limiting in its scope of the application of this most useful and versatile tool. The cause and effect diagram, also known as the Ishikawa diagram (after its inventor), or the fishbone diagram (after its appearance), shows the effect at the head of a central 'spine' with the causes at the ends of the 'ribs' that branch from it. The basic form is shown in Figure 11.5. The principal factors or causes are listed first and then reduced to their sub-causes, and sub-sub-causes if necessary. This process is continued until all the conceivable causes have been included.

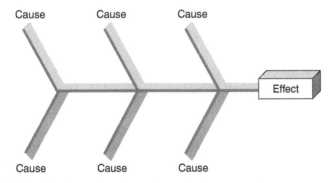

Figure 11.5 Basic form of cause and effect diagram

The factors are then critically analysed in light of their probable contribution to the effect. The factors selected as most likely causes of the effect are then subjected to experimentation to determine the validity of their selection. This analytical process is repeated until the true causes are identified.

Constructing the cause and effect diagram

An essential feature of the cause and effect technique is *brainstorming*, which is used to bring ideas on causes out into the open. A group of people freely exchanging ideas bring originality and enthusiasm to problem solving. Wild ideas are welcomed and safe to offer, as criticism or ridicule is not permitted during a brainstorming session. To obtain the greatest results from the session, all members of the group should participate equally and all ideas offered are recorded for subsequent analysis.

The construction of a cause and effect diagram is best illustrated with an example.

The production manager in a tea-bag manufacturing firm was extremely concerned about the amount of wastage of tea that was taking place. A study group had been set up to investigate the problem but had made little progress, even after several meetings. The lack of progress was attributed to a combination of too much talk, arm-waving and shouting down – typical symptoms of a non-systematic approach. The problem was handed to a newly appointed management trainee who used the following stepwise approach.

Step 1: Identify the effect

This sounds simple enough but, in fact, is often so poorly done that much time is wasted in the later steps of the process. It is vital that the effect or problem is stated in clear, concise terminology. This will help to avoid the situation where the 'causes' are identified and eliminated, only to find that the 'problem' still exists. In the tea-bag company, the effect was defined as 'Waste – unrecovered tea wasted during the tea- bag manufacture'. Effect statements such as this may be arrived at via a number of routes, but the most common are consensus obtained through brainstorming, one of the 'vital few' on a Pareto diagram, and sources outside the production department.

Step 2: Establish goals

The importance of establishing realistic, meaningful goals at the outset of any problem-solving activity cannot be over-emphasized. Problem solving is not a self-perpetuating endeavour. Most people need to know that their efforts are achieving some good in order for them to continue to participate. A goal should, therefore, be stated in some terms of measurement related to the problem and this must include a time limit. In the tea-bag firm, the goal was 'a 50 per cent reduction in waste in 9 months'. This requires, of course, a good understanding of the situation prior to setting the goal. It is necessary to establish the baseline in order to know, for example, when a 50 per cent reduction has been achieved. The tea waste was running at 2 per cent of tea usage at the commencement of the project.

Step 3: Construct the diagram framework

The framework on which the causes are to be listed can be very helpful to the creative thinking process. The authors have found the use of the five 'Ps' of production management very useful in the construction of cause and effect diagrams. The five components of any operational task are the:

* **product,** including services, materials and any intermediates;
* **processes** or methods of transformation;
* **plant,** i.e. the building and equipment;
* **programmes** or timetables for operations;
* **people,** operators, staff and managers.

These are placed on the main ribs of the diagram with the effect at the end of the spine of the diagram (Figure 11.6). The grouping of the sub-causes under the five 'P' headings can be valuable in subsequent analysis of the diagram.

Step 4: Record the causes

It is often difficult to know just where to begin listing causes. In a brainstorming session, the group leader may ask each member, in turn, to suggest a cause. It is essential that the leader should allow only 'causes' to be suggested for it is very easy to slip into an analysis of the possible solutions before all the probable causes have been listed. As suggestions are made, they are written onto the appropriate branch of the diagram. Again, no criticism of any cause is allowed at this stage of the activity. All suggestions are

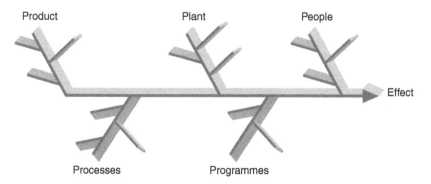

Figure 11.6 Cause and effect analysis and the five 'Ps'

welcomed because even those that eventually prove to be 'false' may serve to provide ideas that lead to the 'true' causes. Figure 11.7 shows the completed cause and effect diagram for the waste in tea-bag manufacture.

Step 5: Incubate and analyse the diagram

It is usually worthwhile to allow a delay at this stage in the process and to let the diagram remain on display for a few days so that everyone involved in the problem may add suggestions. After all the causes have been listed and the cause and effect diagram has 'incubated' for a short period, the group critically analyses it to find the most likely 'true causes'. It should be noted that after the incubation period the members of the group are less likely to remember who made each suggestion. It is, therefore, much easier to criticize the ideas and not the people who suggested them.

If we return to the tea-bag example, the investigation focused on the various stages of manufacture where data could easily be recorded concerning the frequency of faults under the headings already noted. It was agreed that over a 2-week period each incidence of wastage together with an approximate amount would be recorded. Simple clipboards were provided for the task. The break-down of fault frequencies and amount of waste produced led to the information in Table 11.5.

From a Pareto analysis of this data, it was immediately obvious that paper problems were by far the most frequent. It may be seen that two of the seven causes (28 per cent) were together responsible for about 74 per cent of the observed faults. A closer examination of the paper faults showed 'reel changes' to be the most frequent cause. After discussions with the supplier and minor machine modifications, the diameter of the reels of paper was doubled and the frequency of reel changes reduced to approximately one quarter of the original. Prior to this investigation, reel changes were not considered to be a problem – it was accepted as inevitable that a reel would come to an end. Tackling the identified causes in order of descending importance resulted in the tea-bag waste being reduced to 0.75 per cent of usage within 9 months.

Table 11.5 Major categories of causes of tea waste

Category of cause	Percentage wastage
Weights incorrect	1.92
Bag problems	1.88
Dirt	5.95
Machine problems	18.00
Bag formation	4.92
Carton problems	11.23
Paper problems	56.10

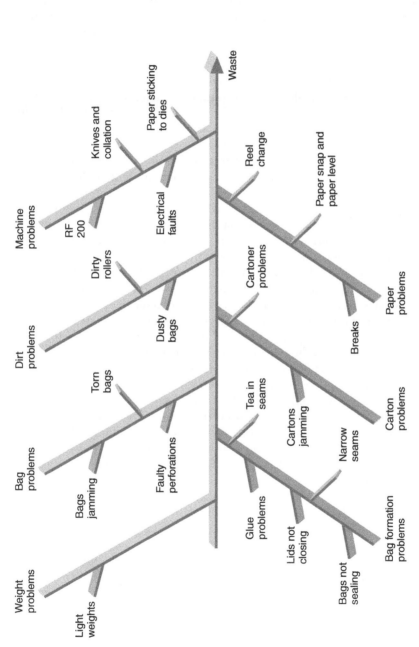

Figure 11.7 Detailed causes of tea wastage

Cause and effect diagrams with addition of cards

The cause and effect diagram is really a picture of a brainstorming session. It organizes free-flowing ideas in a logical pattern. With a little practice it can be used very effectively whenever any group seeks to analyse the cause of any effect. The effect may be a 'problem' or a desirable effect and the technique is equally useful in the identification of factors leading to good results. All too often desirable occurrences are attributed to chance, when in reality they are the result of some variation or change in the process. Stating the desired result as the effect and then seeking its causes can help identify the changes that have decreased the defect rate, lowered the amount of scrap produced or caused some other improvement.

A variation on the cause and effect approach, which was developed at Sumitomo Electric, is the cause and effect diagram with addition of cards (CEDAC).

The effect side of a CEDAC chart is a quantified description of the problem, with an agreed and visual quantified target and continually updated results on the progress of achieving it. The cause side of the CEDAC chart uses two different coloured cards for writing *facts* and *ideas*. This ensures that the facts are collected and organized before solutions are devised.

The basic diagram for CEDAC has the classic fishbone appearance. It is drawn on a large piece of paper, with the effect on the right and causes on the left. A project leader is chosen to be in charge of the CEDAC team, and (s)he sets the improvement target. A method of measuring and plotting the results on the effects side of the chart is devised so that a visual display – perhaps a graph – of the target and the quantified improvements are provided.

The *facts* are gathered and placed on the left of the spines on the cause side of the CEDAC chart (Figure 11.8). The people in the team submitting the fact cards are required to initial them. Improvement *ideas* cards are then generated and placed on the right of the cause spines in Figure 11.8. The ideas are then selected and evaluated for substance and practicality. The test results are recorded on the effect side of the chart. The successful improvement ideas are incorporated into the new standard procedures.

Clearly, the CEDAC programme must start from existing standards and procedures, which must be adhered to if improvements are to be made. CEDAC can be applied to any problem that can be quantified – scrap levels, supplier issues, quality problems, materials usage, sales figures, insurance claims, etc. It is another systematic approach to marshalling the creative resources and knowledge of the people concerned. When they own and can measure the improvement process, they will find the solution.

11.4 Scatter diagrams

Scatter diagrams are used to examine the relationship between two factors to see whether they are related. If they are, then by controlling the independent

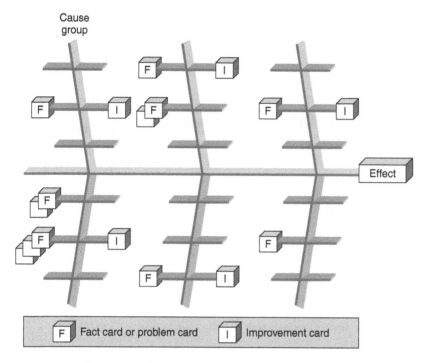

Figure 11.8 The CEDAC diagram with fact and improvement cards

factor, the dependent factor will also be controlled. For example, if the temperature of a process and the purity of a chemical product are related, then by controlling temperature, the quality of the product is determined.

Figure 11.9 shows that when the process temperature is set at A, a lower purity results than when the temperature is set at B. In Figure 11.10 we can see that tensile strength reaches a maximum for a metal treatment time of B, while a shorter or longer length of treatment will result in lower strength.

In both Figures 11.9 and 11.10 there appears to be a relationship between the 'independent factor' on the horizontal axis and the 'dependent factor' on the vertical axis. A statistical hypothesis test could be applied to the data to determine the statistical significance of the relationship, which could then be expressed mathematically. This is often unnecessary, as all that is required is to establish some sort of association. In some cases it appears that two factors are not related. In Figure 11.11, the percentage of defective polypropylene pipework does not seem to be related to the size of granulated polypropylene used in the process.

Scatter diagrams have application in problem solving following cause and effect analyses. After a sub-cause has been selected for analysis, the

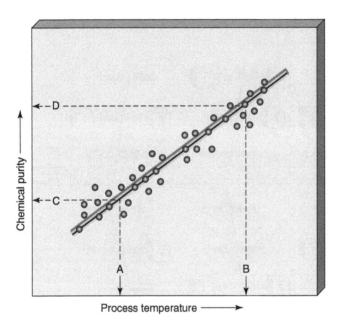

Figure 11.9 Scatter diagram: temperature versus purity

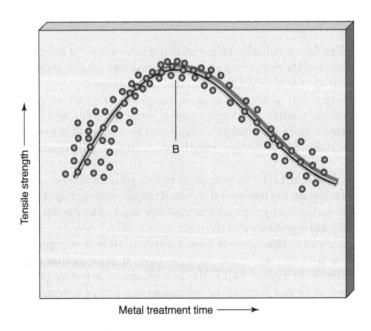

Figure 11.10 Scatter diagram: metal treatment time versus tensile strength

diagram may be helpful in explaining why a process acts the way it does and how it may be controlled.

Simple steps may be followed in setting up a scatter diagram:

1   Select the dependent and independent factors. The dependent factor may be a cause on a cause and effect diagram, a specification, a measure of quality or some other important result or measure. The independent factor is selected because of its potential relationship to the dependent factor.
2   Set up an appropriate recording sheet for data.
3   Choose the values of the independent factor to be observed during the analysis.
4   For the selected values of the independent factor, collect observations for the dependent factor and record on the data sheet.
5   Plot the points on the scatter diagram, using the horizontal axis for the independent factor and the vertical axis for the dependent factor.
6   Analyse the diagram.

This type of analysis is yet another step in the systematic approach to process improvement. It should be noted, however, that the relationship between certain factors is not a simple one and it may be affected by other factors. In these circumstances more sophisticated analysis of variance may be required.

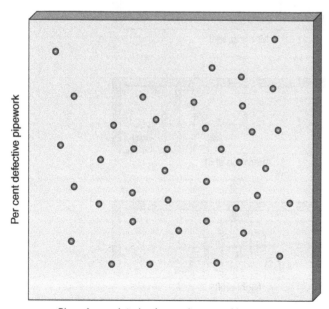

Size of granulated polypropylene used in process

Figure 11.11 Scatter diagram: no relationship between size of granules of polypropylene used and per cent defective pipework produced

11.5 Stratification

This is the sample selection method used when the whole population, or lot, is made up of a complex set of different characteristics, e.g. region, income, age, race, sex, education. In these cases the sample must be very carefully drawn in proportions that represent the makeup of the population.

Stratification often involves simply collecting or dividing a set of data into meaningful groups. It can be used to great effect in combination with other techniques, including histograms and scatter diagrams. If, for example,

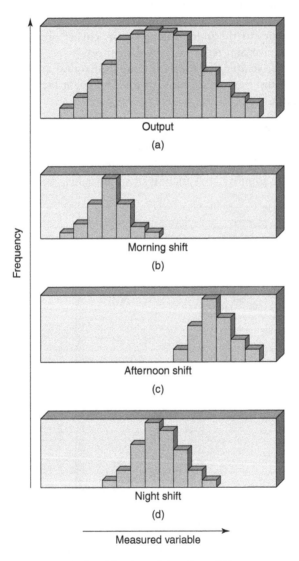

Figure 11.12 Stratification of data into shift teams

three shift teams are responsible for the output described by the histogram (a) in Figure 11.12, 'stratifying' the data into the shift groups might produce histograms (b), (c) and (d), and indicate process adjustments that were taking place at shift changeovers.

Figure 11.13 shows the scatter diagram relationship between advertising investment and revenue generated for all products. In diagram (a) all the data are plotted, and there seems to be no correlation. But if the data are stratified according to product, a correlation is seen to exist.

Of course, the reverse may be true, so the data should be kept together and plotted in different colours or symbols to ensure all possible interpretations are retained.

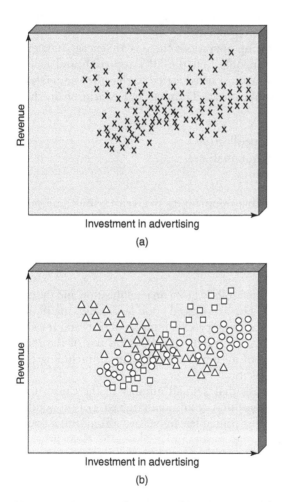

Figure 11.13 Scatter diagrams of investment in advertising versus revenue: (a) without stratification; (b) with stratification by different product

11.6 Summarizing problem solving and improvement

It is clear from the examples presented in this chapter that the principles and techniques of problem solving and improvement may be applied to any human activity, provided that it is regarded as a process. The only way to control process outputs, whether they be artefacts, paperwork, services or communications, is to manage the inputs systematically. Data from the outputs, the process itself, or the inputs, in the form of numbers or information, may then be used to modify and improve the operation.

Presenting data in an efficient and easy to understand manner is as vital in the office as it is on the factory floor and, as we have seen in this chapter, some of the basic tools of SPC and problem solving have a great deal to offer in all areas of management. Data obtained from processes must be analysed quickly so that continual reduction in the variety of ways of doing things will lead to never-ending improvement.

In many non-manufacturing operations there is an 'energy barrier' to be surmounted in convincing people that the SPC approach and techniques have a part to play. Everyone must be educated so that they understand and look for potential SPC applications. Training and education in the basic approach of:

- no process without data collection;
- no data collection without analysis;
- no analysis without action;

will ensure that every possible opportunity is given to use these powerful methods to greatest effect.

Chapter highlights

- Process improvements often follow problem identification and the creation of teams to solve them. The teams need good leadership, the right tools, good data and to take action on process inputs, controls and resources.
- A systematic approach is required to make good use of the facts and techniques, in *all* areas of all types of organization, including those in the service and public sectors.
- Pareto analysis recognizes that a small number of the causes of problems, typically 20 per cent, may result in a large part of the total effect, typically 80 per cent. This principle can be formalized into a procedure for listing the elements, measuring and ranking the elements, creating the cumulative distribution, drawing and interpreting the Pareto curve, and presenting the analysis and conclusions.
- Pareto analysis leads to a distinction between problems that are among the vital few and the trivial many, a procedure that enables effort to be directed towards the areas of highest potential return.

The analysis is simple, but the application requires a discipline that allows effort to be directed to the vital few. It is sometimes called ABC analysis or the 80/20 rule.

- For each effect there are usually a number of causes. Cause and effect analysis provides a simple tool to tap the knowledge of experts by separating the generation of possible causes from their evaluation.
- Brainstorming is used to produce cause and effect diagrams. When constructing the fishbone-shaped diagrams, the evaluation of potential causes of a specified effect should be excluded from discussion.
- Steps in constructing a cause and effect diagram include identifying the effect, establishing the goals, constructing a framework, recording all suggested causes, incubating the ideas prior to a more structured analysis leading to plans for action.
- A variation on the technique is the cause and effect diagram with addition of cards (CEDAC). Here the effect side of the diagram is quantified, with an improvement target, and the causes show *facts* and improvement *ideas*.
- Scatter diagrams are simple tools used to show the relationship between two factors – the independent (controlling) and the dependent (controlled). Choice of the factors and appropriate data recording are vital steps in their use.
- Stratification is a sample selection method used when populations are comprised of different characteristics. It involves collecting or dividing data into meaningful groups. It may be used in conjunction with other techniques to present differences between such groups.
- The principles and techniques of problem solving and improvement may be applied to any human activity regarded as a process. Where barriers to the use of these, perhaps in non-manufacturing areas, are found, training in the basic approach of process data collection, analysis and improvement action may be required.

References and further reading

Crossley, M.L. (2000) *The Desk Reference of Statistical Quality Methods*, ASQ Press, Milwaukee WI, USA.

Ishikawa, K. (1986) *Guide to Quality Control*, Asian Productivity Association, Tokyo, Japan.

Lockyer, K.G., Muhlemann, A.P. and Oakland, J.S. (1992) *Production and Operations Management*, 6th edn, Pitman, London, UK.

Oakland, J.S. (2014) *Total Quality Management and Operational Excellence: Text and Cases*, 4th edn, Routledge, Oxford, UK.

Pyzdek, T. (1990) *Pyzdek's Guide to SPC, Vol. 1: Fundamentals*, ASQC Quality Press, Milwaukee WI, USA.

Sygiyama, T. (1989) *The Improvement Book: Creating the Problem-Free Workplace*, Productivity Press, Cambridge MA, USA.

Discussion questions

1   You are the Production Manager of a small engineering company and have just received the following memo:

MEMORANDUM

To: Production Manager
From: Sales Manager
Subject: *Order Number 2937/AZ*

Joe Brown worked hard to get this order for us to manufacture 10,000 widgets for PQR Ltd. He now tells me that they are about to return the first batch of 1000 because many will not fit into the valve assembly that they tell us they are intended for. I must insist that you give rectification of this faulty batch number one priority, and that you make sure that this does not recur. As you know PQR Ltd are a new customer, and they could put a lot of work our way.

Incidentally I have heard that you have been sending a number of your operators on a training course in the use of the microbang widget gauge for use with that new machine of yours. I cannot help thinking that you should have spent the money on employing more finished product inspectors, rather than on training courses and high technology testing equipment.

(a) Outline how you intend to investigate the causes of the 'faulty' widgets.
(b) Discuss the final paragraph in the memo.

2   You have inherited, unexpectedly, a small engineering business that is both profitable and enjoys a full order book. You wish to be personally involved in this activity where the only area of immediate concern is the high levels of scrap and rework – costing together a sum equivalent to about 15 per cent of the company's total sales. Discuss your method of progressively picking up, analysing and solving this problem over a target period of 12 months. Illustrate any of the techniques you discuss.

3   Discuss in detail the applications of Pareto analysis and cause and effect analysis as aids in solving operations management problems. Give at least two illustrations.

You are responsible for a biscuit production plant, and are concerned about the output from the lines which make chocolate wholemeal biscuits. Output is consistently significantly below target. You suspect that this is because the lines are frequently stopped, so you initiate an in-depth investigation over a typical 2-week period. The table below shows the causes of the stoppages, number of occasions on which each occurred, and the average amount of output lost on each occasion:

Cause	No. of occurrences	Lost production (000s biscuits)
Wrapping		
cellophane wrap breakage	1031	3
cartoner failure	85	100
Enrober		
chocolate too thin	102	1
chocolate too thick	92	3
Preparation		
underweight biscuits	70	25
overweight biscuits	21	25
biscuits misshapen	58	1
Ovens		
biscuits overcooked	87	2
biscuits undercooked	513	1

Use this data and the appropriate techniques to indicate where to concentrate remedial action.

How could *stratification* aid the analysis in this particular case?

4   A company manufactures a range of domestic electrical appliances. Particular concern is being expressed about the warranty claims on one particular product. The customer service department provides the following data relating the claims to the unit/component part of the product which caused the claim:

Unit/component part	Number of claims	Average cost of warranty work (per claim)
Drum	110	48.1
Casing	12842	1.2
Work-top	142	2.7
Pump	246	8.9
Electric motor	798	48.9
Heater unit	621	15.6
Door lock mechanism	18442	0.8
Stabilizer	692	2.9
Power additive unit	7562	1.2
Electric control unit	652	51.9
Switching mechanism	4120	10.2

Discuss what criteria are of importance in identifying those unit/component parts to examine initially. Carry out a full analysis of the data to identify such unit/component parts.

5   The principal causes of accidents, their percentage of occurrence, and the estimated resulting loss of production per annum in the UK is given in the table below:

Accident cause	Percentage of all accidents	Estimated loss of production (£million/annum)
Machinery	16	190
Transport	8	30
Falls from heights >6'	16	100
Tripping	3	10
Striking against objects	9	7
Falling objects	7	20
Handling goods	27	310
Hand tools	7	65
Burns (including chemical)	5	15
Unspecified	2	3

(a) Using the appropriate data draw a Pareto curve and suggest how this may be used most effectively to tackle the problems of accident prevention. How could stratification help in the analysis?

(b) Give three other uses of this type of analysis in non-manufacturing and explain briefly, in each case, how use of the technique aids improvement.

6   The manufacturer of domestic electrical appliances has been examining causes of warranty claims. Ten have been identified and the annual cost of warranty work resulting from these is as follows:

Cause	Annual cost of warranty work (£)
A	1090
B	2130
C	30690
D	620
E	5930
F	970
G	49980
H	1060
I	4980
J	3020

Carry out a Pareto analysis on the above data, and describe how the main causes could be investigated.

7   A mortgage company finds that some 18 per cent of application forms received from customers cannot be processed immediately, owing to the absence of some of the information. A sample of 500 incomplete application forms reveals the following data:

Information missing	Frequency
Applicant's age	92
Daytime telephone number	22
Forenames	39
House owner/occupier	6
Home telephone number	1
Income	50
Signature	6
Occupation	15
Bank account no.	1
Nature of account	10
Postal code	6
Sorting code	85
Credit limit requested	21
Cards existing	5
Date of application	3
Preferred method of payment	42
Others	46

Determine the major causes of missing information, and suggest appropriate techniques to use in form redesign to reduce the incidence of missing information.

8   A company that operates with a 4-week accounting period is experiencing difficulties in keeping up with the preparation and issue of sales invoices during the last week of the accounting period. Data collected over two accounting periods are as follows:

Accounting period 4	Week	1	2	3	4
Number of sales invoices issued		110	272	241	495
Accounting period 5	Week	1	2	3	4
Number of sales invoices issued		232	207	315	270

Examine any correlation between the week within the period and the demands placed on the invoice department. How would you initiate action to improve this situation?

Worked examples

1 Reactor Mooney off-spec results

A project team looking at improving reactor Mooney control (a measure of viscosity) made a study over 14 production dates of results falling ±5 ML points outside the grade aim. Details of the causes were listed (Table 11.6).

Table 11.6 Reactor Mooney off-spec results over 14 production days

Sample	Cause	Sample	Cause
1	Cat. poison	33	H.C.L. control
2	Cat. poison	34	H.C.L. control
3	Reactor stick	35	Reactor stick
4	Cat. poison	36	Reactor stick
5	Reactor stick	37	Reactor stick
6	Cat. poison	38	Reactor stick
7	H.C.L. control	39	Reactor stick
8	H.C.L. control	40	Reactor stick
9	H.C.L. control	41	Instrument/analyser
10	H.C.L. control	42	H.C.L. control
11	Reactor stick	43	H.C.L. control
12	Reactor stick	44	Feed poison
13	Feed poison	45	Feed poison
14	Feed poison	46	Feed poison
15	Reactor stick	47	Feed poison
16	Reactor stick	48	Reactor stick
17	Reactor stick	49	Reactor stick
18	Reactor stick	50	H.C.L. control
19	H.C.L. control	51	H.C.L. control
20	H.C.L. control	52	H.C.L. control
21	Dirty reactor	53	H.C.L. control
22	Dirty reactor	54	Reactor stick
23	Dirty reactor	55	Reactor stick
24	Reactor stick	56	Feed poison
25	Reactor stick	57	Feed poison
26	Over correction F.109	58	Feed poison
27	Reactor stick	59	Feed poison
28	Reactor stick	60	Refridge problems
29	Instrument/analyser	61	Reactor stick
30	H.C.L. control	62	Reactor stick
31	H.C.L. control	63	Reactor stick
32	H.C.L. control	64	Reactor stick
65	Lab result	73	Reactor stick
66	H.C.L. control	74	Reactor stick
67	H.C.L. control	75	B. No. control
68	H.C.L. control	76	B. No control
69	H.C.L. control	77	H.C.L. control
70	H.C.L. control	78	H.C.L. control
71	Reactor stick	79	Reactor stick
72	Reactor stick	80	Reactor stick

Using a ranking method – Pareto analysis – the team were able to determine the major areas on which to concentrate their efforts.

Steps in the analysis were as follows:

1 Collect data over 14 production days and tabulate (Table 11.6).
2 Calculate the totals of each cause and determine the order of frequency (i.e. which cause occurs most often).
3 Draw up a table in order of frequency of occurrence (Table 11.7).
4 Calculate the percentage of the total off-spec that each cause is responsible for.

$$\text{e.g. Percentage due to reactor sticks} = \frac{32}{80} \times 100 = 40 \text{ per cent.}$$

5 Cumulate the frequency percentages.
6 Plot a Pareto graph showing the percentage due to each cause and the cumulative percentage frequency of the causes from Table 11.7 (Figure 11.14).

2 Ranking in managing product range

Some figures were produced by a small chemical company concerning the company's products, their total volume ($), and direct costs. These are given in Table 11.8. The products were ranked in order of income and contribution for the purpose of Pareto analysis, and the results are given in Table 11.9. To consider either income or contribution in the absence of the other could lead to incorrect conclusions; for example, product 013 which is ranked ninth in income actually makes zero contribution.

Table 11.7 Reactor Mooney off-spec results over 14 production dates: Pareto analysis of reasons

Reasons for Mooney off-spec	Tally	Frequency	Percentage of total	Cumulative percentage
Reactor sticks	⫲⫲ ⫲⫲ ⫲⫲ ⫲⫲ ⫲⫲ ⫲⫲ ⫲⫲ ‖	32	40	40
H.C.L. control	⫲⫲ ⫲⫲ ⫲⫲ ⫲⫲ ⦀⦀	24	30	70
Feed poisons	⫲⫲ ⫲⫲	10	12.5	82.5
Cat. Poisons	⦀⦀	4	5	87.5
Dirty stick reactor	⦀	3	3.75	91.25
B. No. control	‖	2	2.5	93.75
Instruments/analysers	‖	2	2.5	96.25
Over correction F.109	∣	1	1.25	97.5
Refridge problems	∣	1	1.25	98.75
Lab results	∣	1	1.25	100

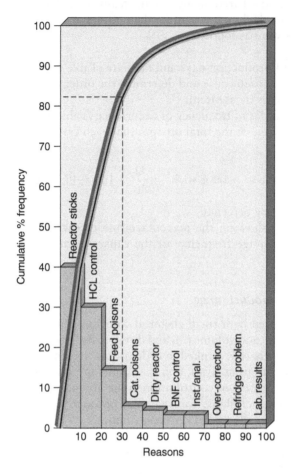

Figure 11.14 Pareto analysis: reasons for off-spec reactor Mooney

Table 11.8 Some products and their total volume, direct costs and contribution

Code number	Description	Total volume ($)	Total direct costs ($)	Total contribution ($)
001	Captine	1040	1066	26
002	BHD-DDB	16240	5075	11165
003	DDB-Sulphur	16000	224	15776
004	Nicotine-Phos	42500	19550	22950
005	Fensome	8800	4800	4000
006	Aldrone	106821	45642	61179
007	DDB	2600	1456	1144
008	Dimox	6400	904	5496
009	DNT	288900	123264	165636

010	Parathone	113400	95410	17990
011	HETB	11700	6200	5500
012	Mepofox	12000	2580	9420
013	Derros-Pyrethene	20800	20800	0
014	Dinosab	37500	9500	28000
015	Maleic Hydrazone	11300	2486	8814
016	Thirene-BHD	63945	44406	19539
017	Dinosin	38800	25463	13337
018	2,4-P	23650	4300	19350
019	Phosphone	13467	6030	7437
020	Chloropicrene	14400	7200	7200

Table 11.9 Income rank/contribution rank table

Code number	Description	Income rank	Contribution rank
001	Captine	20	20
002	BHD-DDB	10	10
003	DDB-Sulphur	11	8
004	Nicotine-Phos	5	4
005	Fensome	17	17
006	Aldrone	3	2
007	DDB	19	18
008	Dimox	18	16
009	DNT	1	1
010	Parathone	2	7
011	HETB	15	15
012	Mepofox	14	11
013	Derros-Pyrethene	9	19
014	Dinosab	7	3
015	Maleic Hydrazone	16	12
016	Thirene-BHD	4	5
017	Dinosin	6	9
018	2,4-P	8	6
019	Phosphone	13	13
020	Chloropicrene	12	14

One way of handling this type of ranked data is to plot an income–contribution rank chart. In this the abscissae are the income ranks, and the ordinates are the contribution ranks. Thus product 010 has an income rank of 2 and a contribution rank of 7. Hence, product 010 is represented by the point (2,7) in Figure 11.15, on which all the points have been plotted in this way.

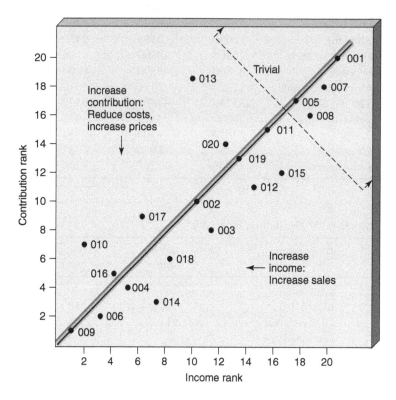

Figure 11.15 Income rank/contribution rank chart

Clearly, those products above the 45° line have income rankings higher than their contribution rankings and may be candidates for cost reduction focus or increases in price. Products below the line are making good contribution and selling more of them would be beneficial. This prior location and focus is likely to deliver more beneficial results than blanket cost reduction programmes or sales campaigns on everything.

3 Process capability in a bank

The process capability indices calculations in Section 10.5 showed that the process was not capable of meeting the requirements and management led an effort to improve transaction efficiency. This began with a flowcharting of the process as shown in Figure 11.16. In addition, a brainstorm session involving the cashiers was used to generate the cause and effect diagram of Figure 11.17. A quality improvement team was formed, further data collected, and the 'vital' areas of incompletely understood procedures and cashier training were tackled. This resulted over a period of 6 months in a reduction in average transaction time and improvement in process capability.

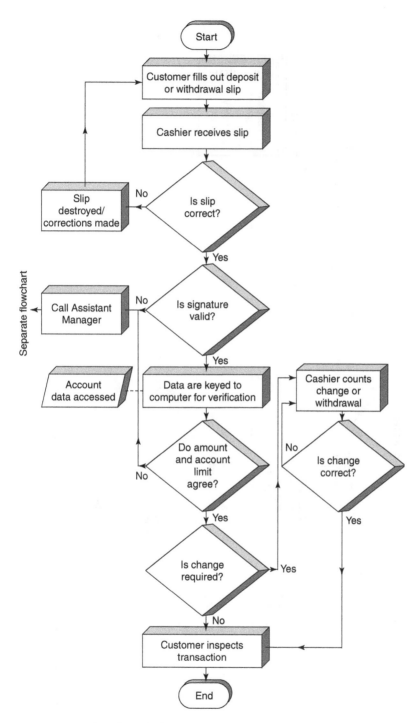

Figure 11.16 Flowchart for bank transactions

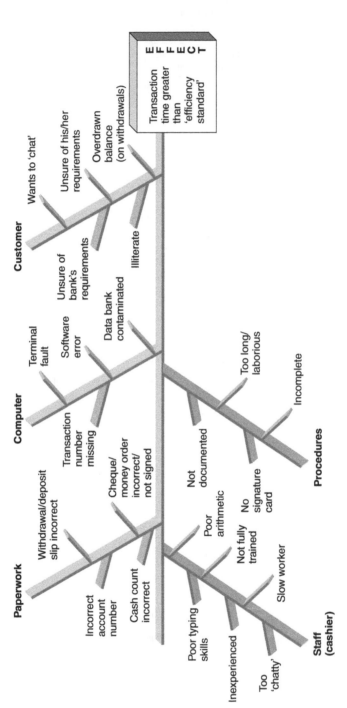

Figure 11.17 Cause and effect diagram for slow transaction times

12 Managing out-of-control processes

Objectives

- To consider the most suitable approach to process trouble-shooting.
- To outline a strategy for process improvement.
- To examine the use of control charts for trouble-shooting and classify out-of-control processes.
- To consider some causes of out-of-control processes.

12.1 Introduction

Historically, the responsibility for trouble-shooting and process improvement, particularly within a manufacturing organization, has rested with a 'technical' department. In recent times, however, these tasks have been carried out increasingly by people who are directly associated with the operation of the process on a day-to-day basis. What is quite clear is that process improvement and trouble-shooting should not become the domain of only research or technical people. In the service sector it very rarely is.

In a manufacturing company, for example, the production people have the responsibility for meeting production targets, which include those associated with the quality of the product. It is unreasonable for them to accept responsibility for process output, efficiency, and cost while delegating elsewhere responsibility for the quality of its output. If problems of low quantity arise during production, whether it be the number of tablets produced per day or the amount of herbicide obtained from a batch reactor, then these problems are tackled without question by production personnel. Why then should problems of – say – defects or excessive process variation leading to out of specification products not fall under the same umbrella?

Problems in process operations are rarely single dimensional. They have at least five dimensions:

- *product* or service, including inputs;
- *plant*, including equipment;
- *programmes*, timetables-schedules;

- *people*, including information;
- *process*, the way things are done.

The indiscriminate involvement of quality/research/technical people in trouble-shooting tends to polarize attention towards the technical aspects, with the corresponding relegation of other vital parameters. In many cases the human, managerial, and even financial dimensions have a significant bearing on the overall problem and its solution. They should not be ignored by taking a problem out of its natural environment and placing it in a 'laboratory'.

The emphasis of any 'trouble-shooting' effort should be directed towards problem prevention with priorities in the areas of:

  (i) maintaining quality of current output;
 (ii) process improvement;
(iii) product development.

Quality Assurance, for example, should not be a department to be ignored when everything is running well, yet saddled with the responsibility for solving quality problems when they arise. Associated with this practice are the dangers of such people being used as scapegoats when explanations to senior managers are required, or being offered as sacrificial lambs when customer complaints are being dealt with. The responsibility for quality must always lie with operators of the process and the role of QA or any other support function is clearly to assist in the meeting of this responsibility. It should not be acceptable for any group within an organization to approach another group with the question, 'We have got a problem, what are *you* going to do about it?' Expert advice may be necessary to tackle particular quality and/or process problems, of course.

Having described Utopia, we must accept that the real world is inevitably less than perfect. The major problem is the one of whether a process has the necessary capabilities required to meet the requirements. It is against this background that the methods in this chapter are presented.

12.2 Process improvement strategy

Process improvement is neither a pure science nor an art. Procedures may be presented but these will nearly always benefit from ingenuity. It is traditional to study cause and effect relationships. However, when faced with a multiplicity of potential causes of problems, all of which involve imperfect data, it is frequently advantageous to begin with studies that identify only blocks or groups as the source of the trouble. The groups may be, for example, a complete filling line or a whole area of a service operation. Thus, the pinpointing of specific causes and effects is postponed.

An important principle to be emphasized at the outset is that initial studies should not aim to discover everything straight away. This is particularly important in situations where more data is obtainable quite easily.

It is impossible to set down everything that should be observed in carrying out a process improvement exercise. One of the most important rules to observe is to be present when data are being collected, at least initially. This provides the opportunity to observe possible sources of error in the acquisition method or the type of measuring equipment itself. Direct observation of data collection may also suggest assignable causes that may be examined at the time. This includes the different effects due to equipment changes, various suppliers, shifts, people skills, etc.

In trouble-shooting and process improvement studies, the planning of data acquisition programmes should assist in detecting the effects of important changes. The opportunity to note possible relationships comes much more readily to the investigator who observes the data collection than the one who sits comfortably in an office chair. The further away the observer is located from the action, the less the information (s)he obtains and the greater the doubt about the value of the information.

Effective methods of planning process investigations have been developed. Many of these began in the chemical, electrical and mechanical engineering industries. The principles and practices are, however, universally applicable. Generally two approaches are available, as discussed in the next two sections.

Effects of single factors

The effects of many single variables (e.g. temperature, voltage, time, speed, concentration) may have been shown to have been important in other, similar studies. The procedures of altering one variable at a time is often successful, particularly in well-equipped 'laboratories' and pilot plants. Frequently, however, the factors that are expected to allow predictions about a new process are found to be grossly inadequate. This is especially common when a process is transferred from the laboratory or pilot plant to full-scale operation. Predicted results may be obtained on some occasions but not on others, even though no known changes have been introduced. In these cases the control chart methods previously described are useful to check on process stability.

Group factors

A trouble-shooting project or any process improvement may begin by an examination of the possible differences in output quality of different people, different equipment, different product or other variables. If differences are established within such a group, experience has shown that careful study of the sources of the variation in performance will often provide important causes of those differences. Hence, the key to making adjustments and improvements is in knowing that actual differences do exist, and being able to pinpoint the sources of the differences.

It is often argued that any change in a product, service, process or plant will be evident to the experienced manager. This is not always the case. It is accepted that many important changes are recognized without resort to analytical studies, but the presence, and certainly the identity, of many economically important factors cannot be recognized without them. Processes are invariably managed by people who combine theory, practical experience and ingenuity. An experienced manager will often recognize a recurring malfunctioning process by characteristic symptoms. As problems become more complex, however, many important changes, particularly gradual ones, cannot be recognized by simple observation and intuition no matter how competent a person may be as an engineer, scientist or psychologist. No process is so simple that data from it will not give added insight into its behaviour. Indeed many processes have unrecognized complex behaviour which can be thoroughly understood only by studying large quantities of data on the product produced or service provided. The manager or supervisor who accepts and learns methods of statistically based investigation, linked with data analytics, to support 'technical' knowledge will be an exceptionally able person in his area.

Discussion of any trouble-shooting investigation between the appropriate people is essential at a very early stage. Properly planned procedures will prevent wastage of time, effort and materials and will avoid embarrassment to those involved. It will also ensure support for implementation of the results of the study. (See also Chapter 14.)

12.3 Use of control charts for trouble-shooting

In some studies, the purpose of the data collection is to provide information on the relationships between variables. In other cases, the purpose is just to find ways to eliminate a serious problem – the data themselves, or a formal analysis of them, are of little or no consequence. The application of control charts to data can be developed in a great variety of situations and provides a simple yet powerful method of presenting and studying results. By this means, sources of assignable causes are often indicated by patterns of trends. The use of control charts always leads to systematic programmes of sampling and measurement. The presentation of results in chart form makes the data more easily assimilated and provides a picture of the process. This is not available from a simple tabulation of the results.

The control chart method is, of course, applicable to sequences of attribute data as well as to variables data, and may well suggest causes of unusual performance. Examination of such charts, as they are plotted, may provide evidence of economically important assignable causes of trouble. The chart does not solve the problem, but it indicates when, and possibly where, to look for a solution.

The applications of control charts that we have met in earlier chapters usually began with evidence that the process was in statistical control. Corrective action of some sort was then indicated when an out-of-control signal was obtained. In many trouble-shooting applications, the initial results show that the process is *not* in statistical control and investigations must begin immediately to discover the special assignable causes of variation.

It must be made quite clear that use of control charts alone will not enable the cause of trouble in a process to be identified. A thorough knowledge of the process and how it is operated is also required. When this is combined with an understanding of control chart principles, then the diagnosis of causes of problems will be possible.

This book cannot hope to provide the intimate knowledge of every process that is required to solve problems. Guidance can only be given on the interpretation of control charts for process improvement and trouble-shooting. There are many and various patterns which develop on control charts when processes are not in control. What follows is an attempt to structure the patterns into various categories. The latter are not definitive, nor is the list exhaustive. The taxonomy is based on the ways in which out-of-control situations may arise, and their effects on various control charts.

When variable data plotted on charts fall outside the control limits there is evidence that the process has changed in some way during the sampling period. This change may take three different basic forms:

- a change in the process mean, with no change in spread or standard deviation;
- a change in the process spread (standard deviation) with no change in the mean;
- a change in both the process mean and standard deviation.

These changes affect the control charts in different ways. The manner of change also causes differences in the appearance of control charts. Hence, for a constant process spread, a maintained drift in process mean will show a different pattern to frequent, but irregular changes in the mean. Therefore the list may be further divided into the following types of changes:

1   Change in process mean (no change in standard deviation):

   (a) sustained shift;
   (b) drift or trend – including cyclical;
   (c) frequent, irregular shifts.

2   Change in process standard deviation (no change in mean):

    (a) sustained changes;
    (b) drift or trends – including cyclical;
    (c) frequent irregular changes.

3   Frequent, irregular changes in process mean and standard deviation.

These change types are shown, together with the corresponding mean, range and cusum charts, in Figures 12.1 to 12.7. The examples are taken from a tablet-making process in which trial control charts were being set up for a sample size of $n = 5$. In all cases, the control limits were calculated using the data that is plotted on the mean and range charts.

Sustained shift in process mean (Figure 12.1)

The process varied as shown in (a). After the first five sample plots, the process mean moved by two standard deviations. The mean chart (b) showed the change quite clearly – the next six points being above the upper action line. The change of one standard deviation, which follows, results in all but one point lying above the warning line. Finally, the out-of-control process moves to a lower mean and the mean chart once again responds immediately. Throughout these changes, the range chart (c) gives no indication of lack of control, confirming that the process spread remained unchanged.

    The cusum chart of means (d) confirms the shifts in process mean.

Drift or trend in process mean (Figure 12.2)

When the process varied according to (a), the mean and range charts ((b) and (c), respectively) responded as expected. The range chart shows an in-control situation since the process spread did not vary. The mean chart response to the change in process mean of *ca.* two standard deviations every ten sample plots is clearly and unmistakably that of a drifting process.

    The cusum chart of means (d) is curved, suggesting a trending process, rather than any step changes.

Frequent, irregular shift in process mean (Figure 12.3)

Figure 12.3a shows a process in which the standard deviation remains constant, but the mean is subjected to what appear to be random changes of between one and two standard deviations every few sample plots. The mean chart (b) is very sensitive to these changes, showing an out-of-control situation and following the pattern of change in process mean. Once again the range chart (c) is in control, as expected.

    The cusum chart of means (d) picks up the changes in process mean.

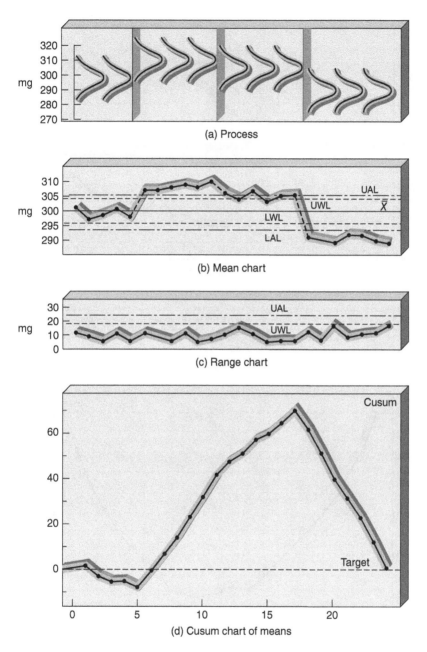

Figure 12.1 Sustained shift in process mean

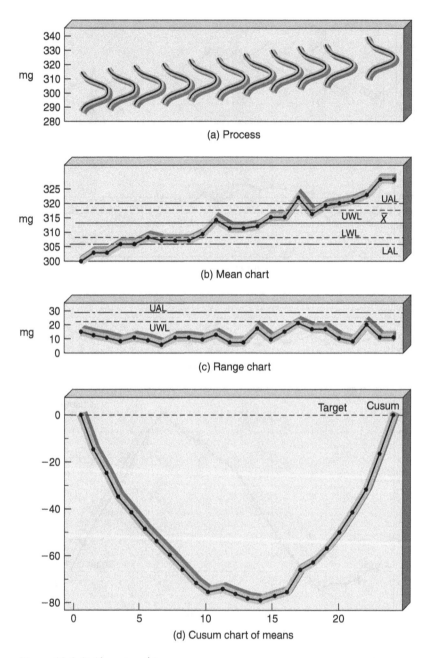

Figure 12.2 Drift or trend in process mean

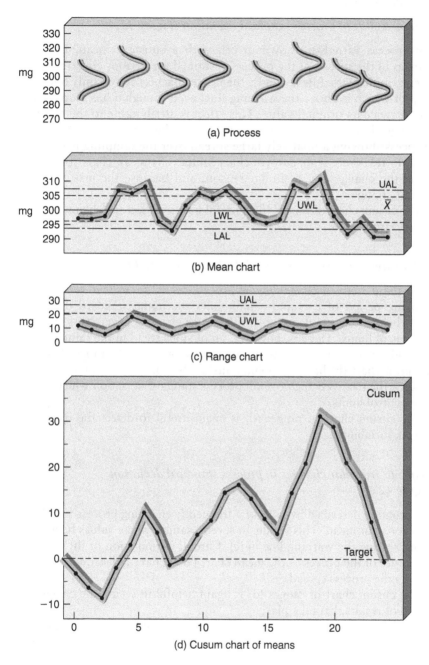

Figure 12.3 Frequent, irregular shift in process mean

Sustained shift in process standard deviation (Figure 12.4)

The process varied as shown in (a), with a constant mean, but with changes in the spread of the process sustained for periods covering six or seven sample plots. Interestingly, the range chart (c) shows only one sample plot which is above the warning line, even though σ has increased to almost twice its original value. This effect is attributable to the fact that the range chart control limits are based upon the data themselves. Hence a process showing a relatively large spread over the sampling period will result in relatively wide control chart limits. The mean chart (b) fails to detect the changes for a similar reason, and because the process mean did not change.

The cusum chart of ranges (d) is useful here to detect the changes in process variation.

Drift or trend in process standard deviation (Figure 12.5)

In (a) the pattern of change in the process results in an increase over the sampling period of two and a half times the initial standard deviation. Nevertheless, the sample points on the range chart (c) never cross either of the control limits. There is, however, an obvious trend in the sample range plot and this would suggest an out-of-control process. The range chart and the mean chart (b) have no points outside the control limits for the same reason – the relatively high overall process standard deviation which causes wide control limits.

The cusum chart of ranges (d) is again useful to detect the increasing process variability.

Frequent, irregular changes in process standard deviation (Figure 12.6)

The situation described by (a) is of a frequently changing process variability with constant mean. This results in several sample range values being near to or crossing the warning line in (c). Careful examination of (b) indicates the nature of the process – the mean chart points have a distribution which mirrors the process spread.

The cusum chart of ranges (d) is again helpful in seeing the changes in spread of results that take place.

The last three examples, in which the process standard deviation alone is changing, demonstrate the need for extremely careful examination of control charts before one may be satisfied that a process is in a state of statistical control. Indications of trends and/or points near the control limits on the range chart may be the result of quite serious changes in variability, even though the control limits are never transgressed.

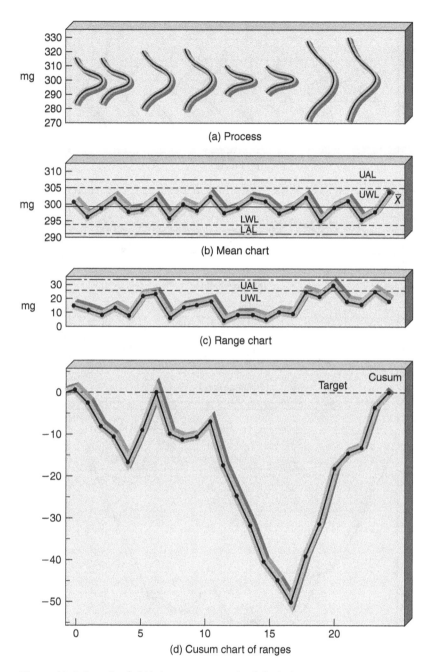

Figure 12.4 Sustained shift in process standard deviation

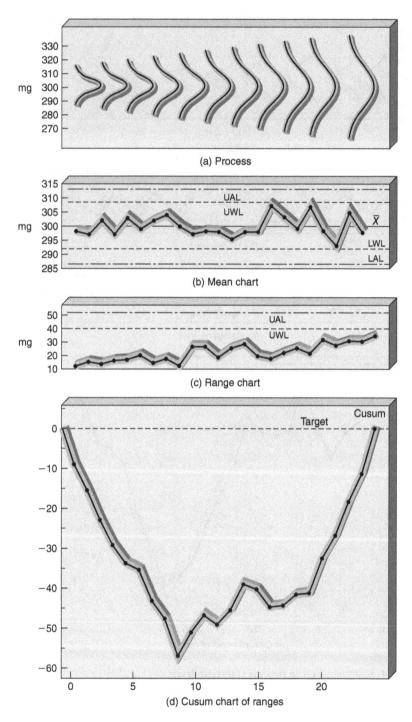

Figure 12.5 Drift or trend in process standard deviation

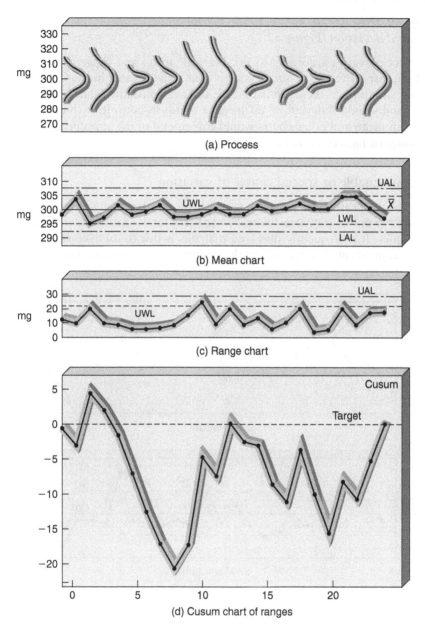

Figure 12.6 Frequent, irregular changes in process standard deviation

Frequent, irregular changes in process mean and standard deviation (Figure 12.7)

The process varies according to (a). Both the mean and range charts ((b) and (c), respectively) are out of control and provide clear indications of a serious situation. In theory, it is possible to have a sustained shift in process mean and standard deviation, or drifts or trends in both. In such cases the resultant mean and range charts would correspond to the appropriate combinations of Figures 12.1, 12.2, 12.4 or 12.5.

12.4 Assignable or special causes of variation

It is worth repeating the point made in Chapter 5, that many processes are found to be out-of-statistical control when first examined using control chart techniques. It is frequently observed that this is due to an excessive number of adjustments being made to the process, based on individual results.

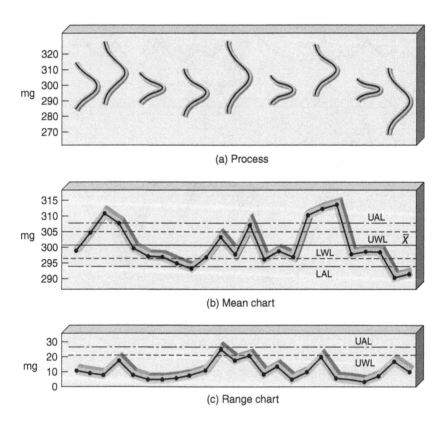

Figure 12.7 Frequent, irregular changes in process mean and standard deviation

This behaviour, commonly known as hunting, causes an overall increase in variability of results from the process, as shown in Figure 12.8.

If the process is initially set at the target value μ_a and an adjustment is made on the basis of a single result A, then the mean of the process will be adjusted to μ_b. Subsequently, a single result at B will result in a second adjustment of the process mean to μ_c. If this behaviour continues, the variability or spread of results from the process will be greatly increased with a detrimental effect on the ability of the process to meet the specified requirements.

Variation cannot be ignored. The simple fact that a measurement, test or analytical method is used to generate data introduces variability. This must be taken into account and the appropriate charts of data used to control processes, instead of reacting to individual results. It is often found that range charts are in control and indicate an inherently capable process. The saw-tooth appearance of the mean chart, however, shows the rapid alteration in the mean of the process. Hence the patterns appear as in Figure 12.3.

When a process is found to be out of control, the first action must be to investigate the assignable or special causes of variation. This may require,

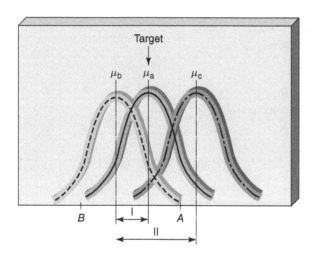

I = First adjustment based on distance of test result A
from target value $(A - \mu_a)$
II = Second adjustment based on distance of test result B
from target value $\mu_a(\mu_a - B)$

Figure 12.8 Increase in process variability due to frequent adjustments based on individual test results

in some cases, the charting of process parameters rather than the product parameters that appear in the specification. For example, it may be that the viscosity of a chemical product is directly affected by the pressure in the reactor vessel, which in turn may be directly affected by reactor temperature. A control chart for pressure, with recorded changes in temperature, may be the first step in breaking into the complexity of the relationship involved. The important point is to ensure that all adjustments to the process are recorded and the relevant data charted.

There can be no compromise on processes that are shown to be not in control. The simple device of changing the charting method and/or the control limits will not bring the process into control. A proper process investigation must take place.

It has been pointed out that there are numerous potential special causes for processes being out-of-control. It is extremely difficult, even dangerous, to try to find an association between types of causes and patterns shown on control charts. There are clearly many causes which could give rise to different patterns in different industries and conditions. It may be useful, however, to list some of the most frequently met types of special causes:

People

- fatigue or illness;
- lack of training/novices;
- unsupervised;
- unaware;
- attitudes/motivation;
- changes/improvements in skill;
- rotation of shifts.

Plant/equipment

- rotation of machines;
- differences in test or measuring devices;
- scheduled preventative maintenance;
- lack of maintenance;
- badly designed equipment;
- worn equipment;
- gradual deterioration of plant/equipment.

Processes/procedures

- unsuitable techniques of operation or test;
- untried/new processes;
- changes in methods, inspection or check.

Materials

- merging or mixing of batches, parts, components, subassemblies, inter-mediates, etc.;
- accumulation of waste products;
- homogeneity;
- changes in supplier/material.

Environment

- gradual deterioration in conditions;
- temperature changes;
- humidity;
- noise;
- dusty atmospheres.

It should be clear from this non-exhaustive list of possible causes of variation that an intimate knowledge of the process is essential for effective process improvement. The control chart, when used carefully, informs us when to look for trouble. This contributes typically 10–20 per cent of the problem. The bulk of the work in making improvements is associated with finding where to look and which causes are operating.

The low yields experienced in some manufacturing industries, e.g. aerospace, are a function of the difficulties in handling large amounts of complex data to make 'accept/reject' decisions on the quality of products such as engine turbine blades and brackets for wing slats and flaps. Traditional SPC is difficult to apply to these situations owing to the number, complexity and possible interactions of the variables present or possible. The huge amount of data generated – often called 'big data' – requires new approaches to handle it. Through a combination of big data computer analytics and SPC, it should be possible now to identify and 'isolate' key variables that contribute mostly to key accept/reject decisions on products. The use of these analytics tools combined with SPC on the key variables should lead to actions to improve control, capability, quality and yield.

For big data applications, machine learning algorithms can be used to develop insights and understand the root causes of underlying problems. Clearly, there are opportunities for using such data in detecting process changes prior to the production of non-conforming products; use of SPC tools in this way enables data-rich sources to achieve their full potential. Multi-stream data presents both opportunities and challenges in terms of minimizing false reject rates, while maintaining the ability to quickly detect problems. The machine learning – big data handling problem differs from traditional SPC applications, however. Noise reduction, data pre-processing and outlier detection are often necessary before any SPC analysis can be done; and the analysis required is more exhaustive than in standard SPC applications.

Chapter highlights

- The responsibility for trouble-shooting and process improvement should not rest with only one group or department, but the shared ownership of the process.

- Problems in process operation are rarely due to single causes, but a combination of factors involving the product (or service), plant, programmes and people.

- The emphasis in any problem-solving effort should be towards prevention, especially with regard to maintaining quality of current output, process improvement and product/service development.

- When faced with a multiplicity of potential causes of problems it is beneficial to begin with studies which identify blocks or groups, such as a whole area of production or service operation, postponing the pinpointing of specific causes and effects until proper data have been collected.

- The planning and direct observation of data collection should help in the identification of assignable causes.

- Generally, two approaches to process investigations are in use; studying the effects of single factors (one variable) or group factors (more than one variable). Discussion with the people involved at an early stage is essential.

- The application of control charts to data provides a simple, widely applicable, powerful method to aid trouble-shooting, and the search for assignable or special causes.

- There are many and various patterns that develop on control charts when processes are not in control. One taxonomy is based on three basic changes: a change in process mean with no change in standard deviation; a change in process standard deviation with no change in mean; a change in both mean and standard deviation.

- The manner of changes, in both mean and standard deviation, may also be differentiated: sustained shift, drift, trend or cyclical, frequent irregular.

- The appearance of control charts for mean and range, and cusum charts should help to identify the different categories of out-of-control processes.

- Many processes are out of control when first examined and this is often due to an excessive number of adjustments to the process, based on individual results, which causes hunting. Special causes such as this must be found through proper investigation.

- The most frequently met causes of out-of-control situations may be categorized under: people, plant/equipment, processes/procedures, materials and environment.

- Machine learning – big data handling problems differ from traditional SPC applications. Noise reduction, data pre-processing and outlier detection are often necessary before any SPC analysis can be done; and the analysis required is more exhaustive than in standard SPC applications. Big data analytical tools are now available which can make these breakthroughs a reality.

References and further reading

Ott, E.R., Schilling, E.G. and Neubauer, D.V. (2005) *Process Quality Control: Troubleshooting and Interpretation of Data*, 4th edn, ASQ Press, Milwaukee WI, USA.

Wheeler, D.J. (1986) *The Japanese Control Chart*, SPC Press, Knoxville TN, USA.

Wheeler, D.J. and Chambers, D.S. (1992) *Understanding Statistical Process Control*, 2nd edn, SPC Press, Knoxville TN, USA.

Discussion questions

1   You are the Operations Manager in a medium-sized manufacturing company which is experiencing quality problems. The Managing Director has asked to see you and you have heard that he is not a happy man; you expect a difficult meeting. Write notes in preparation for your meeting to cover: which people you should see, what information you should collect and how you should present it at the meeting.

2   Explain how you would develop a process improvement study paying particular attention to the planning of data collection.

3   Discuss the 'effects of single factors' and 'group factors' in planning process investigations.

4   Describe, with the aid of sketch diagrams, the patterns you would expect to see on control charts for mean for processes that exhibit the following types of out of control:

    (a) sustained shift in process mean;
    (b) drift/trend in process mean;
    (c) frequent, irregular shift in process mean.

    Assume no change in the process spread or standard deviation.

5   Sketch the cusum charts for mean that you would expect to plot from the process changes listed in question 4.

6   (a) Explain the term 'hunting' and show how this arises when processes are adjusted on the basis of individual results or data points.
    (b) What are the most frequently found assignable or special causes of process change?

13 Designing the statistical process control system

Objectives

- To examine the links between statistical process control and the quality management system, including procedures for out-of-control processes.
- To look at the role of teamwork in process control and improvement.
- To explore the detail of the never-ending improvement cycle.
- To examine Taguchi methods for cost reduction and quality improvement.

13.1 SPC and the quality management system

For successful statistical process control (SPC) there must be an uncompromising commitment to quality, which must start with the most senior management and flow down through the organization. It is essential to set down a *quality policy* for implementation through a *documented quality management system*. Careful consideration must be given to this system as it forms the backbone of the quality skeleton. The objective of the system is to cause improvement of products and services through reduction of variation in the processes. The focus of the whole workforce from top to bottom should be on the processes. This approach makes it possible to control variation and, more importantly, to prevent non-conforming products and services, while steadily improving standards.

The quality management system should apply to and interact with all activities of the organization. This begins with the identification of the customer requirements and ends with their satisfaction, at every transaction interface, both internally and externally. The activities involved may be classified in several ways – generally as processing, communicating and controlling, but more usefully and specifically as:

(i) marketing;
(ii) market research;
(iii) design;
(iv) specifying;
(v) development;

  (vi) procurement/supply chain;
 (vii) process planning;
(viii) process development and assessment;
  (ix) process operation and control;
   (x) product or service testing or checking;
  (xi) packaging (if required);
 (xii) storage (if required);
(xiii) sales;
 (xiv) distribution/logistics;
  (xv) installation/operation;
 (xvi) technical service;
(xvii) maintenance.

The impact of a good management system, such as one that meets the requirements of the international standard ISO 9000 series, is that of gradually reducing process variability to achieve continuous or never-ending improvement. The requirement to set down defined processes for all aspects of an organization's operations, and to stick to them, will reduce the variations introduced by the numerous different ways often employed for doing things. Go into any factory without a good management system and ask to see the operators' 'black-book' of plant operation and settings. Of course, each shift has a different black-book, each with slightly different settings and ways of operating the process. Is it any different in office work or for salespeople in the field? Do not be fooled by the perceived simplicity of a process into believing that there is only one way of operating it. There are an infinite variety of ways of carrying out the simplest of tasks – one author recalls seeing various course participants finding 14 different methods for converting A4 size paper into A5 (half A4) in a simulation of a production task. The ingenuity of human beings needs to be controlled if these causes of variation are not to multiply together to render processes completely incapable of consistency or repeatability.

The role of the management system then is to define and control processes and methods. Continual system audit and review will ensure that processes are either followed or corrected, thus eliminating assignable or special causes of variation in materials, methods, equipment, information, etc., to ensure a 'could we do this job with more consistency?' approach (Figure 13.1).

The task of measuring, inspecting or checking is taken by many to be the passive one of sorting out the good from the bad, when it should be an active part of the feedback system to prevent errors, defects or non-conformance. Clearly any control system based on detection of poor quality by post-production/operation inspection or checking is unreliable, costly, wasteful and uneconomical. It should be replaced eventually by the strategy of prevention, and the inspection then used to check the *system* of transformation, rather than the product. Inputs, outputs and processes need to be measured for effective quality management. The measurements monitor quality and

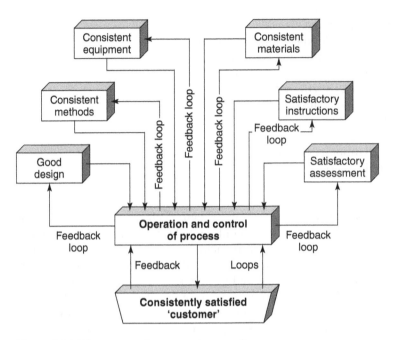

Figure 13.1 The systematic approach to quality management

may be used to determine the extent of improvements and deterioration. Measurement may take the form of simple counting to produce attribute data, or it may involve more sophisticated methods to generate variable data. Processes operated without measurement and feedback are processes about which very little can be known. Conversely, if inputs and outputs can be measured and expressed in numbers, then something is known about the process and control is possible. The first stage in using measurement as part of the process control system is to identify precisely the activities, materials, equipment, etc. that will be measured. This enables everyone concerned with the process to be able to relate to the target values and the focus provided will encourage improvements.

For measurements to be used for quality improvement, they must be accepted by the people involved with the process being measured. The simple self-measurement and plotting, or the 'how-am-I-doing' chart, will gain far more ground in this respect than a policing type of observation and reporting system that is imposed on the process and those who operate it. Similarly, results should not be used to illustrate how bad one operator or group is, unless their performance is entirely under their own control. The emphasis in measuring and displaying data must always be on the assistance that can be given to correct a problem or remove obstacles preventing the process from meeting its requirements first time, every time.

Out-of-control procedures

The rules for interpretation of control charts should be agreed and defined as part of the SPC system design. These largely concern the routes to be followed when an out-of-control (OoC) situation develops. It is important that each process 'operator' responds in the same way to an OoC indication, and it is necessary to get their inputs and those of the supervisory management at the design stage.

Clearly, it may not always be possible to define which corrective actions should be taken, but the intermediate stage of identifying what happened should follow a systematic approach. Recording of information, including any significant 'events', the possible courses of OoC, analysis of causes, and any action taken is a vital part of any SPC system design.

In some processes, the actions needed to remove or prevent causes of OoC are outside the capability or authority of the process 'operators'. In these cases, there must be a mechanism for progressing the preventive actions to be carried out by supervisory management, and their integration into routine procedures.

When improvement actions have been taken on the process, measurements should be used to confirm the desired improvements and checks made to identify any side effects of the actions, whether they be beneficial or detrimental. It may be necessary to recalculate control chart limits when sufficient data are available, following the changes.

Computerized SPC

There are now available many SPC computer software packages that enable the recording, analysis and presentation of data as charts, graphs and summary statistics. Most of the good ones on the market, such as Minitab, will readily produce anything from a Pareto diagram to a cusum chart, and calculate skewness, kurtosis and capability indices. They will draw histograms, normal distributions and plots, scatter diagrams and every type of control chart with decision rules included. In using these powerful aids it is, of course, essential that the principles behind the techniques displayed are thoroughly understood.

13.2 Teamwork and process control/improvement

Teamwork will play a major role in any organization's efforts to make never-ending improvements. The need for teamwork can be seen in many human activities. In most organizations, problems and opportunities for improvement exist between departments. Seldom does a single department own all the means to solve a problem or bring about improvement alone.

Sub-optimization of a process seldom improves the total system performance. Most systems are complex, and input from all the relevant processes is required when changes or improvements are to be made. Teamwork throughout the organization is an essential part of the implementation of SPC. It is necessary in most organizations to move from a state of independence to one of interdependence, through the following stages:

INDEPENDENCE

Little sharing of ideas and information

Exchange of basic information

Exchange of basic ideas

Exchange of feelings and data   TIME

Elimination of fear

Trust

Open communication

INTERDEPENDENCE

The communication becomes more open with each progressive step in a successful relationship. The point at which it increases dramatically is when trust is established. After this point, the barriers that have existed are gone and open communication will proceed. This is critical for never-ending improvement and problem solving, for it allows people to supply good data and all the facts without fear.

Teamwork brings diverse talents, experience, knowledge and skills to any process situation. This allows a variety of problems that are beyond the technical competence of any one individual to be tackled. Teams can deal with problems that cross departmental and divisional boundaries. All of this is more satisfying and morale boosting for people than working alone.

A team will function effectively only if the results of its meetings are communicated and used. Someone should be responsible for taking minutes of meetings. These need not be formal, and simply reflect decisions and action assignments – they may be copied and delivered to the team members on the way out of the door. More formal sets of minutes might be drawn up after the meetings and sent to sponsors, administrators, supervisors or others who need to know what happened. The purpose of minutes is to inform people of decisions made and list actions to be taken. Minutes are an important part of the communication chain with other people or teams involved in the whole process.

Process improvement and 'Kaisen' teams

A process improvement team is a group of people with the appropriate knowledge, skills and experience who are brought together specifically by management to tackle and solve a particular problem, usually on a project basis: they are cross-functional and often multi-disciplinary.

The 'task force' has long been a part of the culture of many organizations at the technological and managerial levels, but process improvement teams go a step further, they expand the traditional definition of 'process' to include the entire production or operating system. This includes paperwork, communication with other units, operating procedures and the process equipment itself. By taking this broader view all process problems can be addressed.

The management of process improvement teams is outside the scope of this book and is dealt with in *Total Quality Management and Operational Excellence*, 4th edition (Oakland, 2014). It is important, however, to stress here the role that SPC techniques themselves can play in the formation and work of teams. For example, the management in one company, which was experiencing a 17 per cent error rate in its invoice generating process, decided to try to draw a flowchart of the process. Two people who were credited with knowledge of the process were charged with the task. They soon found that it was impossible to complete the flowchart, because they did not fully understand the process. Progressively five other people, who were involved in the invoicing, had to be brought to the table in order that the map could be finished to give a complete description of the process. This assembled group were kept together as the process improvement team, since they were the only people who collectively could make improvements. Simple data collection methods, brainstorming, cause and effect and Pareto analysis were then used, together with further process mapping techniques, to reduce the error rate to less than 1 per cent within just 6 months.

The flexibility of the cause and effect (C&E) diagram makes it a standard tool for problem-solving efforts throughout an organization. This simple tool can be applied in manufacturing, service or administrative areas of a company and can be applied to a wide variety of problems from simple to very complex situations.

Again the knowledge gained from the C&E diagram often comes from the *method of construction* not just the completed diagram. A very effective way to develop the C&E diagram is with the use of a team, representative of the various areas of expertise on the effect and processes being studied. The C&E diagram then acts as a collection point for the current knowledge of possible causes, from several areas of experience.

Brainstorming in a team is the most effective method of building the C&E diagram. This activity contributes greatly to the understanding, by all those involved, of a problem situation. The diagram becomes a focal point for the entire team and will help any team develop a course for corrective action.

Process improvement teams usually find their way into an organization as *problem-solving* groups. This is the first stage in the creation of *problem prevention* teams, which operate as common work groups and whose main objective is constant improvement of processes. Such groups may be part of a multi-skilled, flexible workforce, and include 'inspect and repair' tasks as part of the overall process. The so-called 'Kaisen' team operates in this way to eliminate problems at the source by working together and, using very basic tools of SPC where appropriate, to create less and less opportunity for problems and reduce variability. Kaisen teams are usually provided with a 'help line' that, when 'pulled', attracts help from human, technical and material resources from outside the group. These are provided specifically for the purpose of eliminating problems and aiding process control.

13.3 Improvements in the process

To improve a process, it is important first to recognize whether the process control is limited by the common or the special causes of variation. This will determine *who* is responsible for the specific improvement steps, *what resources* are required, and *which statistical tools* will be useful. Figure 13.2, which is a development of the strategy for process improvement presented in Chapter 11, may be useful here. The comparison of actual product quality characteristics with the requirements (inspection) is not a basis for action on the process, since unacceptable products or services can result from either common or special causes. Product or service inspection is useful to sort out good from bad and to perhaps set priorities on which processes to improve.

Any process left to natural forces will suffer from deterioration, wear and breakdown (the second law of thermodynamics: entropy is always increasing!). Therefore, management must help people identify and prevent these natural causes through ongoing improvement of the processes they manage. The organization's culture should encourage communications throughout and promote a participative style of management that allows people to report problems and suggestions for improvement without fear or intimidation, or enquiries aimed at apportioning blame. These should then be addressed with statistical thinking by all members of the organization.

Activities to improve processes may include the assignment of various people in the organization to work on common and special causes. The appropriate people to identify special causes are usually different to those needed to identify common causes. The same is true of those needed to remove causes. Removal of common causes is the responsibility of management, often with the aid of experts in the process such as engineers, chemists and systems analysts. Special causes can frequently be handled at a local level by those working in the process such as supervisors and operators. Without some knowledge of the likely origins of common and special causes it is difficult to efficiently allocate human resources to improve processes.

Most improvements require action by management and, in almost all cases, the removal of special causes will make a fundamental change in the way processes are operated. For example, a special cause of variation in a production process may result when there is a change from one supplier's

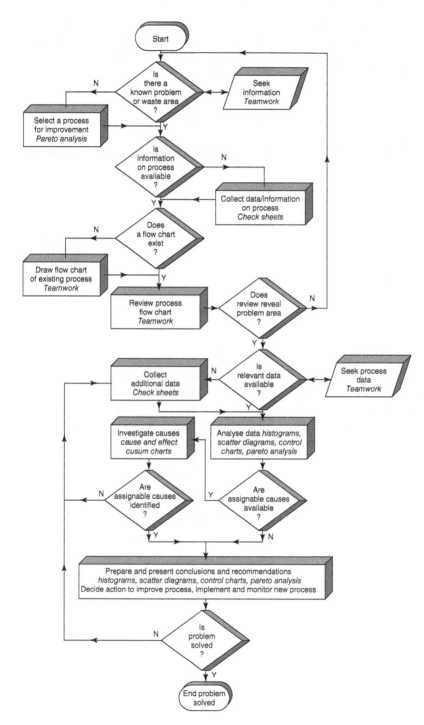

Figure 13.2 The systematic approach to improvement

material to another. To prevent this special cause from occurring in the particular production processes, a change in the way the organization chooses and works with suppliers may be needed. Improvements in conformance are often limited to a policy of single sourcing.

Another area in which the knowledge of common and special causes of variation is vital is in the supervision of people. A mistake often made is the assignment of variation in the process to those working on the process, e.g. operators and staff, rather than to those in charge of the process, i.e. management. Clearly, it is important for a supervisor to know whether problems, mistakes or rejected material are a result of common causes, special causes related to the system, or special causes related to the people under his or her supervision. Again the use of the systematic approach and the appropriate techniques will help the supervisor to accomplish this.

Management need to demonstrate commitment to this approach by providing leadership and the necessary resources. These resources will include training on the job, time to effect the improvements, improvement techniques and a commitment to institute changes for ongoing improvement. This will move the organization from having a reactive management system to having one of prevention. This all requires time and effort by everyone, every day.

Process control charts and improvements

The emphasis placed on never-ending improvement has important implications for the way in which process control charts are applied. They should not be used purely for control, but as an aid in the reduction of variability by those at the point of operation capable of observing and removing special causes of variation. They can be used effectively in the identification and gradual elimination of common causes of variation. In this way the process of continuous improvement may be charted, and adjustments made to the control charts in use to reflect the improvements.

This is shown in Figure 13.3 where progressive reductions in the variability of ash content in a weed killer has led to decreasing sample ranges. If the control limits on the mean and range charts are recalculated periodically or after a step change, their positions will indicate the improvements that have been made over a period of time, and *ensure* that the new level of process capability is maintained. Further improvements can then take place (Figure 13.4). Similarly, attribute or cusum charts may be used, to show a decreasing level of number of errors, or pro-portion of defects and to indicate improvements in capability.

Often in process control situations, action signals are given when the special cause results in a desirable event, such as the reduction of an impurity level, a decrease in error rate or an increase in order intake. Clearly, special causes that result in deterioration of the process must be investigated and eliminated, but those that result in improvements should also be sought out and managed so that they become part of the process operation. Significant

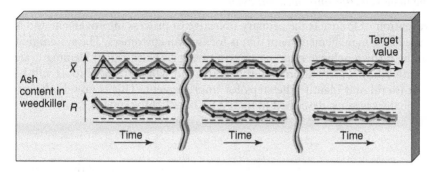

Figure 13.3 Continuous process improvement: reduction in variability

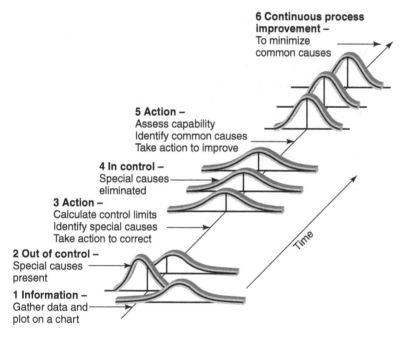

**6 Continuous process
improvement –**
To minimize
common causes

5 Action –
Assess capability
Identify common causes
Take action to improve

4 In control –
Special causes
eliminated

3 Action –
Calculate control limits
Identify special causes
Take action to correct

2 Out of control –
Special causes
present

1 Information –
Gather data and
plot on a chart

Time

Figure 13.4 Process improvement stages

variation between batches of material, operators or differences between suppliers are frequent causes of action signals on control charts. The continuous improvement philosophy requires that these are all investigated and the results used to take another step on the long ladder to perfection. Action signals and special causes of variation should stimulate enthusiasm for solving a problem or understanding an improvement, rather than gloom and despondency.

The never-ending improvement cycle

Prevention of failure is the primary objective of process improvement and is caused by a management team that is focused on customers. The system that will help them achieve ongoing improvement is the so-called Deming cycle (Figure 13.5). This will provide the strategy in which the SPC tools will be most useful and identify the steps for improvement. This is now built into the quality systems described in the ISO 9001: 2015 standard.

Plan

The first phase of the system – plan – helps to focus the effort of the improvement team on SIPOC (Suppliers-Inputs-Process-Outputs-Customers). The following questions should be addressed by the team:

- What are the requirements of the output from the process?
- Who are the customers of the output? Both internal and external customers should be included.
- What are the requirements of the inputs to the process?
- Who are the suppliers of the inputs?
- What are the objectives of the improvement effort? These may include one or all of the following:

    o improve customer satisfaction;
    o eliminate internal difficulties;

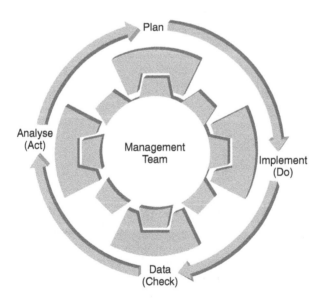

Figure 13.5 The Deming cycle

o   eliminate unnecessary work ('lean' approaches);
o   eliminate failure costs;
o   eliminate non-conforming output.

Every process has many opportunities for improvement, and resources should be directed to ensure that all efforts will have a positive impact on the objectives. When the objectives of the improvement effort are established, output identified and the customers noted, then the team is ready for the implementation stage.

Implement (Do)

The implementation effort will have the purpose of:

*   defining the processes that will be improved;
*   identifying and selecting opportunities for improvement.

The improvement team should accomplish the following steps during implementation:

*   Define the scope of the SIPOC system to be improved and map or flowchart the processes within this system.
*   Identify the key sub-processes that will contribute to the objectives identified in the planning stage.
*   Identify the customer–supplier relationships throughout the key processes.

These steps can be completed by the improvement team through their present knowledge of the SIPOC system. This knowledge will be advanced throughout the improvement effort and, with each cycle, the maps/flowcharts and C&E diagrams should be updated. The following stages will help the team make improvements on the selected process:

*   Identify and select the process in the system that will offer the greatest opportunities for improvement. The team may find that a completed process flowchart will facilitate and communicate understanding of the selected process to all team members.
*   Document the steps and actions that are necessary to make improvements. It is often useful to consider what the flowchart would look like if every job was done right the first time, sometimes called 'imagineering'.
*   Define the C&E relationships in the process using a C&E diagram.
*   Identify the important sources of data concerning the process. The team should develop a data collection plan.

- Identify the measurements that will be used for the various parts of the process.
- Identify the largest contributors to variation in the process. The team should use their collective experience and brainstorm the possible causes of variation.

During the next phase of the improvement effort, the team will apply the knowledge and understanding gained from these efforts and gain additional knowledge about the process.

Data (Check)

The data collection phase has the following objectives:

- To collect data from the process as determined in the planning and implementation phases. 'Big data' analytics may be appropriate here.
- Determine the stability of the process using the appropriate control chart method(s).
- If the process is stable, determine the capability of the process.
- Prove or disprove any theories established in the earlier phases.
- If the team observed any unplanned events during data collection, determine the impact these will have on the improvement effort.
- Update the maps/flowcharts and C&E diagrams, so the data collection adds to current knowledge.

Analyse (Act)

The purpose of this phase is to analyse the findings of the prior phases and help plan for the next effort of improvement. During this phase of improvement, the following should be accomplished:

- Determine the action on the process that will be required. This will identify the inputs or combinations of inputs that will need to be improved. These should be noted on an updated map of the process.
- Develop greater understanding of the causes and effects.
- Ensure that the agreed changes have the anticipated impact on the specified objectives.
- Identify the departments and organizations that will be involved in analysis, implementation and management of the recommended changes.
- Determine the objectives for the next round of improvement. Problems and opportunities discovered in this stage should be considered as objectives for future efforts. Pareto charts should be consulted from the earlier work and revised to assist in this process. Business process redesign (BPR) may be required to achieve step changes in performance.

Plan, do, check, act (PDCA), as the cycle is often called, will lead to improvements if it is taken seriously by the team. Gaps can occur, however, in moving from one phase to another unless good facilitation is provided. The team leader plays a vital role here. One of his/her key roles is to ensure that PDCA does not become 'please don't change anything!'

13.4 Taguchi methods

Genichi Taguchi defined a number of methods to simultaneously reduce costs and improve quality. The popularity of his approach is a fitting testimony to the merits of this work. The Taguchi methods may be considered under four main headings:

- total loss function;
- design of products, processes and production;
- reduction in variation;
- statistically planned experiments.

Total loss function

The essence of Taguchi's definition of total loss function is that the smaller the loss generated by a product or service from the time it is transferred to the customer, the more desirable it is. Any variation about a target value for a product or service will result in some loss to the customer and such losses should be minimized. It is clearly reasonable to spend on quality improvements provided that they result in larger savings for either the producer or the customer. Earlier chapters have illustrated ways in which non-conforming products, when assessed and controlled by variables, can be reduced to events that will occur at very low probabilities – such reductions will have a large potential impact on the customer's losses.

Taguchi's loss function is developed by using a statistical method that need not concern us here – but the concept of loss by the customer as a measure of quality performance is clearly a useful one. Figure 13.6 shows that, if set correctly, a specification should be centred at the position at which the customer would like to receive all the product. This implies that the centre of the specification is where the customer's process works best. Product just above and just below one of the limits is to all intents and purposes the same, it does not perform significantly differently in the customer's process and the losses are unlikely to have the profile shown in (a). The cost of non-conformance is more likely to increase continuously as the actual variable produced moves away from the centre – as in (b).

For any product or service we may identify three stages of design – the product (or service) design, the process (or method) design and the production (or operation) design. Each of these overlapping stages has many steps, the outputs of which are often the inputs to other steps. For all the steps, the

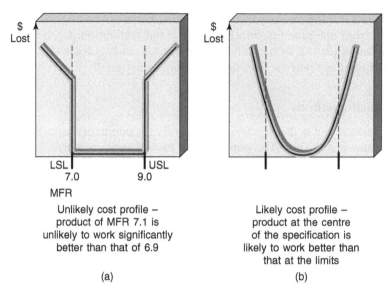

Unlikely cost profile –
product of MFR 7.1 is
unlikely to work significantly
better than that of 6.9

(a)

Likely cost profile –
product at the centre
of the specification is
likely to work better than
that at the limits

(b)

Figure 13.6 Incremental cost ($) of non-conformance

matching of the outputs to the requirements of the inputs of the next step clearly affects the quality and cost of the resultant final product or service. Taguchi's clear classification of these three stages may be used to direct management's effort not only to the three stages but also the separate steps and their various interfaces. Following this model, management is moved to select for study 'narrowed down' subjects, to achieve 'focused' activity, to increase the depth of understanding, and to greatly improve the probability of success towards higher quality levels. This is the essence of the more recently developed approach of 'Advanced Product Quality Planning (APQP'). The authors and their colleagues have extended this to the Services sector with ASQP.

Design or APQP must include consideration of the potential problems that will arise as a consequence of the operating and environmental conditions under which the product or service will be both produced and used. Equally, the costs incurred during production will be determined by the actual manufacturing process. Controls, including SPC techniques, will always cost money but the amount expended can be reduced by careful consideration of control during the initial design of the process. In these, and many other ways, there is a large interplay between the three stages of development.

In this context, Taguchi distinguishes between 'on-line' and 'off-line' quality management. On-line methods are technical aids used for the control of a process or the control of quality during the production of products and services – broadly the subject of this book. Off-line methods use technical aids in the design of products and processes, such as APQP. Too often the off-line methods are based on the evaluation of products and processes rather than their improvement. Effort is directed towards assessing reliability rather than to reviewing the design of both product and process with a view to removing potential imperfections by design. Off-line methods are best directed towards improving the capability of design. A variety of techniques are possible in these APQP activities and include structured teamwork, the use of formal quality/management systems, the auditing of control procedures, the review of control procedures and failure mode and effect analysis (FMEA) applied on a company-wide basis.

Reduction in variation

Reducing the variation of key processes, and hence product parameters about their target values, is the primary objective of a quality improvement programme. The widespread practice of stating specifications in terms of simple upper and lower limits conveys the idea that the customer is equally satisfied with all the values within the specification limits and is suddenly not satisfied when a value slips outside the specification band. The practice of stating a tolerance band may lead to manufacturers aiming to produce and dispatch products whose parameters are just inside the specification band. In any operation, whether mechanical, electrical, chemical, processed food, processed data – as in banking, civil construction, etc. – there will be a multiplicity of activities and hence a multiplicity of sources of variation that all combine to give the total variation.

For variables, the mid-specification or some other target value should be stated along with a specified variability about this value. For those performance characteristics that cannot be measured on a continuous scale it is better to employ a scale such as: excellent, very good, good, fair, unsatisfactory, very poor; rather than a simple pass or fail, good or bad.

Taguchi introduces a three-step approach to assigning nominal values and tolerances for product and process parameters, as defined in the next three sub-sections.

Design system

The application of scientific, engineering and technical knowledge to produce a basic functional prototype design requires a fundamental understanding of both the need of customers and the production possibilities. Trade-offs are not being sought at this stage, but there are requirements for a clear definition of the customer's real needs, possibly classified as critical, important and desirable, and an equally clear definition of the supplier's known capabilities to respond to these needs, possibly distinguishing between the use of existing technology and the development of new techniques.

Parameter design

This entails a study of the whole process system design aimed at achieving the most robust operational settings – those that will lead to least variations of inputs.

Process developments tend to move through cycles. The most revolutionary developments tend to start life as either totally unexpected results (fortunately observed and understood) or success in achieving expected results, but often only after considerable, and sometimes frustrating, effort. Development moves on through further cycles of attempting to increase the reproducibility of the processes and outputs, and includes the optimization of the process conditions to those that are most robust to variations in all the inputs. An ideal process would accommodate wide variations in the inputs with relatively small impacts on the variations in the outputs. Some processes and the environments in which they are carried out are less prone to multiple variations than others. Types of cereal and domestic animals have been bred to produce cross-breeds that can tolerate wide variations in climate, handling, soil, feeding, etc. Machines have been designed to allow for a wide range of the physical dimensions of the operators (motor cars, for example). Industrial techniques for the processing of food will accommodate wide variations in the raw materials with the least influence on the taste of the final product. The textile industry constantly handles, at one end, the wide variations that exist among natural and man-made fibres and, at the other end, garment designs that allow a limited range of sizes to be acceptable to the highly variable geometry of the human form. Specifying the conditions under which such robustness can be achieved is the object of parameter design.

Tolerance design

A knowledge of the nominal settings advanced by parameter design enables tolerance design to begin. This requires a trade-off between the costs of production or operation and the losses acceptable to the customer arising from performance variation. It is at this stage that the tolerance design of cars or clothes ceases to allow for all versions of the human form, and that either blandness or artificial flavours may begin to dominate the taste of processed food.

These three steps pass from the original concept of the potential for a process or product, through the development of the most robust conditions of operation, to the compromise involved when setting 'commercial' tolerances – and focus on the need to consider actual or potential variations at all stages. When considering variations within an existing process it is clearly beneficial to similarly examine their contributions from the three points of view.

Statistically planned experiments

Experimentation is necessary under various circumstances and in particular in order to establish the optimum conditions that give the most robust process – to assess the parameter design. 'Accuracy' and 'precision', as defined in Chapter 5, may now be regarded as 'normal settings' (target or optimum values of the various parameters of both processes and products) and 'noise' (both the random variation and the 'room' for adjustment around the nominal setting). If there is a problem it will not normally be an unachievable nominal setting but unacceptable noise. Noise is recognized as the combination of the random variations and the ability to detect and adjust for drifts of the nominal setting. Experimentation should, therefore, be directed towards maintaining the nominal setting and assessing the associated noise under various experimental conditions. Some of the steps in such research will already be familiar to the reader. These include grouping data together, in order to reduce the effect on the observations of the random component of the noise and exposing more readily the effectiveness of the control mechanism, the identification of special causes, the search for their origins and the evaluation of individual components of some of the sources of random variation. So called 'big data' analytics have a key role to play here, of course.

Noise is divided into three classes, outer, inner and between. Outer noise includes those variations whose sources lie outside the management's controls, such as variations in the environment that influence the process (for example, ambient temperature fluctuations). Inner noise arises from sources that are within management's control but not the subject of the normal routine for process control, such as the condition or age of a machine. Between noise is that tolerated as a part of the control techniques in use – this is the 'room' needed to detect change and correct for it. Trade-off between these different types of noise is sometimes necessary. Taguchi quoted the case of a tile manufacturer who had invested in a large and expensive kiln for baking tiles, and in which the heat transfer through the oven and the resultant temperature cycle variation gave rise to an unacceptable degree of product variation. Whilst a re-design of the oven was not impossible both cost and time made this solution unavailable – the kiln gave rise to 'outer' noise. Effort had to be directed, therefore, towards finding other sources of variation, either 'inner' or 'between', and, by reducing the noise they contributed, bringing the total noise to an acceptable level. It was only at some much later date, when specifying the requirements of a new kiln, that the problem of the outer noise became available and could be addressed.

In many processes, the number of variables that can be the subject of experimentation is vast, and each variable will be the subject of a number of sources of noise within each of the three classes. So the possible combinations for experimentation is seemingly endless. The 'statistically planned experiment' is a system directed towards minimizing the amount of experimentation to yield the maximum of results and in doing this to take account of both accuracy and precision – nominal settings and noise. Taguchi recognized that in any ongoing industrial process the list of the major sources of variation and the critical parameters that are affected by 'noise' are already known. So the combination of useful experiments may be reduced to a manageable number by making use of this inherent knowledge. Experimentation can be used to identify:

- the design parameters that have a large impact on the product's parameters and/or performance;
- the design parameters that have no influence on the product or process performance characteristics;
- the setting of design parameters at levels that minimize the noise within the performance characteristics;
- the setting of design parameters that will reduce variation without adversely affecting cost.

As with nearly all the techniques and facets of SPC, the 'design of experiments' is not new; Tippet used these techniques in the textile industry many years ago. Along with the other quality gurus, Taguchi enlarged the world's view of the applications of established techniques. His major contributions were in emphasizing the cost of quality by use of the total loss function and the sub-division of complex 'problem solving' into manageable component parts. The authors hope that this book, now in its seventh edition, will make a similar, modest, contribution towards the understanding and adoption of under-utilized process management principles.

13.5 Summarizing improvement

Improving products or service quality is achieved through improvements in the processes that produce the product or the service. Each activity and each job is part of a process which can be improved. Improvement is derived from people learning and the approaches presented above provide a 'road map' for progress to be made. The main thrust of the approach is a team with common objectives – using the improvement cycle, defining current knowledge, building on that knowledge, and making changes in the process. Integrated into the cycle are methods and tools that will enhance the learning process.

When this strategy is employed, the quality of products and services is improved, job satisfaction is enhanced, communications are strengthened, productivity is increased, costs are lowered, market share rises, new jobs are provided and additional profits flow. In other words, process improvement as a business strategy provides rewards to everyone involved: customers receive value for their money, employees gain job security, and owners or shareholders are rewarded with a healthy organization capable of paying real dividends. This strategy will be the common thread in all companies that remain competitive in world markets throughout the twenty-first century.

Chapter highlights

- For successful SPC there must be management commitment to quality, a quality policy and a documented quality management system.
- The main objective of the system is to cause improvements through reduction in variation in processes. The system should apply to and interact with all activities of the organization.
- The role of the management system is to define and control processes and methods. The system audit and review will ensure the processes are followed or changed.
- Measurement is an essential part of management and SPC systems. The activities, materials, equipment, etc., to be measured must be identified precisely. The measurements must be accepted by the people involved and, in their use, the emphasis must be on providing assistance to solve problems.
- The rules for interpretation of control charts and routes to be followed when out-of-control (OoC) situations develop should be agreed and defined as part of the SPC system design.
- Teamwork plays a vital role in continuous improvement. In most organizations it means moving from 'independence' to 'interdependence.' Inputs from all relevant processes are required to make changes to complex systems. Good communication mechanisms are essential for successful SPC teamwork and meetings must be managed.
- A process improvement team is a group brought together by management to tackle a particular problem. Process maps/flowcharts, C&E diagrams, and brainstorming are useful in building the team around the process, both in manufacturing and service organizations. Problem-solving groups will eventually give way to problem prevention teams.
- All processes deteriorate with time. Process improvement requires an understanding of who is responsible, what resources are required, and which SPC tools will be used. This requires action by management.
- Control charts should not only be used for control, but as an aid to reducing variability. The progressive identification and elimination of causes of variation may be charted and the limits adjusted accordingly to reflect the improvements.
- Never-ending improvement takes place in the Deming cycle of plan, implement (do), record data (check), analyse (act) (PDCA).
- The Japanese engineer Taguchi has defined a number of methods to reduce costs and improve quality. His methods appear under four headings: the total loss function; design of products processes and production (APQP/ASQP); reduction in variation; and statistically planned

experiments (note also the role of 'big data' analytics). Taguchi's main contribution was to enlarge people's views of the applications of some established techniques.

• Improvements, based on teamwork and the techniques of SPC, will lead to quality products and services, lower costs, better communications and job satisfaction, increased productivity, market share and profits and higher employment.

References and further readings

Oakland, J.S. (2014) *Total Quality Management and Operational Excellence: Text and Cases*, 4th edn, Routledge, Oxford, UK.

Pitt, H. (1993) *SPC for the Rest of Us: A Personal Guide to Statistical Process Control*, Addison-Wesley, Reading MA, USA.

Pyzdek, T. (1992) *Pyzdek's Guide to SPC, Vol. 2: Applications and Special Topics*, ASQC Quality Press, Milwaukee WI, USA.

Roy, R. (1990) *A Primer on the Taguchi Method*, Van Nostrand Reinhold, New York, USA.

Stapenhurst, T. (2005) *Marketing Statistical Process Control: A Handbook for Performance Improvement using SPC cases*, ASQ Press, Milwaukee WI, USA.

Taguchi, G. (1986) *Introduction to Quality Engineering*, Asian Productivity Association, Tokyo, Japan.

Thompson, J.R. and Koronachi, J. (1993) *Statistical Process Control for Quality Improvement*, Kluwer, The Netherlands.

Wheeler, D.J. (1986) *The Japanese Control Chart*, SPC Press, Knoxville TN, USA.

Discussion questions

1   Explain how a documented management system can help to reduce process variation. Give reasons why the system and SPC techniques should be introduced together for maximum beneficial effect.

2   What is the role of teamwork in process improvement? How can the simple techniques of problem identification and solving help teams to improve processes?

3   Discuss in detail the 'never-ending improvement cycle' and link this to the activities of a team facilitator.

4   What are the major headings of Taguchi's approach to reducing costs and improving quality in manufacturing? Under each of these headings, give a brief summary of Taguchi's thinking. Explain how this approach could be applied in a service environment.

5   Reducing the variation of key processes should be a major objective of any improvement activity. Outline a three-step approach to assigning nominal target values and tolerances for variables (product or process parameters) and explain how this will help to achieve this objective.

14 Six-sigma process quality

Objectives

- To introduce the six-sigma approach to process quality, explain what it is and why it delivers high levels of performance.
- To explain the six-sigma improvement model – DMAIC (Define, Measure, Analyse, Improve, Control).
- To show the role of design of experiments in six-sigma.
- To explain the building blocks of a six-sigma organization and culture.
- To show how to ensure the financial success of six-sigma projects.
- To demonstrate the links between six-sigma, TQM, SPC and the EFQM Excellence Model®.

14.1 Introduction

Motorola, one of the world's leading manufacturers and suppliers of semiconductors and electronic equipment systems for civil and military applications, introduced the concept of *six-sigma process quality* to enhance the reliability and quality of their products, and cut product cycle times and expenditure on test/repair. Motorola used the following statement to explain:

> Sigma is a statistical unit of measurement that describes the distribution about the mean of any process or procedure. A process or procedure that can achieve plus or minus *six-sigma* capability can be expected to have a defect rate of no more than a few parts per million, even allowing for some shift in the mean. In statistical terms, this approaches *zero defects*.

The approach was championed by Motorola's chief executive officer at the time, Bob Galvin, to help improve competitiveness. The six-sigma approach became widely publicized when Motorola won the US Baldrige National Quality Award in 1988.

Other early adopters included Allied Signal, Honeywell, ABB, Kodak and Polaroid. These were followed by Johnson and Johnson and perhaps most famously General Electric (GE) under the leadership of Jack Welch.

Six-sigma is a disciplined approach for improving performance by focusing on producing better products and services faster and cheaper. The emphasis is on improving the capability of processes through rigorous data gathering, analysis and action, and:

- enhancing value for the customer;
- eliminating costs that add no value (waste).

Unlike simple cost-cutting programmes six-sigma delivers cost savings while retaining or even improving value to the customers.

Why six-sigma?

In a process in which the characteristic of interest is a variable, defects are usually defined as the values that fall outside the specification limits (LSL–USL). Assuming and using a normal distribution of the variable, the percentage and/or parts per million defects can be found (Appendix A or Table 14.1). For example, in a centred process with a specification set at $\bar{x} \pm 3\sigma$ there will be 0.27 per cent or 2700 ppm defects. This may be referred to as 'an unshifted ±3 sigma process' and the quality called '±3 sigma quality'. In an 'unshifted ±6 sigma process', the specification range is $\bar{x} \pm 6\sigma$ and it produces only 0.002 ppm defects.

It is difficult in the real world, however, to control a process so that the mean is always set at the nominal target value – in the centre of the specification. Some shift in the process mean is expected. Figure 14.1 shows a centred process (normally distributed) within specification limits:

$$\text{LSL} = \bar{x} - 6\sigma; \ \text{USL} = \bar{x} + 6\sigma, \text{ with an } \textit{allowed shift} \text{ in mean of } \pm 1.5\sigma.$$

Table 14.1 Percentage of the population inside and outside the interval $\bar{x} \pm a\sigma$ of a normal population, with ppm

Interval	% Inside interval	Outside each interval (tail)		Outside the spec. interval ppm
		%	ppm	
$\bar{x} \pm \sigma$	68.27	15.865	158,655	317,310
$\bar{x} \pm 1.5\sigma$	86.64	6.6806	66,806	133,612
$\bar{x} \pm 2\sigma$	95.45	2.275	22,750	45,500
$\bar{x} \pm 3\sigma$	99.73	0.135	1350	2700
$\bar{x} \pm 4\sigma$	99.99367	0.00315	31.5	63.0
$\bar{x} \pm 4.5\sigma$	99.99932	0.00034	3.4	6.8
$\bar{x} \pm 5\sigma$	99.999943	0.0000285	0.285	0.570
$\bar{x} \pm 6\sigma$	99.9999998	0.0000001	0.001	0.002

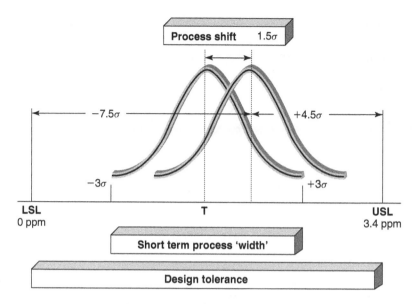

Figure 14.1 Normal distribution with a process shift of ±1.5σ. The effect of the shift is demonstrated for a specification width of ±6σ

The ppm defects produced by such a 'shifted process' are the sum of the ppm outside each specification limit, which can be obtained from the normal distribution or Table 14.1. For the example given in Figure 14.1, a ±6σ process with a maximum allowed process shift of ±1.5σ, the defect rate will be 3.4 ppm $(\bar{x} + 4.5\sigma)$. The ppm outside $\bar{x} - 7.5\sigma$ is negligible.

Similarly, the defect rate for a ±3 sigma process with a process shift of ±1.5σ will be 66,810 ppm:

$$\bar{x} - 1.5\sigma \equiv 66,806 \, \text{ppm}$$
$$\bar{x} + 4.5\sigma \equiv 3.4 \, \text{ppm}$$

Figure 14.2 shows the levels of improvement necessary to move from a ±3 sigma process to a ±6 sigma process, with a 1.5 sigma allowed shift. This feature is not as obvious when the linear measures of process capability *Cp/Cpk* are used:

$$\pm 6 \text{ sigma process} \equiv Cp / Cpk = 2$$
$$\pm 3 \text{ sigma process} \equiv Cp / Cpk = 1$$

This leads to comparative sigma performance, as shown in Table 14.2.

The means of achieving six-sigma capability are, of course, the key. At Motorola this included millions of dollars spent on a company-wide education

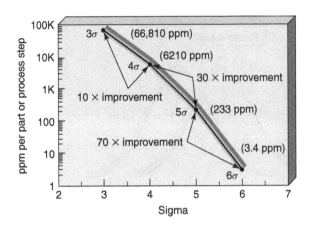

Figure 14.2 The effect of increasing sigma capability on ppm defect levels

Table 14.2 Comparative sigma performance

Sigma	Parts per million out of specification	Percentage out of specification	Comparative position
6	3.4	0.00034	World class
5	233	0.0233	Industry best in class
4	6210	0.621	Industry average
3	66,807	6.6807	Lagging industry standards
2	308,537	30.8537	Non-comparative
1	690,000	69	Out of business!

programme, documented quality systems linked to quality goals, formal processes for planning and achieving continuous improvements, individual QA organizations acting as the customer's advocate in all areas of the business, a Corporate Quality Council for co-ordination, promotion, rigorous measurement and review of the various quality systems/programmes to facilitate achievement of the policy.

14.2 The six-sigma improvement model

There are five fundamental phases or stages in applying the six-sigma approach to improving performance in a process: Define, Measure, Analyse, Improve, and Control (DMAIC). These form an improvement cycle grounded in Deming's original Plan, Do, Check, Act (PDCA), (Figure 14.3). In the six-sigma approach, DMAIC provides a breakthrough strategy and disciplined methods of using rigorous data gathering and statistically based analysis to identify sources of errors and ways of eliminating them.

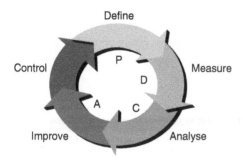

Figure 14.3 The six-sigma improvement model: DMAIC

It has become increasingly common in so-called 'six-sigma organizations', for people to refer to 'DMAIC Projects'. These revolve around the three major strategies for processes we have met in this book:

Process design/re-design

Process management

Process improvement

to bring about rapid bottom-line achievements.

Table 14.3 shows the outline of the DMAIC steps and Figures 14.4(a)–(e) give the detail in process chevron form for each of the steps.

Table 14.3 The DMAIC steps

D	Define the scope and goals of the improvement project in terms of customer requirements and the process that delivers these requirements – inputs, outputs, controls and resources.
M	Measure the current process performance – input, output and process – and calculate the short- and longer-term process capability – the sigma value.
A	Analyse the gap between the current and desired performance, prioritize problems and identify root causes of problems. Benchmarking the process outputs, products or services, against recognized benchmark standards of performance may also be carried out.
I	Generate the improvement solutions to fix the problems and prevent them from reoccurring so that the required financial and other performance goals are met.
C	This phase involves implementing the improved process in a way that 'holds the gains'. Standards of operation will be documented in systems such as ISO9000 and standards of performance will be established using techniques such as statistical process control (SPC).

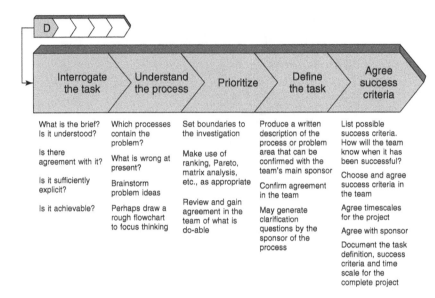

Figure 14.4(a) Dmaic: Define the scope

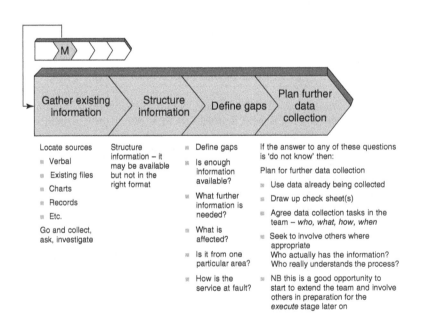

Figure 14.4(b) dMaic: Measure current performance

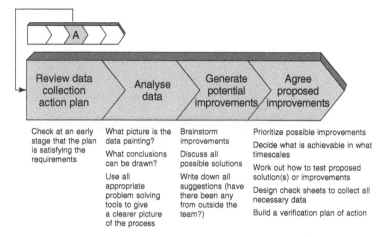

Figure 14.4(c) dmAic: Analyse the gaps

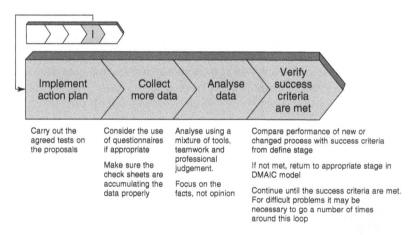

Figure 14.4(d) dmaIc: Improvement solutions

14.3 Six-sigma and the role of Design of Experiments

Design of Experiments (DoE) provides methods for testing and optimizing the performance of a process, product, service or solution. It draws heavily on statistical techniques, such as tests of significance, analysis of variance (ANOVA), correlation, simple (linear) regression and multiple regression. As we have seen in Chapter 13 (Taguchi methods), DoE uses 'experiments' to learn about the behaviour of products or processes under varying conditions, and allows planning and control of variables during an efficient number of experiments.

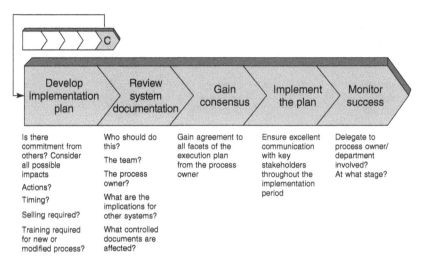

Develop implementation plan	Review system documentation	Gain consensus	Implement the plan	Monitor success
Is there commitment from others? Consider all possible impacts Actions? Timing? Selling required? Training required for new or modified process?	Who should do this? The team? The process owner? What are the implications for other systems? What controlled documents are affected?	Gain agreement to all facets of the execution plan from the process owner	Ensure excellent communication with key stakeholders throughout the implementation period	Delegate to process owner/ department involved? At what stage?

Figure 14.4(e) dmaiC: Control – execute the solution

Design of Experiments supports six-sigma approaches in the following:

- Assessing 'Voice of the Customer' systems to find the best combination of methods to produce valid process feedback.
- Assessing factors to isolate the 'vital' root causes of problems or defects.
- Piloting or testing combinations of possible solutions to find optimal improvement strategies.
- Evaluating product or service designs to identify potential problems and reduce defects.
- Conducting experiments in service environments – often through 'real-world' tests.

The basic steps in DoE are:

- Identify the factors to be evaluated.
- Define the 'levels' of the factors to be tested.
- Create an array of experimental combinations.
- Conduct the experiments under the prescribed conditions.
- Evaluate the results and conclusions.

In identifying the factors to be evaluated, important considerations include what you want to learn from the experiments and what the likely influences are on the process, product or service. As factors are selected it is important to balance the benefit of obtaining additional data by testing more factors with the increased cost and complexity.

When defining the 'levels' of the factors, it must be borne in mind that variable factors, such as time, speed, weight, may be examined at an infinite number of levels and it is important to choose how many different levels are to be examined. Of course, attribute or discrete factors may be examined at only two levels – on/off type indicators – and are more limiting in terms of experimentation.

When creating the array of experimental conditions, avoid the 'one-factor-at-a-time' (OFAT) approach where each variable is tested in isolation. DoE is based on examining arrays of conditions to obtain representative data for all factors. Possible combinations can be generated by statistical software tools or found in tables; their use avoids having to test every possible permutation.

When conducting the experiments, the prescribed conditions should be adhered to. It is important to avoid letting other, untested factors, influence the experimental results.

In evaluating the results, observing patterns and drawing conclusions from DoE data, tools such as ANOVA and multiple regression are essential. From the experimental data some clear answers may be readily forthcoming, but additional questions may arise that require additional experiments. The use of so-called 'big data' analytics, in conjunction with design of experiments and six-sigma approaches presents opportunities to crack age-old problems of multiple and complex variables interactions that result in high product rejection rates and/or low yields.

14.4 Building a six-sigma organization and culture

Six-sigma approaches question many aspects of business, including its organization and the cultures created. The goal of most commercial organizations is to make money through the production of saleable goods or services and, in many, the traditional measures used are capacity or throughput based. As people tend to respond to the way they are being measured, the management of an organization tends to get what it measures. Hence, throughput measures may create work-in-progress and finished goods inventory thus draining the business of cash and working capital. Clearly, supreme care is needed when defining what and how to measure.

Six-sigma organizations focus on:

- understanding their customers' requirements;
- identifying and focusing on core/critical processes that add value to customers;
- driving continuous improvement by involving all employees;
- being very responsive to change;
- basing management on factual data and appropriate metrics;
- obtaining outstanding results, both internally and externally.

The key is to identify and eliminate variation in processes. Every process can be viewed as a chain of independent events and, with each event subject to variation, variation accumulates in the finished product or service. Because of this, research suggests that most businesses operate somewhere between the 3 and 4 sigma level. At this level of performance, the real cost of quality is about 25–40 per cent of sales revenue. Companies that adopt a six-sigma strategy can readily reach the five sigma level and reduce the cost of quality to 10 per cent of sales. They often reach a plateau here and to improve to six-sigma performance and 1 per cent cost of quality takes a major rethink.

Properly implemented six-sigma strategies involve:

- leadership involvement and sponsorship;
- whole organization training;
- project selection tools and analysis;
- improvement methods and tools for implementation;
- measurement of financial benefits;
- communication;
- control and sustained improvement.

One highly publicized aspect of the six-sigma movement, especially its application in companies such as General Electric (GE), Motorola, Allied Signal and GE Capital in Europe, is the establishment of process improvement experts, known variously as 'Master Black Belts', 'Black Belts' and 'Green Belts'. In addition to these martial arts related characters, who perform the training, lead teams and do the improvements, are other roles that the organization may consider, depending on the seriousness with which they adopt the six-sigma discipline. These include the:

> Leadership Group or Council/Steering Committee
> Sponsors and/or Champions/Process Owners
> Implementation Leaders or Directors – often Master Black Belts
> Six-sigma Coaches – Master Black Belts or Black Belts
> Team Leaders or Project Leaders – Black Belts or Green Belts
> Team Members – usually Green Belts

Many of these terms will be familiar from TQM and continuous improvement activities. The 'Black Belts' reflect the finely honed skill and discipline associated with the six-sigma approaches and techniques. The different levels of Green, Black and Master Black Belts recognize the depth of training and expertise.

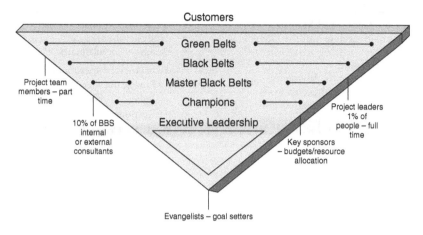

Figure 14.5 A six-sigma company

Mature six-sigma programmes, such as at GE, Johnson & Johnson and Allied Signal, have about 1 per cent of the workforce as full-time Black Belts. There is typically one Master Black Belt to every ten Black Belts or about one to every 1000 employees. A Black Belt typically oversees/ completes 5–7 projects per year, which are led by Green Belts who are not employed full-time on six-sigma projects (Figure 14.5).

The leading exponents of six-sigma have spent millions of dollars on training and support. Typical six-sigma training content is shown in Table 14.4.

14.5 Ensuring the financial success of six-sigma projects

Six-sigma approaches typically are not looking for incremental or 'virtual' improvements, but breakthroughs. This is where six-sigma has the potential to outperform other improvement initiatives. An intrinsic part of implementation is to connect improvement to bottom-line benefits and projects should not be started unless they plan to deliver significantly to the bottom line.

Estimated cost savings vary from project to project, but reported average results range from $150,000 to $250,000 per project, which typically last 4 months. The average Black Belt will generate $600,000–$1,250,000, benefits per annum, and large savings are claimed by the leading exponents of six-sigma. For example, GE has claimed returns of $1.2 billion from its investment of $450 m.

Table 14.4 Typical Six-Sigma training content

• **Week 1 – Define and Measure**
• Six-sigma overview and the DMAIC roadmap
• Process mapping
• Quality function deployment
• Failure mode and effect analysis
• Organizational effectiveness concepts, such as team development
• Basic statistics and use of Excel/Minitab
• Process capability
• Measurement systems analysis
	• **Week 3 – Improve**
	• Analysis of variance
	• Design of experiments
	  o Factorial experiments
	  o Fractional factorials
	  o Balanced block design
	  o Response surface design
	• **Week 4 – Control**
	  o Control plans
	  o Mistake proofing
	  o Special applications:
	  o discrete parts, continuous processes, administration,
	  o design
	  o Final exercise

• **Week 2 – Analyse**
  o Statistical thinking
  o Hypothesis testing and confidence intervals
  o Correlation analysis
  o Multivariate and regression analysis
	Project reviews every day
	Hands on exercises assigned every day
	Learning applied during 3 week gaps between sessions

Linking strategic objectives with measurement of six-sigma projects

Six-sigma project selection takes on different faces in different organizations. While the overall goal of any six-sigma project should be to improve customer results and business results, some projects will focus on production/service delivery processes, and others will focus on business/commercial processes. Whichever they are, all six-sigma projects must be linked to the highest levels of strategy in the organization and be in direct support of specific business objectives. The projects selected to improve business performance must be agreed upon by both the business and operational leadership, and someone must be assigned to 'own' or be accountable for the projects, as well as someone to execute them.

At the business level, projects should be selected based on the organization's strategic goals and direction. Specific projects should be aimed at improving such things as customer results, non-value add, growth, cost and cash flow. At the operations level, six-sigma projects should still tie to the overall strategic goals and direction but directly involve the process/operational management. Projects at this level then should focus on key operational and technical problems that link to strategic goals and objectives.

When it comes to selecting six-sigma projects, key questions that must be addressed include:

- What is the nature of the projects being considered?
- What is the scope of the projects being considered?
- How many projects should be identified?
- What are the criteria for selecting projects?
- What types of results may be expected from six-sigma projects?

Project selection can rely on a 'top-down' or 'bottom-up' approach. The top-down approach considers a company's major business issues and objectives and then assigns a champion – a senior manager most affected by these business issues – to broadly define the improvement objectives, establish performance measures, and propose strategic improvement projects with specific and measurable goals that can be met in a given time period. Following this, teams identify processes and critical-to-quality characteristics, conduct process baselining and identify opportunities for improvement. This is the favoured approach and the best way to align 'localized' business needs with corporate goals.

A word of warning, the bottom-up approach can result in projects being selected by managers under pressure to make budget reductions, resolve specific quality problems or improve process flow. These projects should be considered as 'areas or opportunities for improvement', as they do not always fit well with the company's strategic business goals. For example, managers may be trying to identify specific areas of waste, supply problems, supplier quality issues, or unclear or impractical 'technical' issues, and then a project is assigned to solve a specific problem. With this approach, it is easy for the operational-level focus to become diffused and disjointed in relation to the higher strategic aims and directions of the business.

At the process level, six-sigma projects should focus on those processes and critical-to-quality characteristics that offer the greatest financial and customer results potential. Each project should address at least one element of the organization's key business objectives, and be properly planned.

Metrics to use in tracking project progress and success

The organization's leadership needs to identify the primary objectives, identify the primary operational objectives for each business unit and baseline the key processes before the right projects can be selected. Problem areas need to be identified and analysed to pinpoint sources of waste and inefficiency. Every six-sigma project should be designed to ultimately benefit the customer and/or improve the company's profitability. But projects may also need to improve yield, scrap downtime and overall capacity.

Successful projects, once completed, should each add at least – say – $150,000 to the organization's bottom line. In other words, projects should be selected based on the potential cash they can return to the company, the amount and types of resources they will require, and the length of time it will take to complete the project. Organizations may choose to dedicate time and money to a series of small projects rather than a few large projects that would require the same investment in money, time and resources.

The key to good project selection is to identify and improve those performance metrics that will deliver financial success and impact the customer base. By analysing the performance of the key metric areas, organizations can better understand their operations and create a baseline to determine:

- how well a current process is working;
- theoretically how well a process should work;
- how much a process can be improved;
- how much a process improvement will affect customer results;
- how much impact will be realized in costs.

Information requirements at project charter stage for prioritizing projects

Prioritizing six-sigma projects should be based on four factors. The first is to determine the project's value to the business. The six-sigma approach should be applied only to projects where improvements will significantly impact the organization's overall financial performance and, in particular, profitability. Projects that do not significantly decrease costs are not worthwhile six-sigma projects. Cost-avoidance projects should not be considered at the onset of a six-sigma initiative, simply because there is far too much 'low-hanging fruit' to provide immediate cash. This applies to virtually all organizations in the 3.5 to 4.5 sigma category that need to focus on getting back the money they are losing today before they focus on what they *might* lose next year.

The second factor to be considered is the resource required. Resources used to raise the sigma level of a process must be offset by significant gains in profits and/or market share.

The third factor to be considered is whether any lost sales are the result of the length of time it takes to get new products to market, or whether there is an eroding customer base because of specific problems with a product or service.

The fourth factor is whether or not a six-sigma project aligns with the overall goals of the business.

Not all six-sigma projects need to have a direct impact on the customer. For example Pande *et al.* (2000) quoted a company whose finance department believed that their role was to track the financial savings generated by six-sigma projects and see that the money was returned to the company's overall bottom line. Although the finance department claimed they were different because they generated pieces of paper instead of components, they finally realized that their profitability was also influenced by such factors as productivity, defect rates, and cycle time. By using six-sigma methodology, the finance department reduced the amount of time it took to close its books each month from 12 working days to two. Decreasing defects and cycle time in the finance department alone saved the company $20 million each year.

This same company's legal department has also benefited by applying six-sigma to the length of time it took to file patent applications. Through process mapping, measuring performance, and identifying sources of errors and unnecessary variations, the company streamlined the process so that a patent application went through a chain of lawyers assigned to handle one aspect of the process, rather than a single lawyer handling the entire patent application. The outcome was that, without adding more lawyers, the company's legal department files more patents in shorter periods of time.

In both cases, it was recognized that even a small improvement would produce great savings for the company – the six-sigma projects were chosen to support the company's goal of becoming more efficient and profitable in all its processes.

Immediate savings versus cost avoidance

As already stated, most organizations have 'low-hanging fruit' – processes that can be easily fixed with an immediate impact on profits. Six-sigma provides an easy avenue to almost immediately increasing profitability by focusing the strategy on those 'cost problems' that will produce immediate results in the form of cash. Rework, scrap and warranty costs drop, quickly taking companies up to about three sigma. But it is at the top of the tree where the bulk of the fruit is hidden, and where companies need to apply the six-sigma strategy in full strength.

The theoretical view of how well a process should work should lead to 'best possible performance' which usually occurs intermittently and for very short periods of time. The logic behind cost avoidance is that if processes function well, even for a short period of time, by using simple process improvements, they should be able to function at the 'best possible performance' level all the time. This does not necessarily involve creating new technologies or significantly re-designing current processes.

Allied Signal found that in its first two years of applying six-sigma nearly 80 per cent of its projects fell into the category of low-hanging fruit – processes that could be easily improved with simple tools such as scatter plots, fish bone diagrams, process maps, cause and effect diagrams, histograms, FMEA, Pareto charts, and elementary control charting. As a result, Allied was able to move quickly through a series of projects that returned significant sums to the bottom line. However, as the relatively simpler processes were improved, Allied began to select projects that focused on harvesting the 'sweet fruit' – the fruit found at the top of the tree and the hardest to reach – and it required more sophisticated tools such as design of experiments and design for six-sigma (Pande *et al.* 2000).

In over 20 years of guiding companies through the implementation of the six-sigma approach, the authors and their colleagues have found that the first six-sigma project is especially important. Projects selected for the 'training phase' should not be those with the biggest and most difficult return potential, but ones that are straightforward and manageable. Management cannot expect a six-sigma project to immediately solve persistent problems that have been entrenched and tolerated for long periods of time. Despite the effectiveness of a disciplined six-sigma strategy, it takes training and practice to gain speed and finesse.

Six-sigma is far more than completing projects, of course. Over time, organizations discover what kinds of measures and metrics are needed to improve quality and deliver real financial benefits. Each new insight needs to be integrated into management's knowledge base, strategies and goals. Ultimately, six-sigma transforms how an organization does business, which, in turn, transforms the essence of its culture. It learns how to focus its energy on specific targets rather than random and nebulous goals.

Establishing a baseline project: a performance measurement framework

In the organization that is to succeed with six-sigma over the long term, performance must be measured by improvements seen by the customer and/or financial success. Involving accounting and finance people to enable the development of financial metrics will help in:

- tracking progress against organizational goals;
- identifying opportunities for improvement in financial performance;
- comparing financial performance against internal standards;
- comparing financial performance against external standards.

The authors have seen many examples of so-called performance measurement systems that frustrated improvement efforts. Various problems include systems that:

- produce irrelevant or misleading information;
- track performance in single, isolated dimensions;
- generate financial measures too late, e.g. quarterly, for mid-course corrections;
- do not take account of the customer perspective, both internal and external;
- distort management's understanding of how effective the organization has been in implementing its strategy;
- promote behaviour that undermines the achievement of the financial strategic objectives.

The measures used should be linked to the processes where the value-adding activities take place. This requires a performance measurement framework (PMF) that provides feedback to people in all areas of business operations and stresses the need to fulfil customer needs.

The critical elements of such a good performance measurement framework are:

- leadership and commitment;
- employee involvement;
- good planning;
- sound implementation strategy;
- measurement and evaluation;
- control and improvement;
- achieving and maintaining standards of excellence.

A performance measurement framework is proposed, based on the strategic planning and process management models outlined in the author's *Total Quality Management and Operational Excellence*, 4th edition (2014).

The PMF has four elements related to: strategy development and goal deployment, process management, individual performance management and review. This reflects an amalgamation of the approaches used by a range of organizations using six-sigma approaches and distinguishes between the 'whats' and the 'hows'.

The key to six-sigma planning and deployment is the identification of a set of critical success factors (CSFs) and associated key performance indicators (KPIs). These factors should be derived from the organization's vision and mission, and represent a balance mix of stakeholders. The strategic financial goals should be clearly communicated to all individuals, and translated into measures of performance at the process/functional level. This approach is in line with the EFQM's Excellence Model® and its 'balanced scorecard' of performance measures: customer, people, society and key performance results.

The key to successful performance measurement at the process level is the identification and translation of customer requirements and strategic objectives into an integrated set of process performance measures. The documentation and management of processes has been found to be vital in this translation process. Even when a functional organization is retained, it is necessary to treat the measurement of performance between departments as the measurement of customer–supplier performance.

Performance measurement at the individual level usually relies on performance appraisal, i.e. formal planned performance reviews, and performance management, namely day-to-day management of individuals. A major drawback with some performance appraisal systems, of course, is the lack of their integration with other aspects of company performance measurement, particularly financial.

Performance review techniques are used by many world-class organizations to identify improvement opportunities, and to motivate performance improvement. These companies typically use a wide range of such techniques and are innovative in baselining performance in their drive for continuous improvement.

The links between performance measurement at the four levels of the framework are based on the need for measurement to be part of a systematic approach to six-sigma. The framework should provide for the development and use of measurement, rather than prescriptive lists of measures that should be used. It is, therefore, applicable in all types of organization.

A number of factors have been found to be critical to the success of six-sigma performance measurement systems. These factors include the level of involvement of the finance and accounting people in the identification of the vital few measures, the developing of a performance measurement framework, the clear communication of strategic objectives, the inclusion of customers and suppliers in the measurement process, and the identification of the key drivers of performance. These factors will need to be taken into account by managers wishing to establish successful six-sigma projects.

14.6 Concluding observations and links with Excellence

Six-sigma is not a new technique, its roots can be found in Total Quality Management (TQM) and Statistical Process Control (SPC) but it is more than TQM or SPC re-badged. It is a framework within which powerful TQM and SPC tools can be allowed to flourish and reach their full improvement potential. With the TQM philosophy, many practitioners promised long-term benefits over 5–10 years, as the programmes began to change hearts and minds. Six-sigma by contrast is about delivering breakthrough benefits in the short term and is distinguished from TQM by the intensity of the intervention and pace of change.

Excellence approaches such as the EFQM Excellence Model®, and six-sigma are complementary vehicles for achieving better organizational performance. The Excellence Model can play a key role in the baselining phase of strategic improvement, whilst the six-sigma breakthrough strategy is a delivery vehicle for achieving excellence through:

1 Committed leadership.
2 Integration with top level strategy.
3 A cadre of change agents – Black Belts.
4 Customer and market focus.
5 Bottom-line impact.
6 Business process focus.
7 Obsession with measurement.
8 Continuous innovation.
9 Organizational learning.
10 Continuous reinforcement.

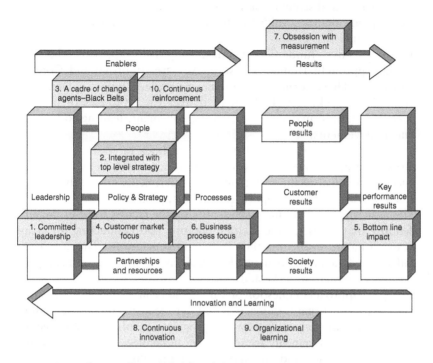

Figure 14.6 The Excellence Model and six-sigma

These are 'mapped' onto the Excellence Model in Figure 14.6. (See also Porter, L. (2002) 'Six Sigma Excellence', *Quality World*, pp. 12–15.)

There is a whole literature and many conferences have been held on the subject of six-sigma and it is not possible here to do justice to the great deal of thought that has gone into the structure of these approaches. As with Taguchi methods, described in the previous chapter, the major contribution of six-sigma has not been in the creation of new technology or methodologies, but in bringing to the attention of senior management the need for a disciplined structured approach and their commitment, if real performance and bottom-line improvements are to be achieved.

Chapter highlights

- Motorola introduced the concept of six-sigma process quality to enhance reliability and quality of products and cut product cycle times and expenditure on test and repair.
- A process that can achieve six-sigma capability (where sigma is the statistical measure of variation) can be expected to have a defect rate of a few parts per million, even allowing for some drift in the process setting.

- Six-sigma is a disciplined approach for improving performance by focusing on enhancing value for the customer and eliminating costs that add no value.
- There are five fundamental phases/stages in applying the six-sigma, approach: Define, Measure, Analyse, Improve and Control (DMAIC). These form an improvement cycle, similar to Deming's Plan, Do, Check, Act (PDCA), to deliver the strategies of process design/re-design, management and improvement, leading to bottom line achievements.
- Design of Experiments (DoE) provides methods for testing and optimizing the performance of a process, product or service. Drawing on known statistical techniques DoE uses experiments efficiently to provide knowledge which supports six-sigma approaches.
- The basic steps of DoE include: identifying the factors to be evaluated, defining the 'levels' of the factors, creating and conducting an array of experiments, evaluating the results and conclusions. The use of so-called 'big data' analytics may allow problems of multiple and complex variables interactions to be addressed.
- Six-sigma approaches question organizational cultures and the measures used. Six-sigma organizations, in addition to focusing on understanding customer requirements, identify core processes, involve all employees in continuous improvement, are responsive to change, base management on fact and metrics, and obtain outstanding results.
- Properly implemented six-sigma strategies involve: leadership involvement and sponsorship, organization-wide training, project selection tools and analysis, improvement methods and tools for implementation, measurement of financial benefits, communication, control and sustained improvement.
- Six-sigma process improvement experts, named after martial arts – Master Black Belts, Black Belts and Green Belts – perform the training, lead teams and carry out the improvements. Mature six-sigma programmes have about 1 per cent of the workforce as Black Belts.
- Improvement breakthroughs are characteristic of six-sigma approaches, which are connected to significant bottom line benefits. In order to deliver these results, strategic objectives must be linked with measurement of six-sigma projects and appropriate information and metrics used in prioritizing and tracking project progress and success. Initial focus should be on immediate savings rather than cost avoidance, to deliver the 'low-hanging fruit' before turning to the 'sweet fruit' higher in the tree.
- A PMF should be used in establishing baseline projects. The PMF should have four elements related to: strategy development and goal deployment; process management; individual performance management; review.

- Six-sigma is not a new technique – its origins may be found in TQM and SPC. It is a framework through which powerful TQM and SPC tools flourish and reach their full potential. It delivers breakthrough benefits in the short term through the intensity and speed of change. The Excellence Model is a useful framework for mapping the key six-sigma breakthrough strategies.

References and further reading

Basu, R. (2002) *Quality Beyond Six Sigma*, Elsevier Butterworth-Heinemann, Oxford, UK.

Breyfogle, F.W. (2003) *Implementing Six-Sigma*, 2nd edn, Wiley-Interscience, New York, USA.

Eckes, G. (2003) *Six Sigma for Everyone*, John Wiley & Sons, New York, USA.

Harry, M. and Schroeder, R. (2000) *Six-Sigma: The Breakthrough Management Strategy Revolutionizing the World's Top Corporations*, Doubleday, New York, USA.

Hayler, R. and Nicholo, M. (2005) *What Is Six Sigma Process Management?* ASQ Press, Milwaukee WI, USA.

Mödl, A. (1992) 'Six-Sigma Process Quality', *Quality Forum*, Vol. 18, No. 3, pp. 145–9.

Oakland, J.S. (2014) *Total Quality Management & Operational Excellence*, 4th edn, Routledge, Oxford, UK.

Pande, P.S., Neuman, R.P. and Cavanagh, R.R. (2000) *The Six-Sigma Way: How GE, Motorola and Other Top Companies Are Honing Their Performance*, McGraw-Hill, New York, USA.

Porter, L. (2002) 'Six Sigma Excellence', Quality World (CQI – London), pp. 12–15.

Wilson, G. (2005) *Six Sigma and the Product Development Cycle*, Elsevier Butterworth-Heinemann, Oxford, UK.

Discussion questions

1   Explain the statistical principles behind six-sigma process quality and why it is associated with *ca* 3 ppm defect rate. Show the effect of increasing sigma capability on 'defects per million opportunities' and how this relates to increased profits.

2   Using process capability indices, such as Cp and Cpk (see Chapter 10) explain the different performance levels of 1 to 6 sigma increasing by integers.

3   Detail the steps of the six-sigma DMAIC methodology (Define, Measure, Analyse, Improve, Control) and indicate the tools and techniques which might be appropriate at each stage.

4   You have been appointed operations director of a manufacturing and service company which has a poor reputation for quality. There have been several attempts to improve this during the previous 10 years,

including quality circles, ISO9000-based quality systems, SPC, TQM and the Excellence Model. These have been at best partially successful and left the organization 'punch-drunk' in terms of waves of management initiatives.

Write a presentation for the board of directors of the company, where you set out the elements of a six-sigma approach to tackling the problems, explaining what will be different to the previous initiatives.

5   What is Design of Experiments (DoE)? What are the basic steps in DoE and how do they link together to support six-sigma approaches?

6   As quality director of a large aircraft manufacturing organization, you are considering the launch of a six-sigma-based continuous improvement programme in the company. Explain in detail the key stages of how you will ensure the financial success of the six-sigma projects that will be part of the way forward.

15 The implementation of statistical process control

Objectives

- To examine the issues involved in the implementation of SPC.
- To outline the benefits to be derived from successful introduction of SPC.
- To provide a methodology for the implementation of SPC.
- To emphasize the link between a good quality management system and SPC.
- To provide some short case study examples of SPC implementation.

15.1 Introduction

The original techniques of statistical quality control (SQC) have been available for over nearly a century; Shewhart's first book on control charts was written in 1924. There is now a vast academic literature on SPC and related subjects such as six-sigma. However, research work carried out by the authors and their colleagues in the Oakland Institute, the Research and Education Division of the Oakland Group, has shown that managers still do not understand variation.

Where SPC is properly used it has been shown that quality-related costs are usually known and low, and that often the use of SPC was specified by a customer, at least initially. Companies using the techniques frequently require their suppliers to use them and generally find SPC to be of considerable benefit.

Where there is low usage of SPC the major reason found is lack of knowledge of variation and its importance, particularly among senior managers. Although they sometimes recognize quality as being an important part of corporate strategy, they do not appear to know what effective steps to take in order to carry out the strategy. Even now in some organizations, quality is seen as an abstract property and not as a measurable and controllable parameter.

It would appear that, as a large majority of companies that have tried SPC are happy with its performance and continue to use it, the point at which resistance occurs is in introducing the techniques. Clearly there is a need to increase knowledge, awareness of the benefits, and an understanding of how SPC, and the reduction/control of variability, should be introduced.

15.2 Successful users of SPC and the benefits derived

In-depth work in organizations that use SPC successfully has given clear evidence that customer driven management systems push suppliers towards the use of process capability assessments and process control charts. It must be recognized, however, that external pressure alone does not necessarily lead to an understanding of either the value or the relevance of the techniques.

Close examination of organizations in which SPC was used incorrectly has shown that there was no real commitment or encouragement from senior management. It was apparent in some of these that lack of knowledge and even positive deceit can lead to unjustifiable claims to either customers or management. No system of quality or process control will survive the lack of full commitment by senior management. The failure to understand or accept this will lead to loss of control of quality and the very high cost associated with it.

Truly successful users of SPC can remain so only when the senior management is both aware of and committed to the continued use and development of the techniques to manage variation. The most commonly occurring influence contributing to the use of SPC was exerted by an enthusiastic member of the management team.

Other themes that recur in successful user organizations are:

- Top management understood variation and the importance of SPC techniques to successful performance improvement.
- All the people involved in the use of the techniques understood what they were being asked to do and why it should help them.
- Training, followed by clear and written instructions on the agreed procedures, was systematically introduced and followed up.

These requirements are, of course, contained within the general principles of good quality management.

The benefits to be derived from the application of statistical methods of process control are many and varied. A major spin-off is the improved or continuing reputation for consistent quality products or service. This leads to a steady or expanding, always healthy, share of the market, or improved effectiveness/efficiency. The improved process consistency derived causes a direct reduction in external failure costs – warranty claims, customer complaints and the intractable 'loss of good will'. The corresponding reduction in costs of internal failure – scrap, rework, wasted time, secondary or low value product, etc., generates a bonus increase in productivity, by reducing the size of the 'hidden plant' which is devoted to producing non-conforming products or services.

The greater degree of process control allows an overall reduction in the checking/inspection/testing efforts, often resulting in a reduction or redeployment of staff. The benefits are not confined to a substantial lowering of total quality-related costs, for additional information such as vendor rating allows more efficient management of areas such as procurement, design, marketing and even accounting.

Two major requirements then appear to be necessary for the successful implementation of SPC, and these are present in all organization that continue to use the techniques successfully and derive the benefits:

1   Real commitment and understanding from senior management.
2   Dedicated and well-informed quality-related manager(s).

It has also been noted by the authors and their colleagues that the intervention of a 'third party' such as a consultant or external trainer has a very positive effect.

15.3 The implementation of SPC

Successful implementation of SPC depends on the approach to the work being structured. This applies to all organizations, whatever their size, technology or product/service range. Unsuccessful SPC implementation programmes usually show weaknesses within either the structure of the project or commitment to it. Any procedure adopted requires commitment from senior management to the objectives of the work and an in-house coordinator to be made available. The selection of a specific project to launch the introduction of SPC should take account of the knowledge available and the improvement of the process being:

- highly desirable;
- measurable;
- possible within a reasonable time period;
- possible by the use of techniques requiring operator level training for their introduction.

The first barrier which usually has to be overcome is that organizations still pay insufficient attention to good training, outside the technological requirements of their processes. With a few notable exceptions, they are often unsympathetic to the devotion of anything beyond minimal effort and time for training in the wider techniques of management. This exacerbates the basic lack of knowledge about processes and derives from lack of real support from the senior management. Lame excuses such as 'the operators will never understand it', 'it seems like a lot of extra work' or 'we lack the necessary facilities' should not be tolerated. A further frequently occurring source of difficulty, related to knowledge and training, is the absence from the management team of a knowledgeable enthusiast.

The impact of the intervention of a third party here can be remarkable. The third party's views will seldom be different from those of some of the management but are simply more willingly received. The expertise of the 'consultant', whilst indispensable, may well be incidental to the wider impact of their presence.

Short case studies in SPC implementation

1   A global pharmaceutical contract manufacturer that turned over $1.5bn with 30 global sites was struggling with process and product variation through lack of validation and verification. Monitoring and controlling processes for the early identification of issues and driving down variation was seen as a critical component in reducing the reputational risk of product failures (e.g. contamination, recalls, etc.). Regulatory bodies were looking for a more proactive approach to product quality control through establishing process capability and using real time process controls. A quick return on investment could be achieved through a reduction in material losses, product deviations, scrapped batches and operator time.

The first stage in the approach was to choose a variable on which to focus. This involved studying yield loss data, deviation analysis and completing a Failure Mode and Effect Analysis (FMEA) for operations. As a critical compliance measure, product weight was selected. Having determined the variable and identified the data source, the next stage was to understand the compliance implications. The project team worked closely with the Quality Department and other key stakeholders to understand the key issues and quickly resolve them.

The IT department were also a critical part of the solution. Their role was three fold:

- installation of the hardware;
- creation of the network link to allow data to be pulled from the validated product weight database;
- development of the software capability to accommodate the particular requirements of this application.

Having installed the capability to see, monitor and analyse the variables, the next stage was to ensure the operators, managers, quality advisors and shift engineers were able to interpret the information presented. A three phased approach was used:

(i)   Initial process awareness sessions were conducted to introduce statistically based concepts, where appropriate, and provide insights into how they could help and seek inputs from the staff;
(ii) Process variation, control and capability introduction training to build on the concepts introduced in the awareness sessions, demonstrating how to use the system and explaining how to interpret the outputs;
(iii) A Standard Operating Procedure (SOP) was developed that included a comprehensive troubleshooting and corrective action guide.

Process verification and validation approaches were successfully introduced on time and within budget. Essential to the success of the project was a close collaborative partnership with the process operators to ensure skills were quickly transferred into the organization, along with the engagement of key stakeholders regarding the new ways of working. The project delivered:

- a workforce more engaged in continuous improvement – operators were immediately able to gain insights into their process, that were previously inadvertently hidden, and thereby identify improvement opportunities;
- clarity on the critical process steps – set-up was shown to have a far greater influence on the process performance than previously imagined – performance by specific parts of the process could also be identified;
- immediate improvements to production control and efficiency – reductions in raw material losses, product deviations and scrapped batches, along with improvements to the production set-ups were realized, delivering a rapid return on investment;
- data driven improvement opportunities – the approach gave the operators the data to challenge the Engineering Team to put in place solutions that resolved some of the root causes and thereby reduced variability, cost and risk;
- a shift in perception of clients and potential clients – the more proactive nature of the approach demonstrated that the company was introducing tools to continuously improve.

The project also established a structured process and the in-house capability for rolling-out process validation and verification across other lines, along with the basis and impetus to implement it globally across the organization.

2 EADS (the company at the time) specialized in aerospace, defence and related products and services. The group included Airbus, Eurocopter and Astrium, the European leader in space programmes. EADS was the major partner in the Eurofighter consortium, developed the A400M military transport aircraft, and held a stake in MBDA, the international leader in missile systems. The group employed over 100,000 people at more than 70 production sites in France, Germany, the UK and Spain as well as in the US and Australia.

Within the Airbus facility at Filton , UK, work was being done on hydraulic pipes to fit within the massive wings of the Airbus A380. The production team was finding it very difficult to bend the pipes to the specified tolerances. The non-conformance rate was running at over 80 per cent, requiring large amounts of rework and also creating high scrap levels.

Improvements to the process had been considered and, although a plan to purchase a new inspection jig had been approved, a project was created to find alternative ways to reduce the non-conformance levels, hence eliminating the need for an expensive capital investment.

The team adopted a structured methodology for problem solving that is represented by the acronym 'DRIVER' (Define, Review, Investigate, Verify, Execute, Reinforce – similar to the DMAIC approach described in fully in Chapter 14 ; see also Oakland's TQM book 2014). After creating a clear definition of the problem the project team collected performance data and ascertained the extent to which the process was incapable of meeting the design tolerances. The next step was to identify and reduce the effects of the significant common causes of variation. Little data existed so the team undertook a major data collection exercise to get the facts.

A number of significant causes were identified, and actions implemented to reduce their effects. However, when the process capability was measured again, it still showed an incapable process. Instead of running a common cause variation reduction method again, the team decided to take a slightly different perspective on the problem and examine the specification to see whether it may be possible to widen the tolerance without affecting quality or operation of any other functions within the wing.

New people were invited into the team from Design Engineering and a study undertaken of true required tolerances. Such work had to be undertaken to ensure that functionality would be not be adversely affected and was compliant with all appropriate design and regulatory standards. As a result, design tolerances were found to be over specified. It was then possible to significantly widen tolerances without any adverse effects on fit or functionality within the wing.

The non-conformance rate was dramatically reduced, giving a saving of €70k per wing set and the proposal for the new inspection jig was abandoned avoiding an investment of *ca* €1m. Work continued to further improve process capability and the principles uncovered in this project were applied to new aircraft in the design phase.

3　A global organization providing specialist transportation services to offshore oil and gas platforms, medical evacuation and civilian search and rescue ran more than 250 aircraft across 30 countries, with major operating units in Australia, Brazil, Norway and the UK. The highly variable nature of the business, with unique contracts and locations across each operating region had created a significant challenge in establishing a meaningful set of metrics to help track performance and drive improvement. This was compounded by a series of acquisitions that had further complicated reporting and a move from regional to centralized operational control, driving the need for common and standardized measures across the business. To enable data-driven improvement in this highly variable project-based business, the company wished to examine the application of SPC methods to tackle the challenges.

The first step was to establish a clear understanding of the management requirement underpinning the Key Performance Indicators (KPIs), especially given the new operating structures with changes in accountability and ownership. At the highest level, all contracts were serviced using the same set of interrelated processes – the key was therefore to find a mechanism to link operational indicators across all aspects of the business (the cause) with financial outcome metrics (the effect) in a way that:

- measured performance against an agreed baseline and targets;
- identified opportunities to focus improvement action;
- enabled data-driven management decisions;
- refined productivity assumptions for future bids and forecasts.

Working closely with the Business Intelligence and functional teams, an initial set of KPIs linking operational performance with financial impact were defined and piloted in Brazil. This acted as a proof of concept to demonstrate what was possible, but also provided the vehicle to refine the metrics and trial the supporting meeting and governance structures, an essential component in driving management decisions and real improvements from the information.

The key to roll-out was sustainability on a global stage – automated KPI dashboards between each region and the global operating centre were built into an online performance hub with live data feeds. This provided a significant change in the way management could access, review and act upon data. An essential component of the project was, therefore, to manage this change, establishing standard ways of reporting and reviewing information, coaching operational staff on the metrics and underlying data, and emphasizing the required behaviours at all levels of the organization.

Live integrated dashboards, with clear ownership of each KPI, were implemented to be fully operational across the organization's global footprint. They provided the essential anchor points for the governance processes, creating visibility to make strategic decisions, identifying the key areas to focus improvement effort and allow the operational teams to share best practice. In the first three months following launch, improvements to crew rostering, training and manpower management yielded $2.3M audited savings p.a.

As is often the case, it was important not to assume that having the right data and information will lead to the right decisions – it is vital to lead the business through the journey to becoming a more data centric organization, demonstrating what is possible and establishing the necessary foundations to ensure the transformation is sustainable.

4   Sonaca Group is a global Belgian company active in the development, manufacturing, and assembly of advanced structures for civil, military aviation and space markets. It has production facilities in China, Europe, North-America and South-America and employs over 2500 people including 350 engineers.

A training and coaching programme was devised and delivered that generated six different six-sigma process improvement projects at the Gossellies plant. The consultant team worked with Quality Director to agree the objectives and format for the programme which involved Green Belt and Yellow Belt training for employees across several production departments.

A quote from the Quality Director: "Six sigma is now officially part of our Quality Toolbox. Six-sigma and the DMAIC methodology help us to shift the limits of the continuous improvement."

Bilingual training sessions and a monthly coaching programme was followed by successful close-out of the projects, and: "The specific benefits of six-sigma techniques were demonstrated through each DMAIC phase:

- Definition of key opportunities for improvement through use of tools in workshops and involving the wider team.
- Visual communication boards to track incidents resulted in the definition of effective quick wins.
- Six-sigma analysis of different factors supported the cost–benefit analysis of proposed innovations.
- A strict evidence-based approach and strong leadership resulted in improved collaboration between departments."

Lessons learned were gathered for direct implementation in the company's new facility which constitutes another significant ongoing benefit for the company.

5   Very short examples of the use of DRIVER tools and techniques in improving finance-related processes

a)   The company operated a payroll process in four regional service centres across the globe. Performance across this global footprint varied in terms of On Time; On Cost; On Quality. External benchmarks were used to define targets for improvement. The focus for the first phase was cost, without degradation in time or quality. The DRIVER methodology was used and in particular an 'assumption busting' tool. The outcome was a revised global process running at 40 per cent per cent lower costs.

b)   The company was a multinational oil and petrochemical company that was operating a very poorly performing invoicing process. The error rates were high – 17 per cent of invoices had errors going out to the customer. This resulted in high levels of customer dissatisfaction, high recovery costs, delays in payments and cash flow problems. An improvement team addressed the problem using the DRIVER approach and tools, including simple process mapping and data collection & analysis techniques. Not only were error rates dramatically reduced – to less than 1 per cent – but they developed a much greater understanding of the process so that it could be redesigned to recover employee and customer satisfaction, reduce costs and improve cash flow.

c) A major international bank set up a project team using the DRIVER methodology and tools to improve productivity and quality of transaction processing operations. Work during the implementation of statistical process control techniques had identified transaction times as a potential area for improvement. Using efficiency standards and activity based sampling, it was discovered that the process was not capable of meeting the requirements. Improvements of 50 per cent in on-time and on-quality measures were achieved by the small team in a few weeks.

d) A multinational energy company was experiencing problems with the quality of its inter-company trading transactions resulting in 40 per cent of the labour time being consumed in error handling and reconciliation activities The use of the DRIVER methodology identified a number of improvements including primary data cleansing, error proofing in coding, changes to policies and better linkage of policies and practices. The identified improvements generated over 50 per cent reduction in rework time.

e) A global company was required to comply with the Sarbanes-Oxley legislation that was created in 2002 following the Enron scandal. The legislation required a trace between key reported data and the processes in the company that created and used the data. In turn, these processes needed to be at high levels of integrity. Many companies adopted an approach characterized by heavy audit activity, but this company used DRIVER methods in adopting an assurance-based approach based on process capability and control.

f) An airline company used DRIVER and improvement techniques, such as cusum charts, on sales and percentage of turnover as profit data. Assignable causes of variation were thus identified, such as the introduction of an "efficiency" bonus scheme. The motivational (or otherwise) impact of managerial decisions and actions often manifest themselves in performance results and the use of a rigorous and disciplined approach was invaluable in identifying the best way forward for the future.

15.4 Proposed methodology for implementation

The conclusions of the authors' and their colleagues' work in helping organizations to improve product consistency and delivery times, and reduce costs by implementing SPC programmes is perhaps best summarized by detailing a proposed methodology. This is given below under the various sub-headings that categorize the essential steps in the process.

Review quality management systems

The 'quality status' of the organization has no bearing on the possibility of help being of value – a company may or may not have 'quality problems', in any event it will always benefit from a review of its quality management systems. The first formal step should be a written outline of the objectives, programme of work, timing and reporting mechanisms. Within this formal approach it is necessary to ensure that the quality policy is defined in writing, that the requirement for a documented system is recognized, that a management representative responsible for quality is appointed. His/her role should be clearly defined, together with any part to be played by a third party. A useful method of formalizing reporting is to prepare on a regular basis a memorandum account of quality-related costs – this monitors progress and acts as a useful focus for management.

Review the requirements and design specifications

Do design specifications exist and do they represent the true customer needs? It is not possible to manufacture a product or carry out the operations to provide a service without a specification – yet written specifications are often absent, out of date, or totally unachievable, particularly in service organizations. The specification should describe in adequate detail what has to be done, how it has to be done, and how checks, inspection or tests will show that it has been done. It will also indicate who is responsible for what, what records shall be kept and the prescribed action when specifications are not met. The format of specifications should also be reviewed and, if necessary, represented as targets with minimum variation, rather than as upper and lower specification limits.

Emphasize the need for process understanding and control

For a variety of reasons the control of quality is still, in some organizations, perceived as being closely related to inspection, inspectors, traceability and heavy administrative costs. It is vital that the organization recognizes that the way to control quality is to understand and control the various processes involved. The inspection of final products can serve as a method of measuring the effectiveness of the control of the processes, but here it is too late to exercise control. Sorting the good from the bad, which is often attempted at final inspection, is a clear admission of the fact that the company does not understand or expect to be able to control its processes.

Process control methods are based on the examination of data at an early stage with a view to rapid and effective feedback. Rapid feedback gives tighter control, saves adding value to poor quality, saves time and reduces the impact on operations scheduling and hence output. Effective feedback can be achieved by the use of statistically based process control methods – other methods will often ignore the difference between common and special causes of variation and consequential action will lead to 'hunting' the process.

Where the quality status of an organization is particularly low and no reliable records are available, it may prove necessary to start the work by data collection from either bought-in goods/services or company products/services. This search for data is, of course, only a preliminary to process control. The increased availability of 'big data', makes it essential for organizations to combine methods of data analytics and SPC techniques in many cases.

In the majority of cases the problems can be solved only by the adoption of better process control techniques. These techniques have been the subject of renewed emphasis throughout the world and new terms are sometimes invented to convey the impression that the techniques are new. In fact, as pointed out earlier, the techniques have been available for decades.

Plan for education and training

This is always required whether it is to launch a new management system or to maintain or improve an existing one. Too often organizations see training as useful and profitable only when it is limited to the technical processes or those of its suppliers and customers. Education often is needed at the top of the organization. The amount of time spent need not be large; for example, with proper preparation and qualified teachers, a short training programme can:

- provide a good introduction for senior managers – enough to enable them to initiate and follow up work within their own organization; or
- provide a good introduction for middle managers – enough to enable them to follow up and encourage work within their domain; or
- put quality managers on the right road – give them the incentive to further their studies either by supervised or unsupervised study; or
- train the people and provide them with an adequate understanding of the techniques so they may use them without a sense of mystique.

Follow-up education and training

For the continued successful use of SPC, all education and training must be followed up during the introductory period. Follow-up can take many forms. Ideally, an in-house expert will provide the lead through the design of implementation programmes. The most satisfactory strategy is to start small and build up a bank of knowledge and experience. Techniques should be introduced alongside existing methods of process control, if they exist. This allows comparisons to be made between the new and old methods. When confidence has been built up from these comparisons, the SPC techniques will almost take over the control of the processes themselves. Improvements in one or two areas of the organization's operations using this approach will quickly establish the techniques as reliable methods for understanding and controlling processes.

The authors and their colleagues have found that another successful formula is the in-house training course plus follow-up projects and workshops. Typically, a short course in SPC is followed within six weeks by a 1- or 2-day workshop. At this, delegates on the initial training course present the results of their project efforts. Specific process control and implementation problems may be discussed. A series of such 'surgery' workshops will add continuity to the follow-up. A wider presence should be encouraged in the follow-up activities, particularly from senior management.

Tackle one process or problem at a time

In many organizations there will be a number of processes or problems all requiring attention and the first application of SPC may well be the use of Pareto analysis in order to decide the order in which to tackle them. It is then important to choose one process or problem and work on it until satisfactory progress has been achieved before passing on to a second. The way to tackle more than one process/problem simultaneously is to engage the interest and commitment of more people, but only provided that everyone involved is competent to tackle their selected area. The co-ordination of these activities then becomes important in selecting the area most in need of improved performance.

Record all observed data in detail

A very common fault in all types of organizations is the failure to record observations properly. This often means that effective analysis of performance is not possible and for subsequent failures, either internal or external, the search for corrective action is frustrated.

The 'inspector's tick' is a frequent feature of many control systems. This actually means that the inspector passed by; it is often assumed that the predetermined observation was carried out and that, although the details are now lost, all was well. Detailed data can be used for performance and trend analysis. Recording detail is also a way of improving the accuracy of records – it is easier to tick off and accept something just outside the 'limits' than it is to deliberately record erroneously a measured parameter.

Measure the capability of processes

Process capability should be assessed and not assumed. The capability of all processes can be measured. This is true both when the results are assessed as attributes and when measured as variables. Once the capability of the process is known, it can be compared with the requirements. Such comparison will show whether the process can achieve the process or service requirements. Where the process is adequate the process capability data can be used to set up control charts for future process control and data recording. Where the process is incapable, the basis is laid for a rational decision concerning the required action – the revision of the requirements or revision of the process.

Make use of the data on the process

This may be cumulated, provide feedback, or refined in some way. Cusum techniques for the identification of either short- or long-term changes can give vital information, not only for process control, but also for fault finding and future planning. The feedback of process data enables remedial action to be planned and taken – this will result in steady improvements over time to both process control and product/ service quality. As the conformance to requirement improves, the data can be refined. This may require either greater precision in measurement or less frequent intervention for collection. The refinement of the data should be directed toward the continuing improvement of the processes and product or service consistency.

A final comment

A good quality management system provides a foundation for the successful application of SPC techniques. It is not possible to 'graft' SPC onto a poor system. Without well-understood ways of operating processes, inspection/ test, and for the recording of data, SPC will lie dormant.

Many organizations would benefit from the implementation of statistical methods of process control and the understanding of variation this brings. The systematic structured approach to their introduction, which is recommended here, provides a powerful spearhead with which to improve conformance to requirements and consistency of products and services. Increased knowledge of process capability will also assist in marketing decisions and product service design.

The importance of the systematic use of statistical methods of process control in all types of activity cannot be over-emphasized. To compete internationally, both in home markets and overseas, or to improve cost effectiveness and efficiency, organizations must continue to adopt a professional approach to the collection, analysis and use of process data.

Acknowledgements

The authors would like to acknowledge the contribution of their colleagues in The Oakland Group to the preparation of this chapter. It is the outcome of many years' collaborative work in helping organizations to overcome the barriers to acceptance of SPC and improve performance.

Chapter highlights

- Research work shows that managers still do not understand variation, in spite of the large number of books and papers written on the subject of SPC and related topics.
- Where SPC is used properly, quality costs are lower; low usage is associated with lack of knowledge of variation and its importance, especially in senior management.
- Successful users of SPC have, typically, committed knowledgeable senior management, people involvement and understanding, training, clear management systems, a systematic approach to SPC introduction, and a dedicated well-informed internal champion.
- The benefits of SPC include: improved or continued reputation for consistent quality products/service, healthy market share or improved efficiency/effectiveness, and reduction in failure costs (internal and external) and appraisal costs.
- Several case studies provide compelling evidence of sustained benefits when a structured approach is taken.
- A stepwise approach to SPC implementation should include the phases: review management systems, review requirements/design specification, emphasize the need for process understanding and control, plan for education and training (with follow-up), tackle one process or problem at a time, record detailed observed data, measure process capabilities and make use of data on the process. 'Big data' makes it essential for methods of data analytics to be combined with SPC techniques.
- A good management system provides a foundation for successful application of SPC techniques. These together will bring a much better understanding of the nature and causes of process variation to deliver improved performance.

References and further reading

Oakland, J.S. (2014) *Total Quality Management & Operational Excellence*, 4th edn, Routledge, Oxford, UK.

Roberts, L. (2006) *SPC for Right Brain Thinkers: Process Control for Non-Statisticians*, ASQ Press, Milwaukee WI, USA.

Appendices

Appendix A

The normal distribution and non-normality

The mathematical equation for the normal curve (alternatively known as the Gaussian distribution) is:

$$y = \frac{1}{\sigma\sqrt{2\pi}} e^{-(x-\bar{x})^2/2\sigma^2},$$

where y = height of curve at any point x along the scale of the variable

σ = standard deviation of the population

$\bar{x}$ = average value of the variable for the distribution

π = ratio of circumference of a circle to its diameter ($\pi = 3.1416$).

If $z = (x - \bar{x})/\sigma$, then the equation becomes:

$$y = \frac{1}{\sigma\sqrt{2\pi}} e^{-z^2/2}.$$

The constant $1/\sqrt{2\pi}$ has been chosen to ensure that the area under this curve is equal to unity, or probability 1.0. This allows the area under the curve between any two values of z to represent the probability that any item chosen at random will fall between the two values of z. The values given in Table A.1 show the proportion of process output beyond a single specification limit that is z standard deviation units away from the process average. It must be remembered, of course, that the process must be in statistical control and the variable must be normally distributed (see Chapters 5 and 6).

Normal probability paper

A convenient way to examine variables data is to plot it in a cumulative form on probability paper. This enables the proportion of items outside a given limit to be read directly from the diagram. It also allows the data to be tested for normality – if it is normal the cumulative frequency plot will be a straight line.

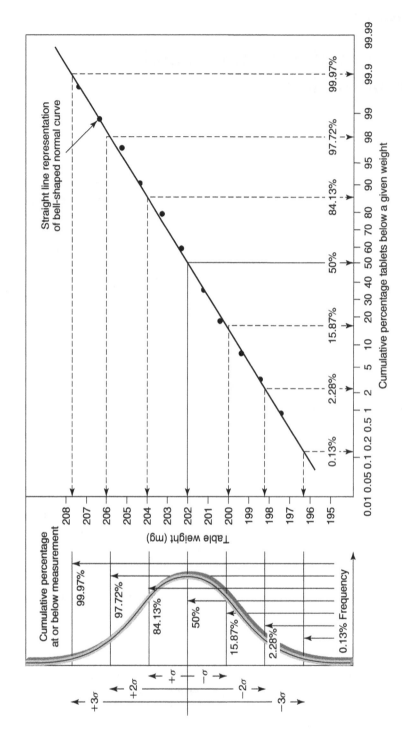

Figure A.1 Probability plot of normally distributed data (tablet weights)

Table A.1 Proportions under the tail of the normal distribution

$Z = (x - \mu)/\sigma$	.00	.01	.02	.03	.04	.05	.06	.07	.08	.09
0.0	.5000	.4960	.4920	.4880	.4840	.4801	.4761	.4721	.4681	.4641
0.1	.4602	.4562	.4522	.4483	.4443	.4404	.4364	.4325	.4286	.4247
0.2	.4207	.4168	.4129	.4090	.4052	.4013	.3974	.3936	.3897	.3859
0.3	.3821	.3783	.3745	.3707	.3669	.3632	.3594	.3557	.3520	.3483
0.4	.3446	.3409	.3372	.3336	.3300	.3264	.3228	.3192	.3156	.3121
0.5	.3085	.3050	.3015	.2981	.2946	.2912	.2877	.2843	.2810	.2776
0.6	.2743	.2709	.2676	.2643	.2611	.2578	.2546	.2514	.2483	.2451
0.7	.2420	.2389	.2358	.2327	.2296	.2266	.2236	.2206	.2177	.2148
0.8	.2119	.2090	.2061	.2033	.2005	.1977	.1949	.1922	.1894	.1867
0.9	.1841	.1814	.1788	.1762	.1736	.1711	.1685	.1660	.1635	.1611
1.0	.1587	.1562	.1539	.1515	.1492	.1469	.1446	.1423	.1401	.1379
1.1	.1357	.1335	.1314	.1292	.1271	.1251	.1230	.1210	.1190	.1170
1.2	.1151	.1131	.1112	.1093	.1075	.1056	.1038	.1020	.1003	.0985
1.3	.0968	.0951	.0934	.0918	.0901	.0885	.0869	.0853	.0838	.0823
1.4	.0808	.0793	.0778	.0764	.0749	.0735	.0721	.0708	.0694	.0681
1.5	.0668	.0655	.0643	.0630	.0618	.0606	.0594	.0582	.0571	.0559
1.6	.0548	.0537	.0526	.0516	.0505	.0495	.0485	.0475	.0465	.0455
1.7	.0446	.0436	.0427	.0418	.0409	.0401	.0392	.0384	.0375	.0367
1.8	.0359	.0351	.0344	.0336	.0329	.0322	.0314	.0307	.0301	.0294
1.9	.0287	.0281	.0274	.0268	.0262	.0256	.0250	.0244	.0239	.0233
2.0	.0228	.0222	.0216	.0211	.0206	.0201	.0197	.0192	.0187	.0183
2.1	.0179	.0174	.0170	.0165	.0161	.0157	.0153	.0150	.0146	.0142
2.2	.0139	.0135	.0132	.0128	.0125	.0122	.0119	.0116	.0113	.0110
2.3	.0107	.0104	.0101	.0099	.0096	.0093	.0091	.0088	.0086	.0084
2.4	.0082	.0079	.0077	.0075	.0073	.0071	.0069	.0067	.0065	.0063
2.5	.0062	.0060	.0058	.0057	.0055	.0053	.0052	.0050	.0049	.0048
2.6	.0046	.0045	.0044	.0042	.0041	.0040	.0039	.0037	.0036	.0035
2.7	.0034	.0033	.0032	.0031	.0030	.0029	.0028	.0028	.0027	.0026
2.8	.0025	.0024	.0024	.0023	.0022	.0021	.0021	.0020	.0019	.0019
2.9	.0018	.0018	.0017	.0016	.0016	.0015	.0015	.0014	.0014	.0013
3.0	.0013									
3.1	.0009									
3.2	.0006									
3.3	.0004									
3.4	.0003									
3.5	.00025									
3.6	.00015									
3.7	.00010									
3.8	.00007									
3.9	.00005									
4.0	.00003									

The type of graph paper shown in Figure A.1 is readily obtainable but the plots and tests can be performed using any good statistical package such as Minitab. The variable is marked along the linear vertical scale, while the horizontal scale shows the percentage of items with variables below that value. The method of performing probability calculations and plots depends upon the number of values available.

Large sample size

Columns 1 and 2 in Table A.2 give a frequency table for weights of tablets. The cumulative total of tablets with the corresponding weights are given in column 3. The cumulative totals are expressed as percentages of $(n + 1)$ in column 4, where n is the total number of tablets. These percentages are plotted against the upper boundaries of the class intervals on probability paper in Figure A.1. The points fall approximately on a straight line indicating that the distribution is normal. From the graph we can read, for example, that about 2 per cent of the tablets in the population weigh 198.0 mg or less. This may be useful information if that weight represents a specification tolerance. We can also read off the median value as 202.0 mg – a value below which half (50 per cent) of the tablet weights will lie. If the distribution is normal, the median is also the mean weight.

It is possible to estimate the standard deviation of the data, using Figure A.1. We know that 68.3 per cent of the data from a normal distribution will lie between the values $\mu \pm \sigma$. Consequently if we read off the tablet weights corresponding to 15.87 and 84.13 per cent of the population, the difference between the two values will be equal to twice the standard deviation (σ).

Table A.2 Tablet weights

Column 1 tablet weights (mg)	Column 2 frequency (f)	Column 3 cumulative (i)	Column 4 percentage $\left(\dfrac{I}{n+1}\right) \times 100$
196.5–197.4	3	3	0.82
197.5–198.4	8	11	3.01
198.5–199.4	18	29	7.92
199.5–200.4	35	64	17.49
200.5–201.4	66	130	35.52
201.5–202.4	89	219	59.84
202.5–203.4	68	287	78.42
203.5–204.4	44	331	90.44
204.5–205.4	24	355	96.99
205.5–206.4	7	362	98.91
206.5–207.4	3	365 (n)	99.73

Table A.3 Lives of light bulbs

Bulb life in hours (ranked in ascending order)	Cumulative number of bulbs failed by a given life (i)	Percentage: $\dfrac{I}{n+1} \times 100$
460	1	9.1
520	2	18.2
550	3	27.3
580	4	36.4
620	5	45.5
640	6	54.5
660	7	63.6
700	8	72.7
740	9	81.8
800	10 (*n*)	90.9

Hence, from Figure A.1:

Weight at 84.13% = 203.85 mg

Weight at 15.87% = 200.15 mg

2σ = 3.70 mg

σ = 1.85 mg.

Small sample size

The procedure for sample sizes of less than 20 is very similar. A sample of 10 light bulbs have lives as shown in Table A.3. Once again the cumulative number failed by a given life is computed (second column) and expressed as a percentage of ($n + 1$) where n is the number of bulbs examined (third column). The results have been plotted on probability paper in Figure A.2. Estimates of mean and standard deviation may be made as before.

Non-normality

There are situations in which the data are not normally distributed. Non-normal distributions are indicated on linear probability paper by non-straight lines. The reasons for this type of data include:

1 The underlying distribution fits a standard statistical model other than normal. Ovality, impurity, flatness and other characteristics bounded by zero often have skew, which can be measured. Kurtosis is another measure of the shape of the distribution being the degree of 'flattening' or 'peaking'.

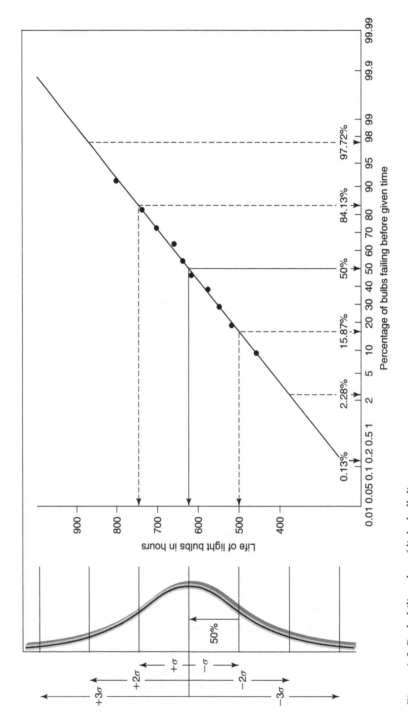

Figure A.2 Probability plot of light bulb lives

2   The underlying distribution is complex and does not fit a standard model. Self-adjusting processes, such as those controlled by software, often exhibit a non-normal pattern. The combination of outputs from several similar processes may not be normally distributed, even if the individual process outputs give normal patterns. Movement of the process mean due to gradual changes, such as tool wear, may also cause non-normality.

3   The underlying distribution is normal, but assignable causes of variation are present causing non-normal patterns. A change in material, operator interference, or damaged equipment are a few of the many examples which may cause this type of behaviour.

Standard probability paper may serve as a diagnostic tool to detect divergences from normality and to help decide future actions:

1   If there is a scattering of the points and no distinct pattern emerges, a technological investigation of the process is called for.

2   If the points make up a particular pattern, various interpretations of the behaviour of the characteristic are possible. Examples are given in Figure A.3. In Figure A.3a, selection of output has taken place to screen out that which is outside the specification. Figure A.3b shows selection to one specification limit or a drifting process. Figure A.3c shows a case where two distinct distribution patterns have been mixed. Two separate analyses should be performed by stratifying the data. If the points make up a smooth curve, as in Figure A.3d, this indicates a distribution other than normal. Interpretation of the pattern may suggest the use of an alternative probability paper.

In some cases, if the data are plotted on logarithmic probability paper, a straight line is obtained. This indicates that the data are taken from a lognormal distribution, which may then be used to estimate the appropriate descriptive parameters. Another type of probability paper is Weibull, used in the study of reliability. Points should be plotted on these papers against the appropriate measurement and cumulative percentage frequency values, in the same way as for normal data. The paper giving the best straight line fit should then be selected. When a satisfactory distribution fit has been achieved, capability indices (see Chapter 10) may be estimated by reading off the values at the points where the best fit line intercepts the 0.13 and 99.87 per cent lines. These values are then used in the formulae:

$$Cp = \frac{\text{USL} - \text{LSL}}{99.87 \text{ percentile} - 0.13 \text{ percentile}},$$

$$Cpk = \text{minimum of } \frac{\text{USL} - \bar{X}}{99.87 \text{ percentile} - \bar{X}} \text{ or } \frac{\bar{X} - \text{LSL}}{\bar{X} - 0.13 \text{ percentile}}.$$

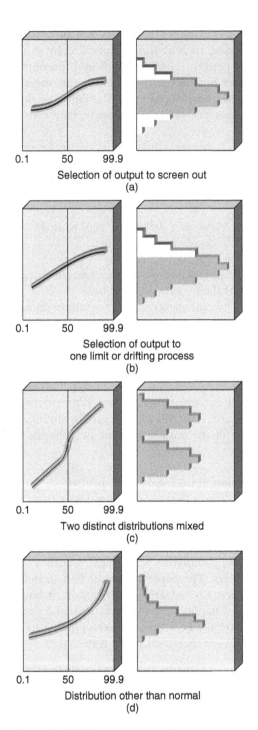

0.1   50   99.9

Selection of output to screen out
(a)

0.1   50   99.9

Selection of output to
one limit or drifting process
(b)

0.1   50   99.9

Two distinct distributions mixed
(c)

0.1   50   99.9

Distribution other than normal
(d)

Figure A.3 Various non-normal patterns on probability paper

Computer methods

There are now many statistical software packages, such as Minitab, that have routine procedures for testing for normality. These will carry out a probability plot and calculate indices for both skewness and kurtosis. As with all indices, these are only meaningful to those who understand them.

Appendix B

Constants used in the design of control charts for mean

Sample size (n)	Hartley's Constant (d_n or d_2)	Constants for mean charts using					
		Sample standard deviation		Sample range		Average sample standard deviation	
		A_1	$2/3\ A_1$	A_2	$2/3\ A_2$	A_3	$2/3\ A_3$
2	1.128	2.12	1.41	1.88	1.25	2.66	1.77
3	1.693	1.73	1.15	1.02	0.68	1.95	1.30
4	2.059	1.50	1.00	0.73	0.49	1.63	1.09
5	2.326	1.34	0.89	0.58	0.39	1.43	0.95
6	2.534	1.20	0.82	0.48	0.32	1.29	0.86
7	2.704	1.13	0.76	0.42	0.28	1.18	0.79
8	2.847	1.06	0.71	0.37	0.25	1.10	0.73
9	2.970	1.00	0.67	0.34	0.20	1.03	0.69
10	3.078	0.95	0.63	0.31	0.21	0.98	0.65
11	3.173	0.90	0.60	0.29	0.19	0.93	0.62
12	3.258	0.87	0.58	0.27	0.18	0.89	0.59

Formulae

$$\sigma = \frac{\bar{R}}{d_n} \text{ or } \frac{\bar{R}}{d_2}$$

Mean charts

$$\text{Action lines} = \bar{\bar{X}} \pm A_1 \sigma \quad \text{Warning lines} = \bar{\bar{X}} \pm 2/3\ A_1 \sigma$$
$$= \bar{\bar{X}} \pm A_2 \bar{R} \qquad\qquad\qquad = \bar{\bar{X}} \pm 2/3\ A_2 \bar{R}$$
$$= \bar{\bar{X}} \pm A_3 \bar{s} \qquad\qquad\qquad = \bar{\bar{X}} \pm 2/3\ A_3 \bar{s}$$

Process capability

$$Cp = \frac{USL - LSL}{\sigma}$$

$$Cpk = \text{minimum of } \frac{USL - \bar{\bar{X}}}{3\sigma} \text{ or } \frac{\bar{\bar{X}} - LSL}{3\sigma}$$

Appendix C

Constants used in the design of control charts for range

Sample size (n)	Constants for use with mean range ($\bar{R}$)				Constants for use with standard deviation (σ)				Constants for use in USA range charts based on $\bar{R}$	
	$D'_{0.999}$	$D'_{0.001}$	$D'_{0.975}$	$D'_{0.025}$	$D_{0.999}$	$D_{0.001}$	$D_{0.975}$	$D_{0.025}$	D_2	D_4
2	0.00	4.12	0.44	2.81	0.00	4.65	0.04	3.17	0	3.27
3	0.04	2.98	0.18	2.17	0.06	5.05	0.30	3.68	0	2.57
4	0.10	2.57	0.29	1.93	0.20	5.30	0.59	3.98	0	2.28
5	0.16	2.34	0.37	1.81	0.37	5.45	0.85	4.20	0	2.11
6	0.21	2.21	0.42	1.72	0.54	5.60	1.06	4.36	0	2.00
7	0.26	2.11	0.46	1.66	0.69	5.70	1.25	4.49	0.08	1.92
8	0.29	2.04	0.50	1.62	0.83	5.80	1.41	4.61	0.14	1.86
9	0.32	1.99	0.52	1.58	0.96	5.90	1.55	4.70	0.18	1.82
10	0.35	1.93	0.54	1.56	1.08	5.95	1.67	4.79	0.22	1.78
11	0.38	1.91	0.56	1.53	1.20	6.05	1.78	4.86	0.26	1.74
12	0.40	1.87	0.58	1.51	1.30	6.10	1.88	4.92	0.28	1.72

Formulae

Action lines: Upper $= D'_{0.001}\,\bar{R}$ or $D_{0.001}\,\sigma$

Lower $= D'_{0.999}\,\bar{R}$ or $D_{0.999}\,\sigma$

Warning lines: Upper $= D'_{0.025}\,\bar{R}$ or $D_{0.025}\,\sigma$

Lower $= D'_{0.975}\,\bar{R}$ or $D_{0.975}\,\sigma$

Control limits (USA): Upper $= D_4\,\bar{R}$

Lower $= D_2\,\bar{R}$

Appendix D

Constants used in the design of control charts for median and range

Sample size (n)	Constants for median charts		Constants for range charts	
	A_4	$2/3\ A_4$	$D^m_{.001}$	$D^m_{.025}$
2	2.22	1.48	3.98	2.53
3	1.27	0.84	2.83	1.79
4	0.83	0.55	2.45	1.55
5	0.71	0.47	2.24	1.42
6	0.56	0.37	2.12	1.34
7	0.52	0.35	2.03	1.29
8	0.44	0.29	1.96	1.24
9	0.42	0.28	1.91	1.21
10	0.37	0.25	1.88	1.18

Formulae

Median chart	Action lines	$= \tilde{\tilde{X}} \pm A_4 \tilde{R}$
	Warning lines	$= \tilde{\tilde{X}} \pm 2/3 A_4 \tilde{R}$
Range chart	Upper action line	$= D^m_{.001} \tilde{R}$
	Upper warning line	$= D^m_{.025} \tilde{R}$

Appendix E

Constants used in the design of control charts for standard deviation

Sample size (n)	C_n	Constants used with $\bar{s}$				Constants used with σ			
		$B'_{.001}$	$B'_{.025}$	$B'_{.975}$	$B'_{.999}$	$B_{.001}$	$B_{.025}$	$B_{.975}$	$B_{.999}$
2	1.253	4.12	2.80	0.04	0.02	3.29	2.24	0.03	0.01
3	1.128	2.96	2.17	0.18	0.04	2.63	1.92	0.16	0.03
4	1.085	2.52	1.91	0.29	0.10	2.32	1.76	0.27	0.09
5	1.064	2.28	1.78	0.37	0.16	2.15	1.67	0.35	0.15
6	1.051	2.13	1.69	0.43	0.22	2.03	1.61	0.41	0.21
7	1.042	2.01	1.61	0.47	0.26	1.92	1.55	0.45	0.25
8	1.036	1.93	1.57	0.51	0.30	1.86	1.51	0.49	0.29
9	1.032	1.87	1.53	0.54	0.34	1.81	1.48	0.52	0.33
10	1.028	1.81	1.49	0.56	0.37	1.76	1.45	0.55	0.36
11	1.025	1.78	1.49	0.58	0.39	1.73	1.45	0.57	0.38
12	1.023	1.73	1.44	0.60	0.42	1.69	1.41	0.59	0.41
13	1.021	1.69	1.42	0.62	0.44	1.66	1.39	0.61	0.43
14	1.019	1.67	1.41	0.63	0.46	1.64	1.38	0.62	0.45
15	1.018	1.64	1.40	0.65	0.47	1.61	1.37	0.63	0.47
16	1.017	1.63	1.38	0.66	0.49	1.60	1.35	0.65	0.48
17	1.016	1.61	1.36	0.67	0.50	1.58	1.34	0.66	0.50
18	1.015	1.59	1.35	0.68	0.52	1.56	1.33	0.67	0.51
19	1.014	1.57	1.34	0.69	0.53	1.55	1.32	0.68	0.52
20	1.013	1.54	1.34	0.69	0.54	1.52	1.32	0.68	0.53
21	1.013	1.52	1.33	0.70	0.55	1.50	1.31	0.69	0.54
22	1.012	1.51	1.32	0.71	0.56	1.49	1.30	0.70	0.56
23	1.011	1.50	1.31	0.72	0.57	1.48	1.30	0.71	0.56
24	1.011	1.49	1.30	0.72	0.58	1.47	1.29	0.71	0.57
25	1.011	1.48	1.30	0.73	0.59	1.46	1.28	0.72	0.58

Formulae

$$\sigma = \bar{s} C_n$$

Standard deviation chart
$$\begin{cases} \text{Upper action line} & = B'_{.001}\ \bar{s} \text{ or } B_{.001}\sigma \\ \text{Upper warning line} & = B'_{.025}\ \bar{s} \text{ or } B_{.025}\sigma \\ \text{Lower warning line} & = B'_{.975}\ \bar{s} \text{ or } B_{.975}\sigma \\ \text{Lower action line} & = B'_{.999}\ \bar{s} \text{ or } B_{.999}\sigma \end{cases}$$

Appendix F
Cumulative Poisson probability curves

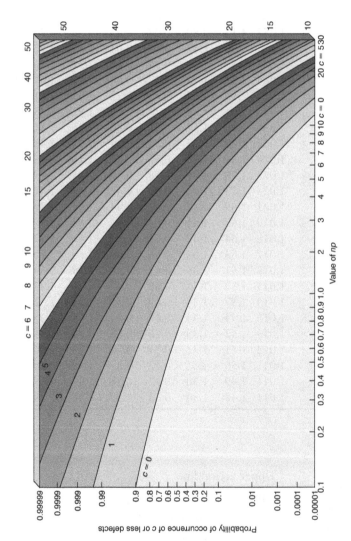

Figure F.1 Cumulative probability curves. For determining probability of occurrence of c or less defects in a sample of n pieces selected from a population in which the fraction defective is p (a modification of chart given by Miss. F. Thorndike, Bell System Technical Journal, October, 1926)

Appendix G

Confidence limits and tests of significance

Confidence limits

When an estimate of the mean of a parameter has been made it is desirable to know not only the estimated mean value, which should be the most likely value, but also how precise the estimate is.

If, for example, 80 results on weights of tablets give a mean $\bar{X} = 250.5$ mg and standard deviation $\sigma = 4.5$ mg, have these values come from a process with mean $\mu = 250.0$ mg? If the process has a mean $\mu = 250.0$, 99.7 per cent of all sample means ($\bar{X}$) should have a value between:

$$\mu \pm 3\sigma / \sqrt{n},$$

i.e.

$$\mu - 3\sigma / \sqrt{n} < \bar{X} < \mu + 3\sigma / \sqrt{n},$$

therefore:

$$\bar{X} - 3\sigma / \sqrt{n} < \mu < \bar{X} + 3\sigma / \sqrt{n},$$

i.e. μ will lie between:

$$\bar{X} \pm 3\sigma \sqrt{n},$$

this is the *confidence interval* at the *confidence coefficient* of 99.7 per cent.

Hence, for the tablet example, the 99.7 per cent interval for μ is:

$$250.5 \pm (3 \times 4.5 / \sqrt{80}) \text{ mg},$$

i.e.

$$249.0–252.0 \text{ mg},$$

which says that we may be 99.7 per cent confident that the true mean of the process lies between 249 and 252 mg, provided that the process was in statistical control at the time of the data collection. A 95 per cent confidence interval may be calculated in a similar way, using the range $\pm 2\sigma / \sqrt{n}$. This is, of course, the basis of the control chart for means.

Difference between two mean values

A problem that frequently arises is to assess the magnitude of the differences between two mean values. The difference between the two observed means is calculated $\bar{X}_1 - \bar{X}_2$, together with the standard error: of the difference. These values are then used to calculate confidence limits for the true difference, $\mu_1 - \mu_2$. If the upper limit is less than zero, μ_2 is greater than μ_1; if the lower limit is greater than zero, μ_1 is greater than μ_2. If the limits are too wide to lead to reliable conclusions, more observations are required.

If we have for sample size n_1, $\bar{X}_1$ and σ_1, and for sample size n_2, $\bar{X}_2$ and σ_2, the standard error of $\bar{X}_1 - \bar{X}_2$ is:

$$SE = \sqrt{\frac{\sigma_1^2}{n_1} + \frac{\sigma_2^2}{n_2}}.$$

When σ_1 and σ_2 are more or less equal:

$$SE = \sigma \sqrt{\frac{1}{n_1} + \frac{1}{n_2}}.$$

The 99.7 per cent confidence limits are, therefore:

$$(\bar{X}_1 - \bar{X}_2) \pm 3\sigma \sqrt{\frac{1}{n_1} + \frac{1}{n_2}}.$$

Tests of significance

A common procedure to aid interpretation of data analysis is to carry out a 'test of significance'. When applying such a test, we calculate the probability p that a certain result would occur if a 'null hypothesis' were true, i.e. that the result does not differ from a particular value. If this probability is equal to or less than a given value, α, the result is said to be significant at the α level. When $p = 0.05$, the result is usually referred to as 'significant' and when $p = 0.01$ as 'highly significant'.

The t-test for means

There are two types of tests for means, the normal test given above and the 'students' t-test. The normal test applies when the standard deviation σ is

Table G.1 Probability points of the *t*-distribution (single sided)

Degrees of freedom (n – 1)	P				
	0.1	0.05	0.025	0.01	0.005
1	3.08	6.31	12.70	31.80	63.70
2	1.89	2.92	4.30	6.96	9.92
3	1.64	2.35	3.18	4.54	5.84
4	1.53	2.13	2.78	3.75	4.60
5	1.48	2.01	2.57	3.36	4.03
6	1.44	1.94	2.45	3.14	3.71
7	1.42	1.89	2.36	3.00	3.50
8	1.40	1.86	2.31	2.90	3.36
9	1.38	1.83	2.26	2.82	3.25
10	1.37	1.81	2.23	2.76	3.17
11	1.36	1.80	2.20	2.72	3.11
12	1.36	1.78	2.18	2.68	3.05
13	1.35	1.77	2.16	2.65	3.01
14	1.34	1.76	2.14	2.62	2.98
15	1.34	1.75	2.13	2.60	2.95
16	1.34	1.75	2.12	2.58	2.92
17	1.33	1.74	2.11	2.57	2.90
18	1.33	1.73	2.10	2.55	2.88
19	1.33	1.73	2.09	2.54	2.86
20	1.32	1.72	2.09	2.53	2.85
21	1.32	1.72	2.08	2.52	2.83
22	1.32	1.72	2.07	2.51	2.82
23	1.32	1.71	2.07	2.50	2.81
24	1.32	1.71	2.06	2.49	2.80
25	1.32	1.71	2.06	2.48	2.79
26	1.32	1.71	2.06	2.48	2.78
27	1.31	1.70	2.05	2.47	2.77
28	1.31	1.70	2.05	2.47	2.76
29	1.31	1.70	2.05	2.46	2.76
30	1.31	1.70	2.04	2.46	2.75
40	1.30	1.68	2.02	2.42	2.70
60	1.30	1.67	2.00	2.39	2.66
120	1.29	1.66	1.98	2.36	2.62
∞	1.28	1.64	1.96	2.33	2.58

known or is based on a large sample, and the *t*-test is used when σ must be estimated from the data and the sample size is small ($n < 30$). The *t*-test is applied to the difference between two means μ_1 and μ_2 and two examples are given below to illustrate the *t*-test method:

1   In the first case μ_1 is known and μ_2 is estimated as $\overline{X}$ The first step is to calculate the t-statistic:

$$t = (\overline{X} - \mu_1)/s/\sqrt{n},$$

where s is the $(n - 1)$ estimate of σ. We then refer to Table G.1 to determine the significance. The following results were obtained for the percentage iron in 10 samples of furnace slag material: 15.3, 15.6, 16.0, 15.4, 16.4, 15.8, 15.7, 15.9, 16.1, 15.7. Do the analyses indicate that the material is significantly different from the declared specification of 16.0 per cent?

$$\overline{X} = \frac{\Sigma X}{n} = \frac{157.9}{10} = 15.79\%,$$

$$s_{(n-1)} = \sqrt{\frac{\Sigma(X_i - \overline{X})^2}{n-1}} = 0.328\%,$$

$$t_{calc} = \frac{\mu_1 - \overline{X}}{s/\sqrt{n}} = \frac{16.0 - 15.79}{0.328/\sqrt{10}}$$

$$= \frac{0.21}{0.1037} = 2.025.$$

Consultation of Table G.1 for $(n - 1) = 9$ (i.e. the 'number of degrees of freedom') gives a tabulated value for $t_{0.05}$ of 1.83, i.e. at the 5 per cent level of significance. Hence, there is only a 5 per cent chance that the calculated value of t will exceed 1.83, if there is no significant difference between the mean of the analyses and the specification. So we many conclude that the mean analysis differs significantly (at 5 per cent level) from the specification. Note, the result is not highly significant, since the tabulated value of $t_{0.01}$, i.e. at the 1 per cent level, is 2.82 and this has not been exceeded.

2   In the second case, results from two sources are being compared. This situation requires the calculation of the t-statistic from the mean of the differences in values and the standard error of the differences. The example should illustrate the method. To check on the analysis of percentage impurity present in a certain product, a manufacturer took 12 samples, halved each of them and had one half tested in his own laboratory (A) and the other half tested by an independent laboratory (B). The results obtained were:

Sample No.	1	2	3	4	5	6
Laboratory A	0.74	0.52	0.32	0.67	0.47	0.77
Laboratory B	0.79	0.50	0.43	0.77	0.67	0.68
Difference, d = A–B	−0.05	+0.02	−0.11	−0.10	−0.20	+0.09

	7	8	9	10	11	12
Laboratory A	0.72	0.80	0.70	0.69	0.94	0.87
Laboratory B	0.91	0.80	0.98	0.67	0.93	0.82
Difference, d = A–B	–0.19	0	–0.28	+0.02	10.01	10.05

Is there any significant difference between the test results from the two laboratories?

Total difference $|\Sigma d| = 0.74$,

Mean difference $|\bar{d}| = \dfrac{|\Sigma d|}{n} = \dfrac{0.74}{12} = 0.062$,

Standard deviation estimate,

$$s_{(n-1)} = \sqrt{\frac{\Sigma(\bar{d} - d_i)^2}{n-1}} = 0.115,$$

$$t_{calc} = \frac{|\bar{d}|}{s/\sqrt{n}} = \frac{0.062}{0.115/\sqrt{12}} = 1.868.$$

From Table G.1 and for $(n-1) = 11$ degrees of freedom, the tabulated value of t is obtained. As we are looking for a difference in means, irrespective of which is greater, the test is said to be *double sided*, and it is necessary to double the probabilities in Table G.1 for the critical values of t. From Table G.1 then:

$$t_{0.025}(11) = 2.20,$$

since

$$1.868 < 2.20,$$

i.e.

$$t_{calc} < t_{0.025}(11),$$

and there is insufficient evidence, at the 5 per cent level, to suggest that the two laboratories differ.

The F-*test for variances*

The F-test is used for comparing two variances. If it is required to compare the values of two variances σ_1^2 and σ_2^2 from estimates s_1^2 and s_2^2, based on $(n_1 - 1)$ and $(n_2 - 1)$ degrees of freedom, respectively, and the alternative to the Null

Table G.2 Critical values of F for variances

Probability point	$n_2 - 1$	Degrees of freedom $n_1 - 1$ (corresponding to greater variance)																		
		1	2	3	4	5	6	7	8	9	10	12	15	20	24	30	40	60	120	∞
0.100	1	39.9	49.5	53.6	55.8	57.2	58.2	58.9	59.4	59.9	60.2	60.7	61.2	61.7	62.0	62.3	62.5	62.8	63.1	63.3
0.050		161	199	216	225	230	234	237	239	241	242	244	246	248	249	250	251	252	253	254
0.025		648	800	864	900	922	937	948	957	963	969	977	985	993	997	1001	1006	1010	1014	1018
0.010		4052	4999	5403	5625	5764	5859	5928	5982	6022	6056	6106	6157	6209	6235	6261	6287	6313	6339	6366
0.100	2	8.53	9.00	9.16	9.24	9.29	9.33	9.35	9.37	9.38	9.39	9.41	9.42	9.44	9.45	9.46	9.47	9.48	9.49	
0.050		18.5	19.0	19.2	19.2	19.3	19.3	19.4	19.4	19.4	19.4	19.4	19.4	19.4	19.5	19.5	19.5	19.5	19.5	19.5
0.025		38.5	39.0	39.2	39.2	39.3	39.3	39.4	39.4	39.4	39.4	39.4	39.4	39.4	39.5	39.5	39.5	39.5	39.5	39.5
0.010		98.5	99.0	99.2	99.2	99.3	99.3	99.4	99.4	99.4	99.4	99.4	99.4	99.4	99.5	99.5	99.5	99.5	99.5	99.5
0.100	3	5.54	5.46	5.39	5.34	5.31	5.28	5.27	5.25	5.24	5.23	5.22	5.20	5.18	5.18	5.17	5.16	5.15	5.14	5.13
0.050		10.1	9.55	9.28	9.12	9.01	8.94	8.89	8.85	8.81	8.79	8.74	8.70	8.66	8.64	8.62	8.59	8.57	8.55	8.53
0.025		17.4	16.0	15.4	15.1	14.9	14.7	14.6	14.5	14.5	14.4	14.3	14.3	14.2	14.1	14.1	14.0	14.0	13.9	13.9
0.010		34.1	30.8	29.5	28.7	28.2	27.9	27.7	27.5	27.3	27.2	27.1	26.9	26.7	26.6	26.5	26.4	26.3	26.2	26.1
0.100	4	4.54	4.32	4.19	4.11	4.05	4.01	3.98	3.95	3.94	3.92	3.90	3.87	3.84	3.83	3.82	3.80	3.79	3.78	3.76
0.050		7.71	6.94	6.59	6.39	6.26	6.16	6.09	6.04	6.00	5.96	5.91	5.86	5.80	5.77	5.75	5.72	5.69	5.66	5.63
0.025		12.2	10.6	10.0	9.60	9.36	9.20	9.07	8.98	8.90	8.84	8.75	8.66	8.56	8.51	8.46	8.41	8.36	8.31	8.26
0.010		21.2	18.0	16.7	16.0	15.5	15.2	15.0	14.8	14.7	14.5	14.4	14.2	14.0	13.9	13.8	13.7	13.7	13.6	13.5
0.100	5	4.06	3.78	3.62	3.52	3.45	3.40	3.37	3.34	3.32	3.30	3.27	3.24	3.21	3.19	3.17	3.16	3.14	3.12	3.10
0.050		6.61	5.79	5.41	5.19	5.05	4.95	4.88	4.82	4.77	4.74	4.68	4.62	4.56	4.53	4.50	4.46	4.43	4.40	4.36
0.025		10.0	8.43	7.76	7.39	7.15	6.98	6.85	6.76	6.68	6.62	6.52	6.43	6.33	6.28	6.23	6.18	6.12	6.07	6.02
0.010		16.3	13.3	12.1	11.4	11.0	10.7	10.5	10.3	10.2	10.1	9.89	9.72	9.55	9.47	9.38	9.29	9.20	9.11	9.02
0.100	6	3.78	3.46	3.29	3.18	3.11	3.05	3.01	2.98	2.96	2.94	2.90	2.87	2.84	2.82	2.80	2.78	2.76	2.74	2.72
0.050		5.99	5.14	4.76	4.53	4.39	4.28	4.21	4.15	4.10	4.06	4.00	3.94	3.87	3.84	3.81	3.77	3.74	3.70	3.67

df_2	α																		
	0.025	8.81	7.26	6.60	6.23	5.99	5.82	5.70	5.60	5.52	5.46	5.37	5.27	5.17	5.12	5.07	5.01	4.96	4.90	4.85
	0.010	13.7	10.9	9.78	9.15	8.75	8.47	8.26	8.10	7.98	7.87	7.72	7.56	7.40	7.31	7.23	7.14	7.06	6.97	6.88
7	0.100	3.59	3.26	3.07	2.96	2.88	2.83	2.78	2.75	2.72	2.70	2.67	2.63	2.59	2.58	2.56	2.54	2.51	2.49	2.47
	0.050	5.59	4.74	4.35	4.12	3.97	3.87	3.79	3.73	3.68	3.64	3.57	3.51	3.44	3.41	3.38	3.34	3.30	3.27	3.23
	0.025	8.07	6.54	5.89	5.52	5.29	5.12	4.99	4.90	4.82	4.76	4.67	4.57	4.47	4.42	4.36	4.31	4.25	4.20	4.14
	0.010	12.2	9.55	8.45	7.85	7.46	7.19	6.99	6.84	6.72	6.62	6.47	6.31	6.16	6.07	5.99	5.91	5.82	5.74	5.65
8	0.100	3.46	3.11	2.92	2.81	2.73	2.67	2.62	2.59	2.56	2.54	2.50	2.46	2.42	2.40	2.38	2.36	2.34	2.32	2.29
	0.050	5.32	4.46	4.07	3.84	3.69	3.58	3.50	3.44	3.39	3.35	3.28	3.22	3.15	3.12	3.08	3.04	3.01	2.97	2.93
	0.025	7.57	6.06	5.42	5.05	4.82	4.65	4.53	4.43	4.36	4.30	4.20	4.10	4.00	3.95	3.89	3.84	3.78	3.73	3.67
	0.010	11.3	8.65	7.59	7.01	6.63	6.37	6.18	6.03	5.91	5.81	5.67	5.52	5.36	5.28	5.20	5.12	5.03	4.95	4.86
9	0.100	3.36	3.01	2.81	2.69	2.61	2.55	2.51	2.47	2.44	2.42	2.38	2.34	2.30	2.28	2.25	2.23	2.21	2.18	2.16
	0.050	5.12	4.26	3.86	3.63	3.48	3.37	3.29	3.23	3.18	3.14	3.07	3.01	2.94	2.90	2.86	2.83	2.79	2.75	2.71
	0.025	7.12	5.71	5.08	4.72	4.48	4.32	4.20	4.10	4.03	3.96	3.87	3.77	3.67	3.61	3.56	3.51	3.45	3.39	3.33
	0.010	10.6	8.02	6.99	6.42	6.06	5.80	5.61	5.47	5.35	5.26	5.11	4.96	4.81	4.73	4.65	4.57	4.48	4.40	4.31
10	0.100	3.28	2.92	2.73	2.61	2.52	2.46	2.41	2.38	2.35	2.32	2.28	2.24	2.20	2.18	2.16	2.13	2.11	2.08	2.06
	0.050	4.96	4.10	3.71	3.48	3.33	3.22	3.14	3.07	3.02	2.98	2.91	2.84	2.77	2.74	2.70	2.66	2.62	2.58	2.54
	0.025	6.94	5.46	4.83	4.47	4.24	4.07	3.95	3.85	3.78	3.72	3.62	3.52	3.42	3.37	3.31	3.26	3.20	3.14	3.08
	0.010	10.0	7.56	6.55	5.99	5.64	5.39	5.20	5.06	4.94	4.85	4.71	4.56	4.41	4.33	4.25	4.17	4.08	4.00	3.91
12	0.100	3.18	2.81	2.61	2.48	2.39	2.33	2.28	2.24	2.21	2.19	2.15	2.10	2.06	2.04	2.01	1.99	1.96	1.93	1.90
	0.050	4.75	3.89	3.49	3.26	3.11	3.00	2.91	2.85	2.80	2.75	2.69	2.62	2.54	2.51	2.47	2.43	2.39	2.34	2.30
	0.025	6.55	5.10	4.47	4.12	3.89	3.73	3.61	3.51	3.44	3.37	3.28	3.18	3.07	3.02	2.96	2.91	2.85	2.79	2.72
	0.010	9.33	6.93	5.95	5.41	5.06	4.82	4.64	4.50	4.39	4.30	4.16	4.01	3.86	3.78	3.70	3.62	3.54	3.45	3.36
15	1.100	3.07	2.70	2.49	2.36	2.27	2.21	2.16	2.12	2.09	2.06	2.02	1.97	1.92	1.90	1.87	1.85	1.82	1.79	1.76
	0.050	4.54	3.68	3.29	3.06	2.90	2.79	2.71	2.64	2.59	2.54	2.48	2.40	2.33	2.29	2.25	2.20	2.16	2.11	2.07
	0.025	6.20	4.77	4.15	3.80	3.58	3.41	3.29	3.20	3.12	3.06	2.96	2.86	2.76	2.70	2.64	2.59	2.52	2.46	2.40
	0.010	8.68	6.36	5.42	4.89	4.56	4.32	4.14	4.00	3.89	3.80	3.67	3.52	3.37	3.29	3.21	3.13	3.05	2.96	2.87

(continued)

Table G.2 (continued)

Probability point	degrees of freedom $n_2 - 1$	Degrees of freedom $n_1 - 1$ (corresponding to greater variance)																		
		1	2	3	4	5	6	7	8	9	10	12	15	20	24	30	40	60	120	∞
0.100	20	2.97	2.59	2.38	2.25	2.16	2.09	2.04	2.00	1.96	1.94	1.89	1.84	1.79	1.77	1.74	1.71	1.68	1.64	1.61
0.050		4.35	3.49	3.10	2.87	2.71	2.60	2.51	2.45	2.39	2.35	2.28	2.20	2.12	2.08	2.04	1.99	1.95	1.90	1.84
0.025		5.87	4.46	3.86	3.51	3.29	3.13	3.01	2.91	2.84	2.77	2.68	2.57	2.46	2.41	2.35	2.29	2.22	2.16	2.09
0.010		8.10	5.85	4.94	4.43	4.10	3.87	3.70	3.56	3.46	3.37	3.23	3.09	2.94	2.86	2.78	2.69	2.61	2.52	2.42
0.100	24	2.93	2.54	2.33	2.19	2.10	2.04	1.98	1.94	1.91	1.88	1.83	1.78	1.73	1.70	1.67	1.64	1.61	1.57	1.53
0.050		4.26	3.40	3.01	2.78	2.62	2.51	2.42	2.36	2.30	2.25	2.18	2.11	2.03	1.98	1.94	1.89	1.84	1.79	1.73
0.025		5.72	4.32	3.72	3.38	3.15	2.99	2.87	2.78	2.70	2.64	2.54	2.44	2.33	2.27	2.21	2.15	2.08	2.01	1.94
0.010		7.82	5.61	4.72	4.22	3.90	3.67	3.50	3.36	3.26	3.17	3.03	2.89	2.74	2.66	2.58	2.49	2.40	2.31	2.21
0.100	30	2.88	2.49	2.28	2.14	2.05	1.98	1.93	1.88	1.85	1.82	1.77	1.72	1.67	1.64	1.61	1.57	1.54	1.50	1.46
0.050		4.17	3.32	2.92	2.69	2.53	2.42	2.33	2.27	2.21	2.16	2.09	2.01	1.93	1.89	1.84	1.79	1.74	1.68	1.62
0.025		5.57	4.18	3.59	3.25	3.03	2.87	2.75	2.65	2.57	2.51	2.41	2.31	2.20	2.14	2.07	2.01	1.94	1.87	1.79
0.010		7.56	5.39	4.51	4.02	3.70	3.47	3.30	3.17	3.07	2.98	2.84	2.70	2.55	2.47	2.39	2.30	2.21	2.11	2.01
0.100	40	2.84	2.44	2.23	2.09	2.00	1.93	1.87	1.83	1.79	1.76	1.71	1.66	1.61	1.57	1.54	1.51	1.47	1.42	1.38
1.050		4.08	3.23	2.84	2.61	2.45	2.34	2.25	2.18	2.12	2.08	2.00	1.92	1.84	1.79	1.74	1.69	1.64	1.58	1.51
0.025		5.42	4.05	3.46	3.13	2.90	2.74	2.62	2.53	2.45	2.39	2.29	2.18	2.07	2.01	1.94	1.88	1.80	1.72	1.64
0.010		7.31	5.18	4.31	3.83	3.51	3.29	3.12	2.99	2.89	2.80	2.66	2.52	2.37	2.29	2.20	2.11	2.02	1.92	1.80
0.100	60	2.79	2.39	2.18	2.04	1.95	1.87	1.82	1.77	1.74	1.71	1.66	1.60	1.54	1.51	1.48	1.44	1.40	1.35	1.29
0.050		4.00	3.15	2.76	2.53	2.37	2.25	2.17	2.10	2.04	1.99	1.92	1.84	1.75	1.70	1.65	1.59	1.53	1.47	1.39
0.025		5.29	3.93	3.34	3.01	2.79	2.63	2.51	2.41	2.33	2.27	2.17	2.06	1.94	1.88	1.82	1.74	1.67	1.58	1.48
0.010		7.08	4.98	4.13	3.65	3.34	3.12	2.95	2.82	2.72	2.63	2.50	2.35	2.20	2.12	2.03	1.94	1.84	1.73	1.60
0.100	120	2.75	2.35	2.13	1.99	1.90	1.82	1.77	1.72	1.68	1.65	1.60	1.54	1.48	1.45	1.41	1.37	1.32	1.26	1.19
0.050		3.92	3.07	2.68	2.45	2.29	2.18	2.09	2.02	1.96	1.91	1.83	1.75	1.66	1.61	1.55	1.50	1.43	1.35	1.25
0.025		5.15	3.80	3.23	2.89	2.67	2.52	2.39	2.30	2.22	2.16	2.05	1.94	1.82	1.76	1.69	1.61	1.53	1.43	1.31
0.01		6.85	4.79	3.95	3.48	3.17	2.96	2.79	2.66	2.56	2.47	2.34	2.19	2.03	1.95	1.86	1.76	1.66	1.53	1.38
0.100	∞	2.71	2.30	2.08	1.94	1.85	1.77	1.72	1.67	1.63	1.60	1.55	1.49	1.42	1.38	1.34	1.30	1.24	1.17	1.00
0.050		3.84	3.00	2.60	2.37	2.21	2.10	2.01	1.94	1.88	1.83	1.75	1.67	1.57	1.52	1.46	1.39	1.32	1.22	1.00
0.025		5.02	3.69	3.12	2.79	2.57	2.41	2.29	2.19	2.11	2.05	1.94	1.83	1.71	1.64	1.57	1.48	1.39	1.27	1.00
0.010		6.63	4.61	3.78	3.32	3.02	2.80	2.64	2.51	2.41	2.32	2.18	2.04	1.88	1.79	1.70	1.59	1.47	1.32	1.00

Hypothesis ($\sigma_1^2 = \sigma_1^2$) is $\sigma_1^2 > \sigma_2^2$, we calculate the ratio $F = s_1^2/s_2^2$ and refer to Table G.2 for the critical values of F, with ($n_1 - 1$) and ($n_2 - 1$) degrees of freedom, where s_1^2 is always the highest variance and n_1 is the corresponding sample size. The levels tabulated in Table G.2 refer to the single upper tail area of the F-distribution. If the alternative to the Null Hypothesis σ_1^2 not equal to σ_2^2, the test is double sided, and we calculate the ratio of the larger estimate to the smaller one and the probabilities in Table G.2 are doubled to give the critical values for this ratio. In each case the calculated values of F must be greater than the tabulated critical values, for significant differences at the appropriate level shown in the probability point column.

For example, in the filling of cans of beans, it is suspected that the variability in the morning is greater than that in the afternoon. From collected data:

$$\text{Morning} \quad n_1 = 40, \quad \bar{X}_1 = 451.78, \quad s_1 = 1.76,$$

$$\text{Afternoon} \quad n_2 = 40, \quad \bar{X}_2 = 450.71, \quad s_2 = 1.55,$$

$$\text{Degrees of freedom} \ (n_1 - 1) = (n_2 - 1) = 39,$$

$$F = \frac{s_1^2}{s_2^2} = \frac{1.76^2}{1.55^2} = \frac{3.098}{2.403} = 1.29$$

(note if $s_1^2 < s_2^2$ the test statistic would have been $F = \dfrac{s_2^2}{s_1^2}$).

If there is a good reason for the variability in the morning to be greater than in the afternoon (e.g. equipment and people 'settling down') then the test will be a one-tail test. For $\alpha = 0.05$, from Table G.2, the critical value for the ratio is $F_{0.05} = 1.70$ by interpolation. Hence, the sample value of s^2/s^2 is not above F, and we accept the Null Hypothesis that $\sigma_1 = \sigma_2$, and the variances are the same in the morning and afternoon.

For confidence limits for the variance ratio, we require both the upper and lower tail areas of the distribution. The lower tail area is given by the reciprocal of the corresponding F-value in the upper tail. Hence, to obtain the 95 per cent confidence limits for the variance ratio, we require the values of $F_{0.975}$ and $F_{0.025}$. For example, if ($n_1 - 1$) = 9 and ($n_2 - 1$) = 15 then:

$$F_{0.975} \ (9,15) = \frac{1}{F_{0.025} \ (15,9)} = \frac{1}{3.77} = 0.27$$

and $F_{0.025} \ (9, 15) = 3.12$.

If s_1^2/s_2^2 exceeds 3.12 or falls short of 0.27, we shall reject the hypothesis that $\sigma_1 = \sigma_2$.

Appendix H

OC curves and ARL curves for $\overline{X}$ and R charts

Operating Characteristic (OC) curves for an R chart (based on upper action line only). Figure H.1 shows, for several different sample sizes, a plot of the probability or chance that the first sample point will fall below the upper action line, following a given increase in process standard deviation. The x axis is the ratio of the new standard deviation (after the change) to the old; the ordinate axis is the probability that this shift will not be detected by the first sample.

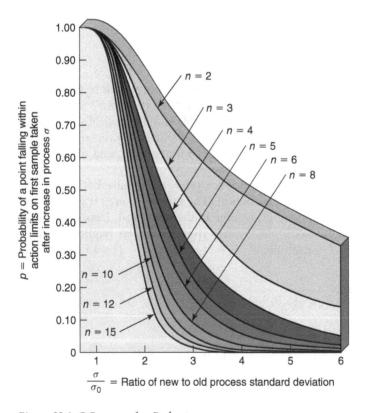

Figure H.1 OC curves for R chart

It is interesting to compare the OC curves for samples of various sizes. For example, when the process standard deviation increases by a factor of 3, the probability of not detecting the shift with the first sample is:

ca. 0.62 for $n = 2$

and

ca. 0.23 for $n = 5$.

The probabilities of *detecting* the change in the first sample are, therefore:

$1 - 0.62 = 0.38$ for $n = 2$

and

$1 - 0.23 = 0.77$ for $n = 5$.

The average run length (ARL) to detection is the *reciprocal of the probability of detection*. In the example of a tripling of the process standard deviation, the ARLs for the two sample sizes will be:

for $n = 2$, ARL $= 1/0.38 = 2.6$

and

for $n = 5$, ARL $= 1/0.77 = 1.3$.

Clearly the R chart for sample size $n = 5$ has a better 'performance' than the one for $n = 2$, in detecting an increase in process variability.

OC curves for an $\overline{X}$ *chart (based on action lines only).* If the process standard deviation remains constant, the OC curve for an $\overline{X}$ chart is relatively easy to construct. The probability that a sample will fall within the control limits or action lines can be obtained from the normal distribution table in Appendix A, assuming the sample size $n \geq 4$ or the parent distribution is normal. This is shown in general by Figure H.2, in which action lines for an $\overline{X}$ chart have been set up when the process was stable at mean μ_0, with standard deviation σ. The $\overline{X}$ chart action lines were set at $\overline{\overline{X}}_0 \pm 3\sigma) / \sqrt{n}$.

If the process mean decreases by $\delta\sigma$ to a new mean μ_1, the distribution of sample means will become centred at $\overline{\overline{X}}_1$, and the probability of the first sample mean falling outside the lower action line will be equal to the shaded proportion under the curve. This can be found from the table in Appendix A.

An example should clarify the method. For the steel rod cutting process, described in Chapters 5 and 6, the process mean $\overline{\overline{X}}_0 = 150.1$ mm and the standard deviation $\sigma = 5.25$ mm. The lower action line on the mean chart, for a sample size $n = 4$,

$$= \overline{\overline{X}}_0 - 3\sigma / \sqrt{n}$$
$$= 150.1 - 3 \times 5.25/\sqrt{4}$$
$$= 142.23 \text{ mm.}$$

If the process mean decreases by one σ value (5.25 mm), the distance between the action line and the new mean of the distribution of sample means $(\overline{X}_1)$ is given by:

$$(3\sigma/\sqrt{n} - \delta\sigma) = 3 \times 5.25/\sqrt{4} - 1 \times 5.25 = 2.625 \text{ mm.}$$

This distance in terms of number of standard errors of the mean (the standard deviation of the distribution) is:

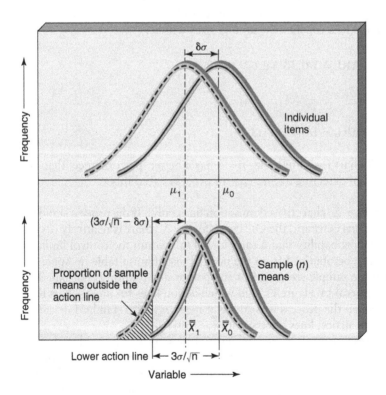

Figure H.2 Determination of OC curves for an X chart

$$\frac{(3\sigma/\sqrt{n} - \delta\sigma)}{\sigma/\sqrt{n}} \text{ standard errors} \qquad\qquad Formula\ A$$

or

$$\frac{2.625}{5.25/\sqrt{4}} = 1 \text{ standard error.}$$

The formula A may be further simplified to:

$$(3 - \delta\sqrt{n}) \text{ standard errors.}$$

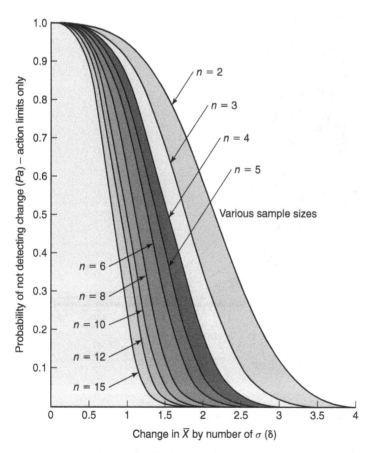

Figure H.3 OC curves for $\overline{X}$ chart

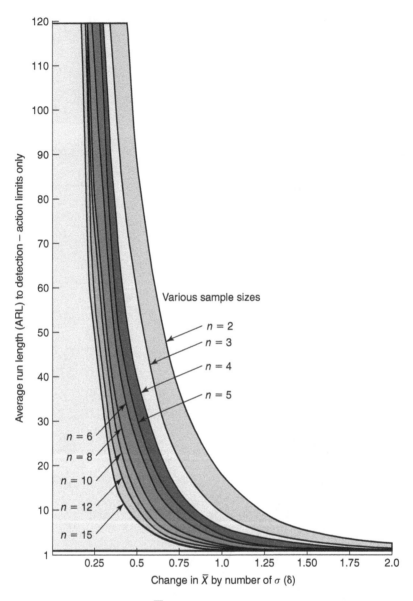

Figure H.4 ARL curves for $\overline{X}$ chart

In the example: $3 - 1 \times \sqrt{4} = 1$ standard error, and the shaded proportion under the distribution of sample means is 0.1587 (from Appendix A).
Hence, the probability of detecting, with the first sample on the means chart
($n = 4$), a change in process mean of one standard deviation is 0.1587. The

probability of *not detecting* the change is $1 - 0.1587 = 0.8413$ and this value may be used to plot a point on the OC curve. The ARL to detection of such a change using this chart, with action lines only, is $1/0.1587 = 6.3$.

Clearly the ARL will depend upon whether or not we incorporate the decision rules based on warning lines, runs and trends. Figures H.3 and H.4 show how the mean chart OC and ARL to action signal (point in zone 3), respectively, vary with the sample size, and these curves may be used to decide which sample size is appropriate, when inspection costs and the magnitude of likely changes have been considered. It is important to consider also the frequency of available data and in certain process industries ARLs in time, rather than points plotted, may be more useful. Alternative types of control charts for variables may be more appropriate in these situations (see Chapter 7).

Appendix I
Autocorrelation

A basic assumption in constructing control charts, such as those for $\overline{X}$ R, moving $\overline{X}$ and moving R, is that the individual data points used are independent of one another. When data are taken in order, there is often a tendency for the observations made close together in time or space to be more alike than those taken further apart. There is often a technological reason for this serial dependence or 'autocorrelation' in the data. For example, physical mixing, residence time or capacitance can produce autocorrelation in continuous processes.

Autocorrelation may be due to shift or day of week effects, or may be due to identifiable causes that are not related to the 'time' order of the data. When groups of batches of material are produced alternatively from two reactors, for example, positive autocorrelation can be explained by the fact that alternate batches are from the same reactor. Trends in data may also produce autocorrelation.

Autocorrelation may be displayed graphically by plotting the data on a scatter diagram, with one axis representing the data in the original order, and the other axis representing the data moved up or down by one or more observations (see Figure I.1).

In most cases, the relationship between the variable and its 'lag' can be summarized by a straight line. The strength of the linear relationship is indicated by the *correlation coefficient*, a number between −1 and 1. The autocorrelation coefficient, often called simply the autocorrelation, is the correlation coefficient of the variable with its lag. Clearly, there is a different autocorrelation for each lag.

If autocorrelated data are plotted on standard control charts, the process may appear to be out of statistical control for mean, when in fact the data represent a stable process. If action is taken on the process, in an attempt to find the incorrectly identified 'assignable' causes, additional variation will be introduced into the process.

When autocorrelation is encountered, there are four procedures to reduce its impact, these are based on *avoidance* and *correction*:

Avoid $\begin{cases} \text{1 Move to 'upstream' measurements to control the process.} \\ \text{2 For continuous processes, sample less often so that the} \\ \quad \text{sample interval is longer than the residence time.} \end{cases}$

Correct $\begin{cases} \text{3 For autocorrelation due to special causes, use stratification} \\ \quad \text{and rational subgrouping to clarify what is really happening.} \\ \text{4 For intrinsic, stable autocorrelation, use knowledge of the} \\ \quad \text{technology to model and 'filter out' the autocorrelation;} \\ \quad \text{control charts may then be applied to the filtered data.} \end{cases}$

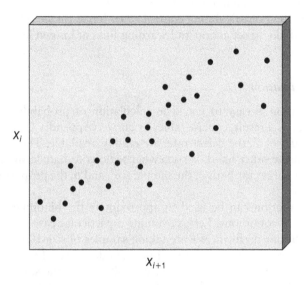

Figure I.1 Scatter plot of autocorrelated data

The mathematics for filtering the data, which can include Laplace transforms, are outside the scope of this book. The reader is referred to the many excellent texts on statistics that deal with these methods.

Appendix J

Approximations to assist in process control of attributes

This appendix is primarily intended for the reader who does not wish to accept the simple method of calculating control chart limits for sampling of attributes, but would like to set action and warning lines at known levels of probability.

The Poisson approximation

The Poisson distribution is easy to use. The calculation of probabilities is relatively simple and, as a result, concise tables or curves (Appendix F) which cover a range of values of $\bar{c}$, the defect rate, are readily available. The binomial distribution, on the other hand, is somewhat tedious to handle since it has to cover different values for both n, the sample size, and p, the proportion defective.

The Poisson distribution can be used to approximate the binomial distribution under certain conditions. Let us examine a particular case and see how the two distributions perform. We are taking samples of size 10 from a pottery process that is producing on average 1 per cent defectives. Expansion of the binomial expression $(0.01 + 0.99)^{10}$ or consultation of the statistical tables will give the following probabilities of finding 0, 1, 2 and 3 defectives:

Number of defectives in sample of 10	Binomial probability of finding that number of defectives
0	0.9044
1	0.0913
2	0.0042
3	0.0001

There is virtually no chance of finding more than three defectives in the sample. The reader may be able to appreciate these figures more easily if we imagine that we have taken 10,000 of these samples of 10. The results should look like this:

Number of defectives in sample of 10	Number of samples out of 10,000 that have that number of defectives
0	9044
1	913
2	42
3	1

We can check the average number of defectives per sample by calculating:

Average number

of defectives per sample $= \dfrac{\text{Total number of defectives}}{\text{Total number of samples}}$,

$$n\bar{p} = \frac{913 + (42 \times 2) + (3 \times 1)}{10,000},$$

$$= \frac{1000}{10,000} = 0.1.$$

Now, in the Poisson distribution we must use the average number of defectives $\bar{c}$ to calculate the probabilities. Hence, in the approximation we let:

$$\bar{c} = n\bar{p} = 0.1,$$

so:

$$e^{-\bar{c}}(\bar{c}^x/x!) = e^{-n\bar{p}}((n\bar{p})^x/x!) = e^{-0.1}(0.1^x/x!),$$

and we find that the probabilities of finding defectives in the sample of 10 are:

Number of defectives in sample of 10	Poisson probability of finding that number of defectives	Number of samples out of 10,000 that have that number of defectives
0	0.9048	9048
1	0.0905	905
2	0.0045	45
3	0.0002	2

The reader will observe the similarity of these results to those obtained using the binomial distribution:

Average number

of defectives per sample $= \dfrac{905 + (45 \times 2) + (2 \times 3)}{10{,}000}$,

$$n\bar{p} = \frac{1001}{10{,}000} = 0.1001.$$

We may now compare the calculations for the standard deviation of these results by the two methods:

Binomial $\sigma = \sqrt{n\bar{p}(1-\bar{p})} = \sqrt{10 \times 0.01 \times 0.99} = 0.315$.

Poisson $\sigma = \sqrt{\bar{c}} = \sqrt{n\bar{p}} = \sqrt{10 \times 0.01} \qquad = 0.316$.

The results are very similar because $(1 - \bar{p})$ is so close to unity that there is hardly any difference between the formulae for σ. This brings us to the conditions under which the approximation holds. The binomial can be approximated by the Poisson when:

$$p \leq 0.10$$
and $\qquad np \leq 5$.

The normal approximation

It is also possible to provide an approximation of the binomial distribution by the normal curve. This applies as the proportion of classified units p approaches 0.5 (50 per cent), which may not be very often in a quality control situation, but may be very common in an activity sampling application. It is, of course, valid in the case of coin tossing where the chance of obtaining a head in an unbiased coin is 1 in 2. The number of heads obtained if 20 coins are tossed have been calculated from the binomial in Table J.1. The results are plotted on a histogram in Figure J.1. The corresponding normal curve has been superimposed on to the histogram. It is clear that, even though the probabilities were derived from a binomial distribution, the results are virtually a normal distribution and that we may use normal tables to calculate probabilities.

An example illustrates the usefulness of this method. Suppose we wish to find the probability of obtaining 14 or more heads when 20 coins are tossed. Using the binomial:

$$P(\geq 14) = P(14) + P(15) + P(16) + P(17) + P(18)$$
$$\text{(there is zero probability of finding more than 18)}$$
$$= 0.0370 + 0.0148 + 0.0046 + 0.0011 + 0.0002$$
$$= 0.0577.$$

Table J.1 Number of heads obtained from coin tossing

Number of heads in tossing 20 coins	Probability (binomial n = 20, p = 0.5)	Frequency of that number of heads if 20 coins are tossed 10,000 times
2	0.0002	2
3	0.0011	11
4	0.0046	46
5	0.0148	148
6	0.0370	370
7	0.0739	739
8	0.1201	1201
9	0.1602	1602
10	0.1762	1762
11	0.1602	1602
12	0.1201	1201
13	0.0739	739
14	0.0370	370
15	0.0148	148
16	0.0046	46
17	0.0011	11
18	0.0002	2

Using the normal tables:

$$\mu = np = 20 \times 0.5 = 10.$$
$$\sigma = \sqrt{np\,(1-p)} = \sqrt{20 \times 0.5 \times 0.5} = 2.24.$$

Since the data must be continuous for the normal curve to operate, the probability of obtaining 14 or more heads is considered to be from 13.5 upward.
The general formulae for the z factor is:

$$z = \frac{x - 0.5 - np}{\sigma}.$$

Now,

$$z = \frac{14 - 0.5 - 10}{2.24} = 1.563,$$

and from the normal tables (Appendix A) the probability of finding 14 or more heads is 0.058.

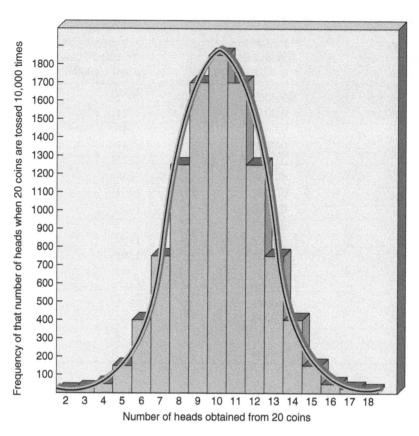

Figure J.1 Coin tossing: the frequency of obtaining heads when tossing 20 coins

The normal curve is an excellent approximation to the binomial when p is close to 0.5 and the sample size n is 10 or more. If n is very large then, even when p is quite small, the binomial distribution becomes quite symmetrical and is well approximated by the normal curve. The nearer p becomes to 0.5, the smaller n may be for the normal approximation to be applied.

Appendix K
Glossary of terms and symbols

A Constants used in the calculation of the control lines for mean, moving mean, median and mid-range control chart, with various suffixes.

Accuracy Associated with the nearness of a process to the target value.

Action limit (line) Line on a control chart beyond which the probability of finding an observation is such that it indicates that a change has occurred to the process and that action should be taken to investigate and/or correct for the change.

Action zone The zones outside the action limits/lines on a control chart where a result is a clear indication of the need for action.

ARL The average run length to detection of a change in a process.

Assignable causes Sources of variation for which an explicit reason exists.

Attribute charts Control charts used to assess the capability and monitor the performance of parameters assessed as attributes or discrete data.

Attribute data Discrete data which can be counted or classified in some meaningful way which does not include measurement.

Average *See* Mean.

B Constants used in the calculation of control chart lines for standard deviation charts.

Bar A bar placed above any mathematical symbol indicates that it is the mean value.

Bar chart A diagram which represents the relative frequency of data.

Binomial distribution A probability distribution for samples of attributes that applies when both the number of conforming and non-conforming items is known.

Brainstorming An activity, normally carried out in groups, in which the participants are encouraged to allow their experience and imagination to run wild, while centred around specific aspects of a problem or effect.

c chart A control chart used for attributes when the sample is constant and only the number of non-conformances is known; c is the symbol which represents the number of non-conformances present in samples of a constant size. c-bar ($\bar{c}$) represents the average value of a series of values of c.

Capable A process that is in statistical control and for which the combination of the degree of random variation and the ability of the control procedure to detect change is consistent with the requirements of the specification.

Cause and effect diagram A graphic display that illustrates the relationship between an effect and its contributory causes.

Central tendency The clustering of a population about some preferred value.

Centre line (CL) A line on a control chart at the value of the process mean.

Checklist A list used to ensure that all steps in a procedure are carried out.

Common causes *See* Random causes.

Conforming Totally in agreement with the specification or requirements.

Continuous data Quantitative data concerning a parameter in which all measured values are possible, even if limited to a specific range.

Control The ability or need to observe/monitor a process, record the data observed, interpret the data recorded and take action on the process if justified.

Control chart A graphical method of recording results in order to readily distinguish between random and assignable causes of variation.

Control limits (lines) Limits or lines set on control charts which separate the zones of stability (no action required), warning (possible problems and the need to seek additional information) and action.

Countable data A form of discrete data where occurrences or events can only be counted (*see also* Attribute data).

Cp A process capability index based on the ratio of the spread of a frequency distribution to the width of the specification.

Cpk A process capability index based on both the centring of a frequency distribution and the ratio of the spread of the distribution to the width of the specification.

Cusum chart A graphic presentation of the cusum score. The cusum chart is particularly sensitive to the detection of small sustained changes.

Cusum score The cumulative sum of the differences between a series of observed values and a predetermined target or average value.

dn or d_2 Symbols that represent Hartley's constant, the relationship between the standard deviation (σ) and the mean range ($\bar{R}$).

D Symbol that represents the constant used to determine the control limits on a range chart, with various suffixes.

Data Facts.

Defect A fault or flaw that is not permitted by the specification requirements.

Defective An item that contains one or more defects and/or is judged to be non-conforming.

Detection The act of discovering.

Deviation The dispersion between two or more data.

Difference chart A control chart for differences from a target value.

Discrete data Data not available on a continuous scale (*see also* Attribute data).

Dispersion The spread or scatter about a central tendency.

DMAIC   Six sigma improvement model – Define, Measure, Analyse, Improve, Control.

DRIVER   General performance improvement model – Define, Review, Investigate, Verify, Execute, Reinforce.

Frequency   How often something occurs.

Frequency distribution   A table or graph that displays how frequently some values occur by comparison with others. Common distributions include normal, binomial and Poisson.

Grand mean   The mean of either a whole population or the mean of a series of samples taken from the population. The grand mean is an estimate of the true mean (*see* Mu).

Histogram   A diagram which represents the relative frequency of data.

Individual   An isolated result or observation.

Individuals plot   A graph showing a set of individual results.

LAL   Lower action limit or line.

LCL   Lower control limit or line.

LSL   Lower specification limit.

LWL   Lower warning limit or line.

$\widetilde{M}_R$ Median of the sample mid-ranges.

Mean   The average of a set of individual results, calculated by adding together all the individual results and dividing by the number of results. Means are represented by a series of symbols and often carry a bar above the symbol which indicates that it is a mean value.

Mean chart   A graph with control lines used to monitor the accuracy of a process, being assessed by a plot of sample means.

Mean range   The mean of a series of sample ranges.

Mean sample size   The average or mean of the sample sizes.

Median   The central value within a population above and below which there are an equal number of members of the population.

Mode   The most frequently occurring value within a population.

Moving mean   A mean value calculated from a series of individual values by moving the sample for calculation of the mean through the series in steps of one individual value and without changing the sample size.

Moving range   A range value calculated from a series of individual values by moving the sample for calculation of the range through the series in steps of one individual value and without changing the sample size.

Mu (μ)   The Greek letter used as the symbol to represent the true mean of a population as opposed to the various estimates of this value that measurement and calculation make possible.

n   The number of individuals within a sample of size n. n-bar ($\bar{n}$) is the average size of a series of samples.

Non-conforming   Not in conformance with the specification/requirements.

Non-conformities   Defects, errors, faults with respect to the specification/requirements.

Normal distribution   Also known as the Gaussian distribution of a continuous variable and sometimes referred to as the 'bell-shaped' distribution. The normal distribution has the characteristic that 68.26 per cent of the population is contained within ± one standard deviation from the mean value, 95.45 per cent within ± two standard deviations from the mean and 99.73 per cent within ± three standard deviations from the mean.

np chart   A control chart used for attributes when the sample size is constant and the number of conforming and non-conforming items within a sample are both known. *n* is the sample size and *p* is the proportion of non-conforming items.

p chart   A control chart used for attributes showing the proportion of non-conforming items in a sample. *p* is the proportion of non-conforming items and *p*-bar ($\bar{p}$) represents the average of a series of values of *p*.

Pareto analysis   A technique of ranking data in order to distinguish between the vital few and the trivial many.

Poisson distribution   A probability distribution for samples of attributes that applies when only the number of non-conformities is known.

Population   The full set of data from which samples may be taken.

Precision   Associated with the scatter about a central tendency.

Prevention   The act of seeking to stop something occurring.

Probability   A measure of the likelihood of an occurrence or incident.

Process   Any activity that converts inputs into outputs.

Process capability   A measure of the capability of a process achieved by assessing the statistical state of control of the process and the amount of random variation present. It may also refer to the tolerance allowed by the specification.

Process capability index   An index of capability (*see Cp* and *Cpk*).

Process control   The management of a process by observation, analysis, interpretation and action designed to limit variation.

Process mean   The average value of an attribute or a variable within a process.

Proportion defective   The ratio of the defectives to the sample size, represented by the symbol *p*. *p*-bar ($\bar{p}$) represents the average of a series of values of *p*.

Quality   Meeting the customer requirements.

R   The range of values in a sample.

R-bar ($\bar{R}$)   The symbol for the mean of a series of sample ranges.

$\tilde{R}$ The median of sample ranges.

Random causes   The contributions to variation that are random in their behaviour, i.e. not structured or assignable.

Range (R)   The difference between the largest and the smallest result in a sample of individuals – an approximate and easy measure of the degree of scatter.

Range chart   A graph with control lines used to monitor the precision of a process, being assessed by a plot of sample ranges.

Run A set of results that appears to lie in an ordered series.

Run chart A graph with control lines used to plot individual results.

Sample A group of individual results, observations or data. A sample is often used for assessment with a view to determining the properties of the whole population or universe from which it is drawn.

Sample size (n) The number of individual results included in a sample, or the size of the sample taken.

Scatter Refers to the dispersion of a distribution.

Scatter diagram The picture that results when simultaneous results for two varying parameters are plotted together, one on the x axis and the other on the y axis.

Shewhart charts The control charts for attributes and variables first proposed by Shewhart. These include mean and range, np, p, c and u charts.

Sigma (σ) The Greek letter used to signify the standard deviation of a population.

Six sigma A disciplined approach for improving performance by focussing on producing better products and services, faster and cheaper.

Skewed distribution A frequency distribution that is not symmetrical about the mean value.

SPC *See* Statistical process control.

Special causes *See* Assignable causes.

Specification The requirement against which the acceptability of the inputs or outputs of a process are to be judged.

Spread Refers to the dispersion of a distribution.

SQC Statistical quality control – similar to SPC but with an emphasis on product quality and less emphasis on process control.

Stable The term used to describe a process when no evidence of assignable causes is present.

Stable zone The central zone between the warning limits on a control chart and within which most of the results are expected to fall.

Standard deviation (σ) A measure of the spread or scatter of a population around its central tendency. Various estimates of the standard deviation are represented by symbols such as σ_n, $\sigma_{(n-1)}$ and s.

Standard error The standard deviation of sample mean values – a measure of their spread or scatter around the grand or process mean, represented by the symbol SE (or $\sigma_{\bar{x}}$).

Statistical control A condition describing a process for which the observed values are scattered about a mean value in such a way as to imply that the origin of the variations is entirely random with no assignable causes of variation and no runs or trends.

Statistical process control The use of statistically based techniques for the control of a process for transforming inputs into outputs.

Statistics The collection and use of data – methods of distilling information from data.

t   The value of a statistic calculated to test the significance of the difference between two means.

T   A symbol used to represent a tolerance limit ($\pm T$).

Tally chart   A simple tool for recording events as they occur or to extract frequencies from existing lists of data.

Target   The objective to be achieved and against which performance will be assessed, often the mid-point of a specification.

Tolerance   The difference between the lowest and/or the highest value stated in the specification and the mid-point of the specification.

Trend   A series of results that show an upward or downward tendency.

u chart   A control chart used for attributes when the sample size is not constant and only the number of non-conformities is known. *u* is the symbol that represents the number of non-conformities found in a single sample and *u*-bar ($\bar{u}$) represents the mean value of *u*.

UAL   Upper action limit or line. *UCL* Upper control limit or line.

Universe   See Population.

USL   Upper specification limit.

UWL   Upper warning limit or line.

V-mask   A device used in conjunction with a cusum chart to identify trends of known significance.

Variable data   Data which is assessed by measurement.

Variance   A measure of spread equal to the standard deviation squared (σ^2).

Variation   The inevitable differences between outputs.

Warning limit (line)   Lines on a control chart, on each side of the central line, and within which most results are expected to fall, but beyond which the probability of finding an observation is such that it should be regarded as a warning of a possible problem.

Warning zone   The zones on a control chart between the warning and the action limits and within which a result suggests the possibility of a change to the process.

x   An individual value of a variable.

X-bar ($\bar{X}$)   The mean value of a sample, sometimes the symbol x-bar ($\bar{x}$) is used.

X-bar-bar ($\bar{\bar{X}}$)   The grand or process mean, sometimes the symbol X-bar ($\bar{X}$) is used.

$\tilde{X}$ The median value of a sample.

$\tilde{\tilde{X}}$ The grand or process median value of a sample.

Z   The standardized normal variate – the number of standard deviations between the mean of a normal distribution and another defined value, such as a target.

Z chart   A control chart for standardized differences from a target value.

Index

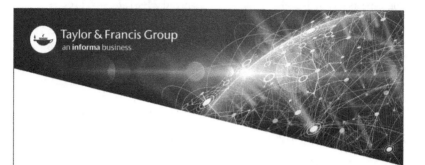